Environmentally Sustainable Production

María del Carmen Valls Martínez ·
José Manuel Santos-Jaén
Editors

Environmentally Sustainable Production

Research for Sustainable Development

Springer

Editors
María del Carmen Valls Martínez
Department of Economics and Business, Mediterranean Research Center on Economics and Sustainable Development (CIMEDES)
University of Almería
Almería, Spain

José Manuel Santos-Jaén
Department of Financial Economics and Accounting
University of Murcia
Murcia, Spain

ISBN 978-3-031-52655-8 ISBN 978-3-031-52656-5 (eBook)
https://doi.org/10.1007/978-3-031-52656-5

This Springer imprint is published by the registered company Springer Nature Switzerland AG
The registered company address is: Gewerbestrasse 11, 6330 Cham, Switzerland

Paper in this product is recyclable.

Introduction

The survival of companies requires obtaining an economic benefit, but today, this should not be the only objective guiding their activities. Indeed, competitiveness, as well as international legal and ethical standards, obliges companies to consider aspects that lead to sustainability through the so-called triple bottom line that expresses the activity of a company in three dimensions: economic, social, and environmental.

Business sustainability involves the development of activities over a long period of time, considering environmental, social, and governance criteria to consolidate the continuity of the company and the protection of people and the planet, in line with the 17 Sustainable Development Goals agreed by the United Nations. This practice is not a mere theory but must be implemented as a business priority necessary for the company's good image in the eyes of investors, the market, and society. However, the active pursuit of sustainability is everyone's responsibility: for companies, the public sector, and citizens in general.

This book aims to provide an overview of what sustainable development implies for both the private and public sectors, with particular emphasis on such vital sectors of the economy as finance and agriculture, highlighting the role to be played by new technologies and revealing generic and specific aspects of corporate social responsibility policies, without forgetting the critical significance of the incorporation of women in positions of maximum organizational responsibility.

The financial sector can guide the actions of companies, public administrations, and the general public, promoting socially responsible investment and channeling funds to sustainable activities. Even for families, for example, by providing mortgage loans with interest rates linked to the home's energy rating. Therefore, the first part of the book is dedicated to the *Financial Sector and Sustainable Development*. Irene Ricote García, head of Sustainability and Rating Agencies at Catalana Occidente Group, a leading worldwide insurer based in Madrid (Spain), writes the first chapter entitled "Current Situation and Perspective of Sustainability in the Financial Sector: A Closer Look at the "E" in ESG," in which she examines the concepts of corporate social responsibility and sustainability, analyzing the world summits and agreements aimed at achieving a more sustainable model. The chapter highlights the importance of the financial system in facilitating the transformation required by the most important national and international agreements and initiatives that have been carried out to date.

The second chapter, written by the reputed professors of the University of Extremadura (Spain), María Mar Miralles-Quirós, José Luis Miralles-Quirós, and Azahara Gil-Corbacho, is entitled "Sustainable Practices and Shareholder Value Creation in FinTech Firms: International Evidence." The chapter analyzes whether the sustainability practices implemented by FinTechs create value for their shareholders in the capital markets, highlighting that not all sustainable activities are rewarded by the markets. In particular, those that add value are related to disclosure that follows GRI standards, assurance of such information, and environmental activities.

The third chapter was written by two great professors from the University of Almería (Spain), María del Mar Gálvez-Rodríguez and Carmen Caba-Pérez, and a professor from the Autonomous University of Asunción (Paraguay), Walter Daniel Ovelar-Fernández. This chapter, which is entitled "Chapter Information Disclosure on the Integration of SDGs into Banking Management: The Mercosur Countries Case," focuses on banks that operate in the Mercosur economic community: Argentina, Brazil, and Paraguay. The chapter addresses the lack of studies about the banking sector's commitment to Sustainable Development Goals in the Latin American region.

The second part of the book, *Environmentally Sustainable Agricultural Production*, is devoted to the farm sector since it is fundamental in all countries, especially the less developed ones. It is also a sector where major environmental sustainability initiatives are being introduced, perhaps because of the greater awareness that care for the environment requires for its continuity, which is severely affected by climate change (droughts, cyclones, water shortages, floods, etc.). The fourth chapter, entitled "Role of Sustainability and Circular Economy in Europe's Common Agricultural Policy," is authored by three professors from the University of Almería (Spain), Francisco José Castillo-Díaz, Luis J. Belmonte-Ureña, and Francisco Camacho-Ferre, and a professor from the Pontificia Universidad Javeriana (Colombia), Juan Fernando Álvarez-Rodríguez. The chapter analyzes the role of sustainability in the European Union's Common Agricultural Policy, which has recently been amended and will apply from 2023 to 2027, focusing on the fight against climate change and the environmental sustainability of agriculture, which must be modernized using the new technologies resulting from digitalization.

The fifth chapter is signed by Ana Isabel García-Agüero, Eduardo Terán-Yépez, Luis J. Belmonte-Ureña, and Francisco Camacho-Ferre, all of them professors at the University of Almería (Spain). The chapter, entitled "The Role of Stakeholders on the Intention to Implement Sustainable Practices: An Exploratory Research in the Agri-Business Spanish Sector," explores the role of stakeholders in the intention to implement sustainable practices within agricultural companies and investigates the multifaceted dynamics that influence their roles in either advancing or impeding sustainable practices. Specifically, eleven stakeholders' roles are analyzed:

top leadership and management, other staff, suppliers, intermediaries, marketplace, wholesalers, customers, end users, governments, laws, and activist groups.

The sixth chapter, entitled "The Role of Family Farming in Socio-Economic Sustainability: An Exploratory Analysis of Rural Development in Southeast Spain," is authored by six professors of the Mediterranean Research Center on Economics and Sustainable Development at the University of Almeríain (Spain), led by Emilio Galdeano-Gómez, a good friend of the lead editor, and his colleagues, Laura Piedra-Muñoz, María del Carmen García-Barranco, Yolanda Sorroche-del-Rey, Jesús Hernández-Rubio, and Javier Sánchez-García. Family-run farms in rural areas play a crucial role and are increasingly recognized as essential for sustainable growth. The chapter analyzes the social and economic impacts of family-based farming, which is achieving eco-labels that reflect their commitment to sustainable practices (waste management, energy optimization, water conservation, etc.) while reinforcing working conditions.

Sepide Mehrabi, Juan Carlos Pérez-Mesa, María del Carmen García-Barranco, and Cynthia Giagnocavo, professors at the University of Almería (Spain), in the seventh chapter, "Innovation Oriented Towards Sustainability in the Value Chain of Agri-Food Cooperatives," examine how producer organizations, in the context of the fruit and vegetable industry, should develop and adapt their business models and strategies for implementation of sustainability-oriented innovation supply chain, by considering all the green supply chain, circular supply chain, sustainable supply chain, and even Bioeconomy.

The public sector is the driving force behind a country's policies, so its importance is crucial in developing and implementing sustainable practices. For this reason, this book's three chapters of third part, *Public Sector and Sustainable Development*, are dedicated to the part closest to the citizens, the municipalities, as well as energy production and consumption. Eighth and Ninth chapters, "Budget Policities Versus Sustainable Development Goal in the Main Spanish Municipalities" and "Analysis of the Influence of Financial Sustainability on the Achievement of SDG 2 in Public Sector," were written by a master's student in Accounting Auditing, Pedro Gil-García, a Finance and Accounting graduate, Natalia Alonso-Morales, and two reputed senior lecturers from the University of Almería (Spain), Arturo Haro-de-Rosario and Alejandro Sáez-Martín. The chapters analyze the reduction of inequalities, corresponding to the Sustainable Development Goals 10 and 2 from the United Nations 2030 Agenda, by identifying the influence of budgetary factors since an adequate budget allocation is essential for sustainable development. Aleksandra Matuszewska-Janica, professor at the Warsaw University of Live Sciences (Poland), authored the tenth chapter, "How Did the Covid-19 Pandemic Affect the Structure of Energy Final Consumption of the Households Across EU Countries? Findings from Eurostat Data." Ensuring accessible, affordable, and clean energy for citizens is deeply reflected in the 7th Sustainable Development Goal and closely linked to responsible consumption and production, and climate action, relating to the 12th and 13th Sustainable Development Goals, respectively.

Fourth part of this book, *Application of New Technologies to Sustainability*, includes two chapters showing how new technologies can be incorporated into sustainable production and environmental care. Chapter eleven, "Integrating Blockchain Technology in Supply Chain Risk Management for Sustainable Development," is signed by Fahim ul Amin, Qingkai Ji, and Azka Amin, professors at Hainan University (China), and Wasim ul Amin, Ph.D. student. Supply chain risk management is a complex procedure focusing on locating, evaluating, and minimizing potential supply chain disruptions. Its significance goes beyond merely ensuring the uninterrupted flow of goods and services since it is crucial in guaranteeing economic resilience and ecological and social equilibrium in a world that is turning toward sustainability. Integrating blockchain technology and supply chain management is no longer just a novel concept in an increasingly globalized world; it is necessary for ethical, transparent, and sustainable business practices.

The twelfth chapter, "The Metaverse as a Solution to Depopulation: Creation of Interconnected Macro-villages Through Digital Twins and Transportation. The Theory of Population Leverage," is authored by Álvaro Bueno-Ferrer, Ph.D. student, and Jaime de Pablo Valenciano, full professor at the University of Almería (Spain). Through digital twins, the metaverse serves as a digital extension of communities. It provides a greater abundance of services available using augmented reality and virtual reality. The scarcity of services and resources in these areas can foster demographic decline, and it is crucial to seek innovative strategies to address the fragility of the most marginalized regions. From the metaverse perspective, digitalization has a greater potential to favor carbon neutrality and green and sustainable development.

Fifth part of the book, *Corporate Social Responsibility*, comprises four chapters, including Environmental, Social, and Governance (ESG) aspects, which are the basis of sustainability. The thirteenth chapter, "Disclosure Practices for Tackling Climate Change in Large Spanish Listed Companies," was written by María Mar Miralles-Quirós, José Luis Miralles-Quirós, and Lorena Leal-Espinosa, scholars from the University of Extremadura (Spain). The problems associated with climate change have increased enormously in recent years. Addressing these problems requires drastic action from the public and private sectors. Large private companies have a crucial role because they strongly influence the economy, society, and the environment. Therefore, it is essential to analyze how these companies address climate change and how they inform their stakeholders, whether there is a real commitment or if they are merely engaging in greenwashing practices.

The co-editor José Manuel Santos-Jaén leads the fourteenth chapter, "The Effect of Audit Committee Characteristics on Corporate Social Responsibility Practices. Evidence from Spain," which is also authored by Mercedes Palacios-Manzano and Ester Gras-Gil, both of them from the University of Murcia (Spain), and Ana León-Gómez, from the University of Málaga (Spain). This chapter investigates the correlation between the attributes of audit committees and the extent of corporate social responsibility initiatives among Spanish companies listed in the Spanish Monitor of Corporate Reputation.

Editors María del Carmen Valls Martínez and José Manuel Santos-Jaén, with the auditor Gema Martín de Almagro Vázquez, authored the fifteenth chapter, "Unveiling Differences in ESG Adoption: A Comparative Analysis of the Big Four Auditors." This chapter examines how auditor tenure and audit fees impact the ESG practices adopted by audited companies. Additionally, the research includes an analysis of the Big Four auditing firms in the USA to assess the role of corporate culture in this context.

The sixteenth chapter, "Does Reducing Carbon Emissions Affect Business Profitability? An Analysis of Family and Non-family Businesses," was written by José L. Gallizo-Larraz, Jordi Moreno-Gené, and Laura Sánchez-Pulido, great professors at the University of Lleida (Spain). Regarding greenhouse gas emissions, this chapter addresses the question of whether it is more profitable to be an environmentally friendly company or one that ignores recommendations to limit GHG emissions. The academic discussion has an additional interest when it applies to familiar companies that have demonstrated their leadership in green practices.

The incorporation of women in positions of maximum organizational responsibility is a hot topic today and is required by more and more countries' legislation. This fact has many implications for corporate policies, including greater sensitivity to corporate social responsibility issues in general and environmental issues in particular. Hence, sixth part of the book deals with *Gender Diversity in Top Management for Greater Sustainability*. The seventeenth chapter, "Carbon Performance and Board Gender Diversity: The Moderating Effect of Patriarchal Attitudes," authored by Sara Corral, a Ph.D. student at the University of León (Spain), analyzes the influence of gender diversity on boards of directors on carbon performance, in European countries, taking into account the moderating impact of prevailing patriarchal attitudes across countries and years.

Finally, the eighteenth chapter, "The Influence of Corporate Governance on the Sustainability of American Company Buildings," is authored by the editor María del Carmen Valls Martínez, José-María Montero, a prestigious full professor at the University of Castilla-La Macha (Spain), and two professors at the International University of Ecuador, María Estefanía Sánchez Pacheco and Fernando José Zambrano Farías. This chapter explores the relationship between the composition and characteristics of the board of directors and companies' adoption of green buildings, showing that companies are more likely to have sustainable buildings when the percentage of women on the board of directors, director compensation and tenure, number of board meetings, and board size increase.

María del Carmen Valls Martínez
José Manuel Santos-Jaén

Contents

Editors and Contributors

About the Editors

María del Carmen Valls Martínez is a Ph.D. in Economics and Business. In 1995, she joined the Economics and Business Department at the University of Almería (Spain), where she is an associate professor in Financial Economics and Accounting and leads the *Ethics, Gender and Sustainability Research Group*. Her research interests include Financial Operations, Ethical Banking, Gender Economics Studies, Health Economics and Sustainability. She has published numerous books and book chapters in reputable publishers such as Springer, Dykinson, Pirámide, McGraw-Hill, etc. She is the author of a large number of articles in prestigious scientific journals, such as the *Journal of Cleaner Production, Corporate Social Responsibility and Environmental Management, Risk Management, Physica A: Statistical Mechanics and its Applications, European Research on Management and Business Economics, Mathematics, Plos One, Women's Studies International Forum, International Journal of Environmental Research and Public Health, Frontiers in Psychology, Oeconomia Copernicana.* She is also an editor at *Plos One, Frontiers in Public Health* and *Mathematics.*

José Manuel Santos-Jaén is a Ph.D. associate professor of Accounting and Finance at the University of Murcia, Spain. He is a graduate in Business Administration and Management and a graduate in Law. He has completed three masters (MBA, Financial Management and Auditing). He finished his Ph.D. in 2011. He is currently teaching financial statement analysis, financial accounting, and corporate accounting. His research is focused on several aspects of Business and Management, such as Firm Performance, Innovation, Sustainability or Corporate Social Responsibility, among others. He is strongly oriented to a quantitative, survey-based methodology, mainly

Structural Equation Modeling (PLS-SEM). He has published in international journals, including *Journal of Cleaner Production, Corporate Social Responsibility and Environment, IEEE Transaction on Engineering Management, Sustainable Production and Consumption, The International Journal of Management Education,* and *Total Quality Management and Business Excellence.* He has also presented the progress of his research at national and international congress.

Contributors

Natalia Alonso-Morales University of Almeria, Almeria, Spain

Juan Fernando Álvarez-Rodríguez Faculty of Environmental and Rural Studies, Pontificia Universidad Javeriana, Bogota, Colombia

Azka Amin International Business School, Hainan University, Haikou, Hainan, China

Luis J. Belmonte-Ureña Department of Economics and Business, University of Almeria (ceiA3), Almeria, Spain

Álvaro Bueno-Ferrer Department of Economics and Business, University of Almeria, Almeria, Spain

Carmen Caba-Pérez University of Almería, Almería, Spain

Francisco Camacho-Ferre Department of Agronomy and Sustainable Protected Agriculture Research Network, University of Almeria (ceiA3), Almeria, Spain

Francisco José Castillo-Díaz Department of Agronomy, University of Almeria, Almería, Spain

Sara Corral Facultad de Ciencias Económicas y Empresariales, Universidad de León, León, Spain

Jaime de Pablo Valenciano Department of Economics and Business, University of Almeria, Almeria, Spain

María del Mar Gálvez-Rodríguez University of Almería, Almería, Spain

Emilio Galdeano-Gómez Mediterranean Research Center on Economics and Sustainable Development (CIMEDES), University of Almería, Almería, Spain

José L. Gallizo-Larraz Department of Economics and Business, University of Lleida, Lleida, Spain

Ana Isabel García-Agüero Department of Economics and Business, University of Almeria (ceiA3), Almeria, Spain

María del Carmen García-Barranco Department of Economics and Business, University of Almería (Agrifood Campus of International Excellence, ceiA3; Centre for Research on Intensive Mediterranean Agrosystems and Agri-Food Biotechnology, CIAMBITAL), Almería, Spain

Cynthia Giagnocavo Department of Economics and Business, University of Almería (Agrifood Campus of International Excellence, ceiA3; Centre for Research on Intensive Mediterranean Agrosystems and Agri-Food Biotechnology, CIAMBITAL), Almería, Spain

Azahara Gil-Corbacho Faculty of Law, University of Extremadura, Cáceres, Spain

Pedro Gil-García University of Almeria, Almeria, Spain

Ester Gras-Gil Department of Accounting and Finance, University of Murcia, Murcia, Spain

Arturo Haro-de-Rosario University of Almeria, Almeria, Spain

Jesús Hernández-Rubio Mediterranean Research Center on Economics and Sustainable Development (CIMEDES), University of Almería, Almería, Spain

Qingkai Ji International Business School, Hainan University, Haikou, Hainan, China

Lorena Leal-Espinosa Faculty of Economic and Business Sciences, University of Extremadura, Badajoz, Spain

Ana León-Gómez Department of Accounting and Management, University of Malaga, Malaga, Spain

Gema Martín de Almagro Vázquez Sector 3 SAP, Murcia, Spain

Aleksandra Matuszewska-Janica Institute of Economics and Finance, Warsaw University of Life Sciences, Warsaw, Poland

Sepide Mehrabi Department of Economics and Business, University of Almería (Agrifood Campus of International Excellence, ceiA3; Centre for Research on Intensive Mediterranean Agrosystems and Agri-Food Biotechnology, CIAMBITAL), Almería, Spain

José Luis Miralles-Quirós Faculty of Economic and Business Sciences, University of Extremadura, Badajoz, Spain

María Mar Miralles-Quirós Faculty of Economic and Business Sciences, University of Extremadura, Badajoz, Spain

José-María Montero Department of Political Economy and Public Finance, Economic and Business Statistics, and Economic Policy, University of Castilla-La Mancha, Toledo, Spain

Jordi Moreno-Gené Department of Economics and Business, University of Lleida, Lleida, Spain

Walter Daniel Ovelar-Fernández Autonomous University of Asunción, Asunción, Paraguay

Mercedes Palacios-Manzano Department of Accounting and Finance, University of Murcia, Murcia, Spain

Juan Carlos Pérez-Mesa Department of Economics and Business, University of Almería (Agrifood Campus of International Excellence, ceiA3; Centre for Research on Intensive Mediterranean Agrosystems and Agri-Food Biotechnology, CIAMBITAL), Almería, Spain

Laura Piedra-Muñoz Mediterranean Research Center on Economics and Sustainable Development (CIMEDES), University of Almería, Almería, Spain

Irene Ricote García Madrid, Spain

Alejandro Sáez-Martín University of Almeria, Almeria, Spain

Javier Sánchez-García Mediterranean Research Center on Economics and Sustainable Development (CIMEDES), University of Almería, Almería, Spain

María Estefanía Sánchez Pacheco International University of Ecuador, Guayaquil, Ecuador

Laura Sánchez-Pulido Department of Economics and Business, University of Lleida, Lleida, Spain

José Manuel Santos-Jaén Department of Accounting and Finance, University of Murcia, Murcia, Spain

Yolanda Sorroche-del-Rey Mediterranean Research Center on Economics and Sustainable Development (CIMEDES), University of Almería, Almería, Spain

Eduardo Terán-Yépez Department of Economics and Business, CIMEDES Research Center, University of Almeria (ceiA3), Almeria, Spain

Fahim ul Amin International Business School, Hainan University, Haikou, Hainan, China;
School of Economics and Management, North China Electric Power University, Beijing, China

Wasim ul Amin School of Economics and Management, North China Electric Power University, Beijing, China

María del Carmen Valls Martínez Mediterranean Research Center on Economics and Sustainable Development, Economics and Business Department, University of Almería, Almería, Spain

Fernando José Zambrano Farías International University of Ecuador, Guayaquil, Ecuador

Financial Sector and Sustainable Development

Current Situation and Perspective of Sustainability in the Financial Sector: A Closer Look at the "E" in ESG

Irene Ricote García

1 Introduction. From CSR to Sustainability

There is a permanent debate about a possible definition of corporate social responsibility (CSR) without academics or professionals recognizing a single, clear and simple definition of it, sometimes generating a certain degree of confusion (Aguilera et al., 2007; Aguinis & Glavas, 2012; Sheehy, 2015).

In general, there has been an evolution in the interpretation of CSR over the decades. In its origins, CSR was conceived as a philanthropic activity. That is, the term CSR referred to those actions carried out by companies to contribute to social well-being, exceeding the companies' own interests and their legal or regulatory obligations (McWilliams & Siegel, 2001). These were then initiatives of a voluntary nature and within the social sphere, leaving aside everything related to the environmental sphere of companies.

Since the 1990s, public opinion has been specially sensitive towards aspects related to the environment and the impact of business activity, presenting globalization as the cause of its deterioration. The classic industrial model associated with companies has historically been associated to this effect.

Awareness of certain global problems, such as the climate change or the deterioration of natural resources, has caused the environmental dimension to gain relevance in business (Auld et al., 2008). This has led to the emergence of corporate environmental responsibility (CER).

The CER encompases the voluntary integration, by companies, of environmental issues in their business operations and their interactions with stakeholders (Rahman & Post, 2012).

The link between CER and CSR can be misleading. Sometimes it is considered that environmental responsibility has its own and independent identity, leaving

I. . Ricote García (✉)
Madrid, Spain
e-mail: iricote@ucm.es

M. d. C. Valls Martínez and J. M. Santos-Jaén (eds.), *Environmentally Sustainable Production*, https://doi.org/10.1007/978-3-031-52656-5_1

social responsibility as a specific scope towards people (excluding environmental aspects). Other interpretations consider that CSR should be considered under a broader approach, which integrates CER as an integral and very important part of CSR (Bansal & Roth, 2000; Gunningham, 2009; Trumpp et al., 2015) and that CSR implicitly includes social and environmental components.

In 1987, the term sustainable development or sustainability was used for the first time as we know it today. It appears in a report prepared by several countries for the UN named Our Common Future (World Commission on Environment & Development, 1987). The commission that prepared the report was headed by Dr. Gro Harlem Brundtland. For that reason it is best known as Brundtland report.

The Brundtland report defines sustainable development as "*development that meets the needs of the present without compromising the ability of future generations to meet their own needs*" (World Commission on Environment & Development, 1987, p. 41).

The report showed that the pace of development being followed as a globalized society is destroying the environment, on the one hand, and increasing the poverty and vulnerability of millions of people, on the other. It also showed that environmental protection should stop being a national or regional task to acquire a global perspective and that development should stop being a problem exclusive to underdeveloped countries. The final conclusion of the report was that it was necessary to evolve towards a more comprehensive approach that took into account in a balanced way the social, environmental and economic dimensions.

Thus, beyond proposing a new definition of CSR, the term sustainable development or sustainability comes to complete the existing definition of CSR, highlighting a fundamental factor for the understanding and implementation of this responsibility: the time. Sustainability aims to ensure that the pace of economic growth and the speed at which the planet can supply it remain balanced, thus avoiding an irreversible impact on its capacity and maintenance.

In 1992, the United Nations Conference on Environment and Development (UNCED), also known as the Earth Summit or Rio Summit, took place in Rio de Janeiro (United Nations, 1992b). It identified three high-risk natural processes, desertification, biodiversity loss and climate change, and recognized that integrating environmental and social concerns along with economic ones is vital to maintaining human life on the planet. Since this summit, the fundamental principle of sustainable development or sustainability has been handled internationally at both a scientific and political level.

This evolution has caused companies in the last decade to abandon the concept of CSR, originally linked basically to philanthropy, and to incorporate a much broader concept, sustainability, as a lever for value creation in the core business of companies.

The acronym ESG (Environmental, Social ans Governance) refers to the different thematic areas that companies focus on to be more sustainable (Fig. 1).

This study has two objectives. First, to analyze the major international cooperation conventions and agreements developed to address the global environmental challenges that the world faces today. Second, to identify the fundamental role that sustainable finance plays in contributing to the fulfillment of previous international

Fig. 1 Relationship between social responsibility, environmental responsibility and sustainability. *Source* Own elaboration

agreements and the efforts made at European, international and national level to develop a package of measures and legislative tools that substantially change the way in which the financial system supports sustainability.

For this purpose, the remainder of this chapter is organized as follows: Sect. 2 reviews the major international sustainability summits and agreements. Section 3 describes the relationship between the financial sector and sustainability. Section 4 analyzes the main sustainability initiatives at international, European and national levels and its interoperability. Finally, Sect. 5 draws the conclusions of the study.

2 Major International Sustainability Summits and Agreements

To understand the current relevance of sustainability concept, and specifically its environmental component, it is interesting to see the main international sustainability summits held and the global agreements agreed upon in them in a global effort to reorient the world towards a more sustainable model.

As mentioned previously, the 1992 Rio de Janeiro Earth Summit was a relevant meeting on sustainable development in which three high-risk natural processes were identified: climate change, loss of biodiversity and desertification (United Nations, 1992b). To address each of these environmental risks, three fundamental agreements were reached: The United Nations Framework Convention on Climate Change (UNFCCC), the Convention on Biological Diversity (CBD) and The United Nations Convention to Combat Desertification (UNCCD) (United Nations, 1992a, 1992c, 1996). These agreements include specific commitments to eradicate each of the three environmental threats. All three have been ratified by more than 190 countries, in addition to the EU, and therefore are ranked as international treaties, which makes them legally binding for the signatory parties.

To supervise the application of these agreements, the Conference of the Parties (COP) were created. The COPs are top-level conferences organized by the UN that bring together all the parties to the three agreements. There is a type of COP for each of the agreements. Each of them examines the application of its corresponding agreement and the commitments assumed by the parties to comply with them. Of the three types of COPs, the best known is climate change one, hold on an annual basis, of which the most famous is COP21, held in Paris in 2015 and leading to the adoption of the Paris Agreement (United Nations, 2015).

The Paris Agreement is an international treaty that established the first global framework to combat climate change starting off in 2020. This agreement materializes in several specific goals, among which are: preventing the planet's temperature from increasing above 2 °C in comparison to pre-industrial levels, increase the economy's ability to adapt to the adverse effects of climate change, and ultimately to provide the financing needed to achieve climate-resilient, low-emission development.

Although less known than the climate change COPs, there are also biodiversity COPs and desertification COPs. Biodiversity COPs are based on the UN Convention on Biological Diversity (CBD) (United Nations, 1992a) and, unlike climate change COPs, are held every two years. The most notable biodiversity COP is COP15 in 2022 as it finished with the approval of the Post-2020 Global Biodiversity Framework (United Nations, 2022). This Framework defines global commitments to stop the loss of animal and plant species worldwide in the period 2020–2030. In effect, the Global Framework for Biological Diversity is for biodiversity, what the Paris agreement is for climate. It is for this reason that it is known as the "Paris Agreement, but of nature." Its objectives include, among others, conserving at least 30% of the world's terrestrial, marine and coastal areas or restoring at least 20% of each of the degraded freshwater, marine and terrestrial ecosystems.

Finally, and although less known than the other COPs, there are the COPs on desertification, held every two years. These COPs arise from the United Nations Convention to Combat Desertification (UNCCD), which constitutes, together with the United Nations Framework Convention on Climate Change (UNFCCC) and the Convention on Biological Diversity (CBD), the triptych of multilateral environmental agreements, created at the 1992 Earth Summit (United Nations, 1992a, 1992b, 1992c, 1996). The COPs on desertification prioritize on the problem of drought and rapid land degradation, especially on the African continent. These COPs seek to address measures that reinforce the commitment to achieve neutrality in land degradation by 2030, which implies a stable maintenance or an increase in the quality of land.

As it can be seen, as the need for an adequate international framework to address environmental threats has become clear, the international response has evolved.

The Paris Agreement and the Global Biodiversity Framework represent the recognition, with a degree of consensus never before achieved, that it is necessary to develop global actions that contribute to mitigating climate change and stopping the loss of biodiversity.

As it will be explained later, compliance with these agreements goes beyond countries and governments. Given the magnitude of the funds needed, the governments of the countries have called the companies and investors, as resource managers, to action.

3 Relationship Between the Financial Sector and Sustainability

The financial sector has been paying attention to the social and ethical impact of its actions for a long time. Banking and insurance companies have allocated resources to these purposes, as part of the philanthropic movement that emerged within the concept of Corporate Social Responsibility (CSR). However, the degree of ambition of the international agreements mentioned above (Paris Agreement and Post-2020 Global Biodiversity Framework) gives financial entities a renewed and important leading role.

Through their investments, financial companies have the ability to channel resources towards more sustainable sectors, companies and projects. In addition, they can operate sustainably and design and make available to the market products and services that promote sustainability.

At the same time, sustainability matters introduce risks that can have a negative impact on the performance of companies and the stability of the financial system in general. The intensification of environmental disasters (e.g. heat waves, floods, temperature rises or droughts) can lead to an increase in insurance claims or in defaults by people exposed to them, causing economic damage for insurers and banks. These risks, derived from the physical impacts of environmental phenomena, are known as physical risks.

Additional to physical risks, financial companies are exposed to other types of risks that arise from the process of change or transition of society towards a more sustainable model. These risks are known as transition risks and derive from politics and legal, technology, market and reputacional changes. Examples of transitional risks are the emergence of stricter legislation to preserve the environment that introduces strict limits on greenhouse gas emissions by companies or changing customer behavior changes in consumer preferences, who reduce their demand for products that harm the environment.

According to the above, a biunivocal relationship emerges between finances and sustainability. On the one hand, the financial sector, through its business decisions, has the capacity to promote and accelerate the change that a sustainable economy requires, reorienting capital flows towards sustainable investments and activities. On the other hand, sustainability must be considered by the financial sector because it introduces physical and transition risks that must be managed. This double paradigm gives rise to the birth of sustainable finance.

Sustainable finance considers how finance (investing and lending) interacts with economic, social and environmental issues (Scholtens, 2006). In the allocation role, finance can assist in making strategic decisions on the trade-offs between sustainable goals. Their importance lies in the fact that they allow economic development to continue without having a negative impact on the environment or society. Sustainable finance is therefore the tool to align the economy with sustainability.

To facilitate sustainable finance, different initiatives have been developed at international, European and national levels, which are subject to analysis in the following section.

4 Main Sustainability Initiatives Developed at International, European and National Level

In order to integrate sustainability into finance, and ensure that the financial sector supports the transition to a more sustainable world, different national, international and, above all, European initiatives have been developed. As of today, the vast majority of initiatives have been focused on the E—for environmental—of the ESG acronym. However, the second wave of initiatives is expected to be focused on developing the S and G.

Due to its current relevance, this section addresses the most important regulations where E is covered either exclusively or jointly with the other components of ESG.

4.1 International Initiatives: Guidelines on Reporting Climate and Nature-Related Information and International Sustainability Standards (IFRS S)

Since the adoption of the Paris Agreement (United Nations, 2015) and the Post-2020 Global biodiversity framework (United Nations, 2022), financial companies began to become aware of the need to act quickly to address the risks derived from climate change and the loss of biodiversity and their possible impact on financial stability and their businesses. In parallel, investors, in order to make informed investment decisions about companies, need to have reliable and up-to-date data on the impact of such risks on companies. Thus, these risks, and all the management instruments to manage them (governance, strategy, metrics, etc.) become something necessary for companies to communicate.

Given this need to have reliable non-financial information related to the management and impact on organizations of climate and biodiversity, the international initiatives of the Taskforce on Climate-Related Financial Disclosures (TCFD) and Taskforce on Nature-Related Financial Disclosures (TNFD) were designed.

TCFD is a working group created in 2015 by the Financial Stability Board (FSB), an international body in charge of supervising the proper functioning of the financial system by promoting supervisory and regulatory policies (Financial Stability Board, n.d.). The TCFD's mission was to develop a set of climate-related disclosure recommendations that financial companies could use when developing their reports to better inform investors. The set of TCFD recommendations were published in 2017 and are known as the TCFD framework (Taskforce on Climate-related Financial Disclosures, 2017).

The TCFD recommendations are structured around four core disclosure elements: climate change governance, climate change strategy, climate change risk management, and climate change metrics and targets.

The first element is based on the fact that, since climate change is another risk for companies, they must take it into account when making their decisions and must integrate it as a central part of their governance. Thus, the TCFD recommends that companies report on whether and how their Board of Directors includes climate change in their decisions and the functions of the Management Committee in relation to climate change.

The second element, strategy, establishes that organizations must disclose their risks and opportunities associated with climate change and the real and potential impacts of them on their business. Companies must also describe their resilience to different possible future climate scenarios, for example, in the hyphotesis of a rise in global temperatures of 2 °C or less.

The third element, climate change management, states that companies must disclose their processes for identifying, assessing and managing climate-related risks, and how these are integrated into their risk management system.

Finally, the fourth element recommends that companies disclose the metrics used to measure and manage climate risks, as well as the targets set to ensure that the increase in temperatures remains below 2 °C, in line with the objectives of the Paris Agreement.

Despite the TCFD recommendations are voluntary, they have gained relevance and turned into the generally accepted framework for companies to explain their approach towards climate-related risks.

A few years after the creation of the TCFD, it was seen that there was also a relevant degree of misinformation in the corporate sphere regarding the protection of nature. To tackle this, the Task Force on Nature-related Financial Disclosures (TNFD) was created in 2021. In reality, the TNFD aims to do for nature what the TCFD does for climate. Thus, the TNFD developed a framework of recommendations that describes what information companies should report regarding their relationship with nature and the impact on their business. This framework was published in September 2023 (Taskforce on Nature-related Financial Disclosures, 2023).

Just as the TCFD, the disclosure recommendations of TNFD are structured around four core elements that represent main areas of how companies operate: governance of nature, strategy regarding nature, nature risk management, and metrics and targets related to nature.

Through the use of both TNFD and TCFD recommendations frameworks, companies' reports are able to offer a holistic view to investors of how climate change and nature loss can affect the results of them and the actions that are taking to address these.

However, following both frameworks is voluntary for companies and in many cases the information reported by companies is insufficient, inconsistent and incomparable. To achieve the objectives agreed in the Paris Agreement and the Global Biodiversity Framework requires a higher level of commitment from companies. The need for mandatory disclosure of sustainability information by companies has led to the creation of the first international sustainability standards by the IFRS Foundation.

IFRS Foundation is a not-for-profit, public interest organisation established to develop high-quality, understandable, enforceable and globally accepted disclosure standards focused on the needs of investors and the financial markets (IFRS Foundation, n.d.). These standards must be followed by companies with accountability obligations. That is, those listed on stock exchanges and financial institutions, such as banks.

The standards are developed by two standard-setting boards, IASB ans ISSB. International Accounting Standards Board (IASB) was created in 2001 and its aim is develops accounting disclosure standards, denominated International Financial Reporting Standards (IFRS standards). IFRS standards establish the requirements to be disclosed by companies for the preparation of their financial statements.

Increasing needs for non-financial information in the markets led the IFRS Foundation to announce the creation of the International Sustainability Standards Board (ISSB) in 2021. ISSB is responsible for developing international disclosure standards for sustainability (referred to as IFRS S standards).

In June 2023 ISSB published the first two IFRS S standards, S1 and S2 (International Sustainability Standards Board, 2023a, 2023b). IFRS S1 establishes the general requirements for disclosure financial information related to sustainability and IFRS S2 sets out specific climate-related disclosures.

Companies must use IFRS S1 and S2 to prepare and report its sustainability-related financial disclosures on or after 1 January 2024. The TCFD framework is now being subsumed into the IFRS S2 about climate change and companies that apply this standard will meet the TCFD recommendations and so do not need to apply the TCFD in addition to it.

It is expected that after IFRS S1 and S2, ISSB will prepare S3 and subsequent ones to address other sustainability issues, such as the loss of nature included in the TNFD.

As explained in Sect. 4.2, in parallel to the IFRS S, the EU has published the Corporate Sustainability Reporting Directive (CSRD), which establishes that companies must present their non-financial information according to common European standards, in order to homogenize to a greater extent the sustainability information published by companies within the European Union.

4.2 *European Initiatives: EU Taxonomy, CSRD, and SFDR as Pivotal Triad of Regulations*

The ambitious agenda defined in the Paris agreement and the Global biodiversity framework has led the European Union to launch a European sustainable finance framework. Its main objective is to facilitate the channeling of capital towards sustainable investments and activities that help the European Union achieve the objectives agreed in previous international agreements.

The key pieces of this European framework are two. On the one hand, the Sustainable Finance Action Plan that the European Commission (EC) published in March 2018 (European Commission, 2018) and which has formed the basis of a significant number of regulatory initiatives in this matter. On the other hand, the Renewed Sustainable Finance Strategy that the EC published in June 2021 (European Commission, 2021) as a complement to the 2018 Action plan and as a logical consequence of the developments in sustainability since the moment of its publication.

Among all the initiatives that are based on the two previous documents, action plan and renewed strategy, this study refers to those especially relevant to the financial sector: the Taxonomy Regulation, the Sustainable Finance Disclosure Regulation (SFDR) and the Corporate Sustainability Reporting Directive (CSRD) (Regulation (EU) 2019/2088, 2019; Regulation (EU) 2020/852, 2020; Directive (EU) 2022/2464, 2022). Each of these initiatives responds to the three targets of the European Commission's Action Plan (refer to Table 1).

Table 1 Action Plan of the European Commission and its link with European sustainability initiatives

Targets of the EC's Action Plan	Key challenges	Actions of the EC	Regulation to comply with the actions
Target 1. Reorienting capital flows towards a more sustainable economy	No common definition of "sustainable activity"	Establish a unified EU classification system that provides clarity on which activities are sustainable	Taxonomy
Target 2. Mainstreaming sustainability into risk management	Investors often disregard sustainability factors or underestimate their impact	Clarify institutional investor duties to consider ESG issues in their investment decision process	SFDR
Target 3. Fostering transparency and long-termism in economic and financial activities	Too little information on corporate sustainability-related activities	Enhancing non-financial information disclosure	CSRD

Source Own elaboration

Taxonomy Regulation

Regulation (EU) 2020/852 on the establishment of a framework to facilitate sustainable investment, known as the Taxonomy Regulation or Green Taxonomy, is a European regulation that is part of the EC's Action Plan and came into force in July 2020. Its aim is to establish the criteria to determine wheter an economic activity is environmentally sustainable.

According to Taxonomy, an activity is sustainable if it contributes significantly to at least one of six environmental objectives, which are the following: (1) climate change mitigation, (2) climate change adaptation, (3) sustainable use and protection of water and marine resources, (4) transition to a circular economy, (5) pollution prevention and control, and (6) protection and restoration of biodiversity and ecosystems (Regulation (EU) 2020/852, 2020, p. 17).

Technical Screening Criteria (TSC) define the specific requirements for an activity to be considered as significantly contributing to a sustainability objective (Regulation (EU) 2020/852, 2020, pp. 17–22). These TSCs are being elaborated in secondary legislation called Delegated Acts (DAs). Currently, only the TSCs for objectives (1) and (2) have been developed (Commission Delegated Regulation (EU) 2021/2139, 2021). The TSCs for objectives (3) to (6) are still pending.

In addition, to be sustainable, the activity cannot cause significant harm to any of the other Taxonomy objectives, while respecting basic human rights and labour standards. For each activity, the TSC lay out thresholds to define compliance with do no significant harm (DNSH) to any of the other objectives.

Any undertaking subject to the Non-Financial Reporting Directive (NFRD) needs to disclose how, and to what extent, its activities are associated with activities that are considered as environmentally sustainable (Regulation (EU) 2020/852, 2020, p. 17).

One of the most important purposes of the Taxonomy is to avoid greenwashing phenomenon. The greenwashing phenomenon is defined as "*the intersection of two companies behaviours: poor environmental performance and positive communication about environmental performance*" (de Freitas Netto et al., 2020, p. 2). That is, greenwashing is the act used by certain companies to present themselves and their products as environmentally friendly entities, without being so.

Taxonomy allows investors and consumers to differentiate which activities (negatively) affect the environment, encouraging the financial sector to flow capital towards activities that truly respond to the needs of sustainable development.

Sustainable Finance Disclosure Regulation (SFDR)

Regulation (EU) 2019/2088 on Sustainability-related Disclosures in the Financial Services Sector, known as SFDR for its acronym in English, is a European regulation that is part of the EC's Action Plan and came into force in March 2021. It aims to help investors by providing more transparency on the degree to which financial products include environmental and social characteristics.

Its difference from the Taxonomy Regulation is that the SFDR Regulation does not define sustainable economic activities (which is the responsibility of the Taxonomy Regulation) but sustainable investments.

The scope of application of the SFDR is very extensive. It affects all financial entities that market investment products in any country in the euro area. It applies to financial advisors, as well as financial market participants including asset managers, investment companies and credit institutions providing portfolio management, as well as some asset owners (Regulation (EU) 2020/852, 2020, p. 8). This criteria brings pension funds, insurance companies and other actors within the scope of the SFDR regulation.

According to the SFDR, entities must regularly report on how they integrate sustainability risks into investment decision-making, how they integrate sustainability into remuneration policy, and the negative impacts on the environment or society that arise from their investment decisions (known as principal adverse incidents or PAIs) (Regulation (EU) 2020/852, 2020, pp. 9–10).

In addition to the information required at the entity level, the SFDR establishes disclosure obligations at the product level. In this sense, the SFDR establishes a classification of financial products into three categories, according to their degree of sustainability.

(i) Financial products with a sustainable investment objective (Regulation (EU) 2020/852, 2020, p. 12). Known as "article 9" or "dark green" products, they are the most sustainable category. They include an explicit sustainable investment objective applicable to the product and are in line with the European Commission's environmental strategy.
(ii) Financial products that promote environmental or social characteristics (Regulation (EU) 2020/852, 2020, p. 11). Denominated "Article 8" or "light green", they are products that, without including sustainable investment objectives, explicitly integrate environmental and/or social considerations in their management into their strategy. They promote objectives not recognized by the regulations of the European Commission but, nevertheless, they can contribute to achieving sustainability in a certain way.
(iii) Rest of products (Regulation (EU) 2020/852, 2020, p. 10). These are products that are not sustainable because they cannot fit into any of the previous categories. They are known as "grey products" or "article 6" products.

For all these financial products, the SFDR requires including in the pre-contractual information an explanation of whether and how sustainability risks are integrated, as well as whether negative impacts (Principal Adverse Impacts, or PAIs) on the environment and society at the product level are taken into consideration.

Furthermore, for Article 8 and 9 products, additional information must be provided on how the social or environmental characteristics are promoted in the first case, or how the sustainable objective is achieved in the second. These products also carry a requirement to disclose additional information in periodic reports.

The main objective of the SFDR is to catalog financial investment products according to their level of sustainability. That is, making the sustainability profile of these products more understandable through specific standards that establish what and how it should be communicated in this regard. This allows investors to know if they are investing in truly sustainable products and thus avoid greenwashing.

Corporate Sustainability Reporting Directive (CSRD)
Directive (EU) 2022/2464, known as Corporate Sustainability Reporting Directive or CSRD for its acronym in English, establishes new information requirements on sustainability reporting in the European Union. In fact, it expands on the relatively limited sustainability reporting disclosure requirements set out in the Non-financial Reporting Directive (NFRD for its acronym in English), which the CSRD replaces (Directive 2014/95/EU, 2014).

Although the CSRD came into force in January 2023, it is the typology of each company that determines the date on which they are obliged to report in accordance with this directive. Among the first obligated companies are those listed on the stock exchange, which must report under CSRD in 2025 (financial year 2024).

The final objective of the CSRD is to equate the non-financial information reporting with the financial information reporting, allowing to access reliable and comparable data on sustainability information. To do this, it imposes on companies the use of new standards and indicators that serve to standarise non-financial reporting, named European Sustainability Reporting Standards (ESRS). These ESRS are being elaborated by the European Commission in secondary legislation called Delegated Acts (DAs). Currently, a first set of 12 general ESRS has been developed, common for all sectors, and covering environmental, social and governance aspects (Commission Delegated Regulation (EU) of 31.7.2023 Supplementing Directive 2013/34/EU, 2023; Annex I to the Commission Delegated Regulation (EU) Supplementing Directive 2013/34/EU, 2023; Annex II to the Commission Delegated Regulation (EU) Supplementing Directive 2013/34/EU, 2023). As next steps, the Commission is expected to publish a second set of sector-specific ESRS.

Compliance with the ESRS Standards depends on the "Double Materiality" analysis. Materiality analysis is a process by which companies must identify which issues are of most concern to their stakeholders (e.g. climate change, business conduct, workers in the value chain, etc.). If an issue is important (material) to the stakeholders, the company must disclose information about it.

According to the double materiality concept, an issue can be material (and therefore, information about it must be disclosed) from two approaches or perspectives: either because it may materially affect the company's financial results (financial perspective) or because the company may affect the issue with its activity (impact perspective) (Raith, 2023).

Companies will need to evaluate whether ESRS 1 to 12 address topics that are material to their business, and will only need to disclose information about those ESRS that meet this characteristic (except ESRS 2—General Disclosures—which is mandatory disclosure). This means that if the issue of water was not material for a company due to the type of activity it carries out, this company would not have to disclose information about the ESRS E3—Water and marine resources—(refer to Table 2).

To guarantee the reliability and credibility of the information reported under the CSRD and each of its developmental ESRS, it is required to be ensured by an independent third-party expert. In the first phase, only limited verification of the information will be required. This means that the independent third-party expert expresses his

Table 2 Structure of the general European Sustainability Reporting Standards (ESRS)

Topics	Standards	
Cross-cutting	ESRS 1	General Requirements
	ESRS 2	General Disclosures
Environment	ESRS E1	Climate
	ESRS E2	Pollution
	ESRS E3	Water and marine resources
	ESRS E4	Biodiversity and ecosystems
	ESRS E5	Resource use and circular economy
Social	ESRS S1	Own workforce
	ESRS S2	Workers in the value chain
	ESRS S3	Affected communities
	ESRS S4	Consumers and end users
Governance	ESRS G1	Business conduct

Source Own elaboration

opinion on the information in negative terms, simply stating that no material errors have been found in the information. In a second phase, it is expected that the assurance can progress towards reasonable opinion, or, in other words, that the independent third-party expert expresses in positive terms that the information is prepared and presented appropriately, according to the requirements of the ESRS.

The ultimate goal of all of the above is to create a common reporting language for companies, which serves to increase transparency, protect against greenwashing and direct capital towards sustainable companies.

4.3 Spanish Initiatives: Climate Change and Energy Transition Law and the Transposition of the CSRD into Spanish Law

With the aim of guaranteeing compliance with the objectives of the Paris Agreement, of which Spain has been a Party since 2016, in 2021 the Spanish Law on Climate Change and Energy Transition was approved (Law 7/2021, 2021).

The final objective of this Law is to guide the decarbonization of the Spanish economy until reaching greenhouse gas emissiones neutrality in 2050. To this end, obligations are established for both the financial and business sector as well as for supervisors.

With respect to the financial and business sector, article 32 of the Law obliges companies to publish an annual report on the financial impact of climate change risks in their business, including the measures adopted to address such financial risks. Likewise, companies must publish specific decarbonization objectives aligned with the Paris Agreement. The content of this report must be determined by Royal

Decree within a period of two years from the approval of the Law. Nevertheless, the Royal Decree has deviated from its original deadline and is yet undergoing the approval process.

Regarding the obligations of the supervisors, the Law establishes that the Bank of Spain, the National Securities Market Commission (CNMV for its acronym in Spanish) and the General Directorate of Insurance and Pension Funds (DGSFP), within the scope of their respective powers, will prepare jointly, every two years, a report on the degree of alignment of the Spanish financial system with the climate goals of the Paris Agreement and European Union regulations.

Additionally, Spain is working on the transposition of the Corporate Sustainability Reporting Directive (CSRD) into Spanish law. As mentioned in Sect. 4.2, the CSRD is an EU directive whose objective is to establish which activities are environmentally sustainable (Regulation (EU) 2020/852 of the European Parliament of the Council, of 18 June 2020, on the Establishment of a Framework to Facilitate Sustainable Investment, 2020). Unlike EU regulations, which are directly applicable by member states, EU directives must be transposed by them, acquiring the status of law in their national systems. The deadline for transposition of the CSRD ends on July 6, 2024. Therefore, Spain must publish a Sustainability Corporate Information Law in its legal system before this date.

4.4 Connection Among International, European and National Sustainability Initiatives

All of the above initiatives were born to increase transparency in financial markets, protect against greenwashing and direct capital towards companies, sectors and activities that are truly sustainable.

In addition to this, there is a relationship between them derived from their contents. As they have different scopes (international, European or national), some of the initiatives have the same purpose or require the similar obligations as others, with the only difference of the different geographical area of application.

This causes some companies to have to simultaneously comply with different standards (national, European and international) that deal with the same issue but have different requirements and enforcement dates.

When this happens, the optimal solution is that by complying with one of the initiatives it is not necessary to fulfill both. This is the case of IFRS S and CSRD. In fact, IFRS S and CSRD are competing standards. Both have the same purpose (they define sustainability disclosure standards for companies) but with different scopes, international and European, respectively. Nevertheless CSRS have a high degree of interoperability with IFRS S requirements with the aim of ensuring that reporting under CSRD does not conflict or create a double reporting burden for companies if they wish to also comply with IFRS S. Therefore, companies that comply with the CSRD, implicitly comply with IFRS S.

At the same time, the CSRD is interconnected with the SFDR and the Taxonomy. On the one hand, the CSRD ensures that companies' sustainability reports include the information necessary to enable market participants to comply with the reporting requirements of the SFDR. On the other hand, the CSRD includes the requirement to report in line with the information of the Taxonomy.

Spanish regulation is also connected to international and European initiatives. On the one hand, the European CSRD directive must be transposed into Spanish Law within the period set in this.

On the other hand, the Spanish law on climate change and energy transition requires companies to publish annually a climate report whose content is similar to the information requested by the CSRD (in ESRS E1) and IFRS S (in IFRS S2). However, no official body has yet issued comments on the possibility of only having to comply with one of these initiatives (Fig. 2).

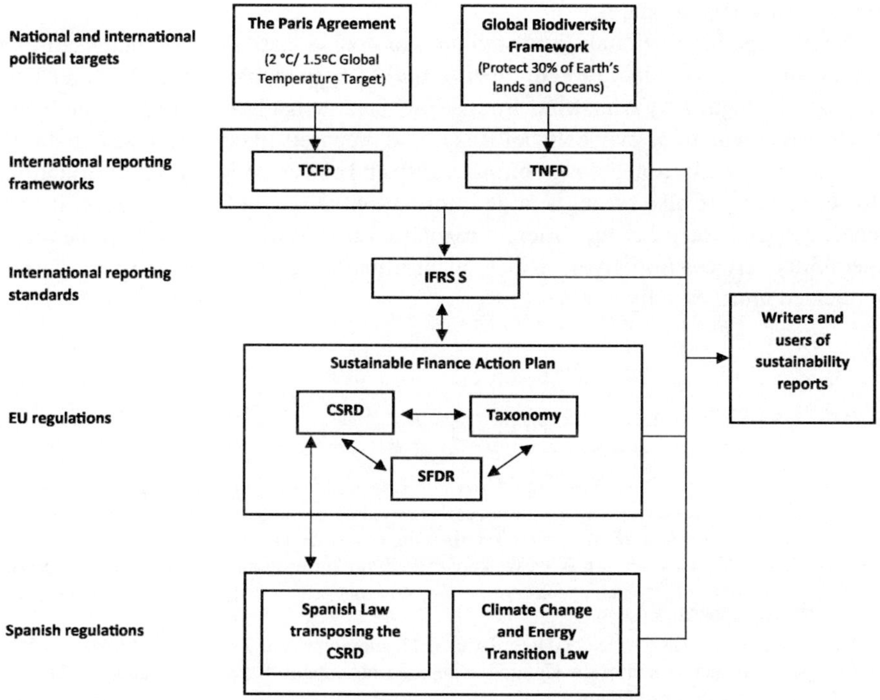

Fig. 2 Architecture of international, European and Spanish initiatives in the preparation of sustainability information. *Source* Own elaboration

5 Conclusion

Since the twentieth century, the humanity stanpoint has shifted towards the alleviation of the environmental situation in which the planet is inmersed, as the sustainability summits and conventions reflect. The Paris Agreement and the Post-2020 Global Biodiversity Framework mark historic milestones in these international negotiations, establishing global objectives to avoid climate change and combat biodiversity loss, respectively.

To comply with both agreements, on the one hand, the involvement of the financial sector is required, since it acts as a vector of transformation, redirecting capital towards sustainable investments and activities that help achieve the climate and nature global objectives. On the other hand, regulatory measures are necessary in the corporate disclosures on sustainability to ensure that the information is consistent and reliable and allow the financial market to identify those activities that really contribute to sustainable development.

All players, international, European and, of course, national, are immersed in a process of regulatory definition in order to build a sustainable financial future. This has caused a regulatory whirlwind around ESG factors that has focused especially on E. The new regulations may help both users and preparers of corporate sustainability information benefit from greater comparability and consistency of ESG information. However, they are also posing a large information workload for companies. In this sense, it is necessary that the different regulatory bodies take into account the interoperability between initiatives, so that the information demands for companies is not duplicated unnecessarily.

References

Aguilera, R. V., Rupp, D. E., Williams, C. A., & Ganapathi, I. (2007). Putting the S back in corporate social responsibility: A multilevel theory of social change in organizations. *Academy of Management Review, 32*(3), 836–863. https://doi.org/10.5465/AMR.2007.25275678

Aguinis, H., & Glavas, A. (2012). What we know and don't know about corporate social responsibility: A review and research agenda. *Journal of Management, 38*(4), 932–968. https://doi.org/10.1177/0149206311436079

Annex I to the Commission Delegated Regulation (EU) supplementing Directive 2013/34/EU of the European Parliament and of the Council as regards sustainability reporting standards, Official Journal of the European Union.

Annex II to the Commission Delegated Regulation (EU) supplementing Directive 2013/34/EU of the European Parliament and of the Council as regards sustainability reporting standards, Official Journal of the European Union.

Auld, G., Bernstein, S., & Cashore, B. (2008). The new corporate social responsibility. *Annual Review of Environment and Resources, 33*, 413–435. https://doi.org/10.1146/annurev.environ.32.053006.141106

Bansal, P., & Roth, K. (2000). Why companies go green: A model of ecological responsiveness. *Source: The Academy of Management Journal, 43*(4). https://www.jstor.org/stable/1556363?seq=1&cid=pdf

Commission Delegated Regulation (EU) of 31.7.2023 supplementing Directive 2013/34/EU of the European Parliament and of the Council as regards sustainability reporting standards—European Sustainability Reporting Standards (ESRS), Official Journal of the European Union (2023).

Commission Delegated Regulation (EU) 2021/2139 of 4 June 2021 supplementing Regulation (EU) 2020/852 of the European Parliament and of the Council by establishing the technical screening criteria for determining the conditions under which an economic activity qualifies as contributing substantially to climate change mitigation or climate change adaptation and for determining whether that economic activity causes no significant harm to any of the other environmental objectives, Official Journal of the European Union (2021).

de Freitas Netto, S. V., Sobral, M. F. F., Ribeiro, A. R. B., & da Luz Soares, G. R. (2020). Concepts and forms of greenwashing: A systematic review. In *Environmental Sciences Europe, 32*(1). Springer. https://doi.org/10.1186/s12302-020-0300-3

Directive 2014/95/EU of the European Parliament and of the Council of 22 October 2014 amending Directive 2013/34/EU as regards disclosure of non-financial and diversity information by certain large undertakings and groups, Official Journal of the European Union (2014).

Directive (EU) 2022/2464 of the European Parliament and of the Council of 14 December 2022 as regards corporate sustainability reporting, Official Journal of the European Union (2022).

European Commission. (2018). Action Plan: Financing sustainable growth. In *EUR-Lex—52018DC0097—ES.*

European Commission. (2021). Strategy for financing the transition to a sustainable economy. *EUR-Lex—52021DC0390—ES.*

Financial Stability Board. (n.d.). https://www.fsb.org/

Gunningham, N. (2009). Shaping corporate environmental performance: A review. *Environmental Policy and Governance, 19*(4), 215–231. https://doi.org/10.1002/eet.510

IFRS Foundation. (n.d.). https://www.ifrs.org/

International Sustainability Standards Board. (2023a). Climate-related Disclosures IFRS S2 IFRS ® Sustainability Disclosure Standard International Sustainability Standards Board. https://www.ifrs.org/content/dam/ifrs/publications/pdf-standards-issb/english/2023/issued/part-a/issb-2023-a-ifrs-s2-climate-related-disclosures.pdf?bypass=on.

International Sustainability Standards Board. (2023b). General Requirements for Disclosure of Sustainability-related Financial Information IFRS S1 IFRS ® Sustainability Disclosure Standard International Sustainability Standards Board. https://www.ifrs.org/content/dam/ifrs/publications/pdf-standards-issb/english/2023/issued/part-a/issb-2023-a-ifrs-s1-general-requirements-for-disclosure-of-sustainability-related-financial-information.pdf?bypass=on

McWilliams, A., & Siegel, D. (2001). Corporate social responsibility: A theory of the firm perspective. *Academy of Management Review, 26*(1), 117–127. https://doi.org/10.5465/AMR.2001.4011987

Rahman, N., & Post, C. (2012). Measurement issues in environmental corporate social responsibility (ECSR): Toward a transparent, reliable, and construct valid instrument. *Journal of Business Ethics, 105*(3), 307–319. https://doi.org/10.1007/s10551-011-0967-x

Raith, D. (2023). The contest for materiality. What counts as CSR? *Journal of Applied Accounting Research, 24*(1), 134–148. https://doi.org/10.1108/JAAR-04-2022-0093

Regulation (EU) 2019/2088 of the European Parliament and of the Council of 27 November 2019 on sustainability-related disclosures in the financial services sector, Official Journal of the European Union (2019).

Regulation (EU) 2020/852 of the European Parliament of the Council, of 18 June 2020, on the establishment of a framework to facilitate sustainable investment, Official Journal of the European Union (2020).

Scholtens, B. (2006). Finance as a driver of corporate social responsibility. *Journal of Business Ethics, 68*(1), 19–33. https://doi.org/10.1007/s10551-006-9037-1

Sheehy, B. (2015). Defining CSR: Problems and solutions. *Journal of Business Ethics, 131*(3), 625–648. https://doi.org/10.1007/s10551-014-2281-x

Taskforce on Climate-Related Financial Disclosures. (2017). *Recommendations of the Taskforce on climate-related financial disclosures.*

Taskforce on Nature-Related Financial Disclosures. (2023). Recommendations of the Taskforce on nature-related financial disclosures. https://tnfd.global/Tnfd-Publications/

Trumpp, C., Endrikat, J., Zopf, C., & Guenther, E. (2015). Definition, conceptualization, and measurement of corporate environmental performance: A critical examination of a multidimensional construct. *Journal of Business Ethics, 126*(2), 185–204. https://doi.org/10.1007/s10551-013-1931-8

United Nations. (1992a). Convention on Biological Diversity. https://www.cbd.int/Convention/Text/

United Nations. (1992b). Report of the United Nations Conference on Environment and Development. https://digitallibrary.un.org/Record/168679

United Nations. (1992c). United Nations Framework Convention on Climate Change. https://unfccc.int/process-and-meetings/the-convention/history-of-the-convention/convention-documents

United Nations. (1996). United Nations Convention to Combat Desertification. https://www.unccd.int/resource/convention-text

United Nations. (2015). Paris Agreement. https://unfccc.int/Sites/Default/Files/English_paris_agreement.Pdf

United Nations. (2022). Kunming-Montreal Global Biodiversity Framework. https://www.cbd.int/Article/Cop15-Final-Text-Kunming-Montreal-Gbf-221222. https://www.cbd.int/conferences/2021-2022/cop-15/documentsasdocument:CBD/COP/15/L25

World Commission on Environment and Development. (1987). *Report of the World Commission on Environment and Development: Our Common Future Towards Sustainable Development 2. Part II. Common Challenges Population and Human Resources 4.*

Irene Ricote García is Head of Sustainability and Rating Agencies at Catalana Occidente Group, a worldwide leading insurer based in Madrid, Spain. Her scope of work encompasses steering and implementing the Group sustainability strategy. Thanks to her profound knowledge of the regulatory framework and extensive experience in various roles within the insurance sector, such as non-life insurance actuary, she has been able to lead the transformation of the Group to a recognized better performer.

From the Sustainability standpoint, she is the General Secretary of the Sustainability Committee in the organization, coordinating non-life and life insurance business lines. In relation to the Rating Agencies role, she manages the interactions with credit rating entities (Moody's, S&P, Fitch, amongst others) and with recently emerged ESG rating agencies.

Academically, she holds a double degree in Law and Business Management and a master's degree in Actuarial and Financial Sciences, which entitled her to be a member of the Association of Actuaries (with no. 4082). Her academic titles were issued by the Complutense University of Madrid, where she defended her master's thesis on IFRS17 implementation and adaptation challenges over actuarial models, published in the university repository with honors. Also in this institution, she presented her Solvency II thesis as part of the closing of the double degree in Law and Business Management, also with honors.

She is currently a member of sustainability workshops in UNESPA (the Spanish association of insurers and reinsurers) and is helping to define sectorial statements and standpoints, whose adaptation and implementation to the Group business will lead in the near future.

Sustainable Practices and Shareholder Value Creation in FinTech Firms: International Evidence

María Mar Miralles-Quirós, José Luis Miralles-Quirós, and Azahara Gil-Corbacho

1 Introduction

FinTech, as the name suggests, is the fusion of finance and technology. However, the concept of a FinTech firm is more difficult to specify. Following Goldstein et al. (2019), we can indicate that these are young start-ups or large established companies in the technology field that are breaking into the financial business by providing new products and services to financial users such as innovations in payments (peer-to-peer systems, cryptocurrencies), lending and equity investments (crowdfunding) or digital financial advice and wealth management (robo-advising), among others.

These innovative companies have caused major changes in the classical financial system, as they are fulfilling the same economic function as traditional financial institutions more efficiently and with lower operating costs (Gil-Corbacho et al., 2023; Romanova and Kudinska, 2016; Sánchez, 2022; Zveryakov et al., 2019). In addition, the loss of trust in traditional financial firms as a consequence of the 2008 global financial crisis contributed to the development and expansion of this new type of business, known as the FinTech revolution (Gomber et al., 2017).

In recent years, investments in FinTech companies have been growing not only in the U.S. but also in Europe and the Asia–Pacific (Baltgailis & Simakhova, 2022; KPMG, 2022). Specifically, global investments in financial technology have increased from $55.3 million in 2019 to $210 million in 2021 (Atayah et al., 2023).

M. M. Miralles-Quirós (✉) · J. L. Miralles-Quirós
Faculty of Economic and Business Sciences, University of Extremadura, Av. Elvas S/N, 06006 Badajoz, Spain
e-mail: marmiralles@unex.es

J. L. Miralles-Quirós
e-mail: miralles@unex.es

A. Gil-Corbacho
Faculty of Law, University of Extremadura, Av. de Las Letras S/N, 10004 Cáceres, Spain
e-mail: azaharagc@unex.es

M. d. C. Valls Martínez and J. M. Santos-Jaén (eds.), *Environmentally Sustainable Production*, https://doi.org/10.1007/978-3-031-52656-5_2

Therefore, we can state that FinTech companies have become an essential segment of the economy and stock markets globally. Although the Financial Stability Board (FSB) stated in 2017 that this type of business could generate global instability by being more vulnerable to economic shocks than traditional financial institutions (Financial Stability Board, 2017; Ng & Kwok, 2017), FinTech firms currently present significant opportunities and challenges for sustainable economic development. For instance, Arner et al. (2020) highlight the multiple possibilities of the FinTech industry in financial inclusion. On the other hand, Hoang et al. (2022) highlight the important role that FinTech companies can play in environmental sustainability, as they can be key to revitalizing investments in renewable energy or financing green projects.

However, to the best of our knowledge empirical research on the sustainability practices of these types of companies is scarce. Moreover, these studies focus on FinTech companies listed on the U.S. stock market and do not specifically analyze the value relevance of such practices. To fill this gap in the literature, this chapter aims to analyze the contribution to sustainable development of an international sample of FinTech companies and whether these sustainability practices generate value. Specifically, we analyze the score given by the Refinitive Eikon database to these types of companies in relation to their sustainable activities related to the environment, society and corporate governance (ESG). In addition, we also consider it essential to analyze not only the valuation of these activities but also how they are being disclosed by the company to its stakeholders. Therefore, we analyze whether these companies prepare and disclose sustainability reports, whether they contribute to making them comparable with other companies in the sector by using the indicators developed by the Global Reporting Initiative (GRI) generally accepted and followed by financial stakeholders and, finally, we analyze whether the information provided has a higher level of transparency through verification by an external auditor.

Our overall results reflect that not all sustainability activities considered generate value. In particular, our findings show that the practices that generate value are disclosure of information using GRI standards, verification of such information and environmental activities. These findings are relevant for academics, the managers and shareholders of these companies, but also for policymakers and citizens as well.

The structure of this chapter is as described below. Section two contains the theoretical background and a brief review of the scarce existing literature. Section three describes the database and the methodology used to carry out the empirical analysis. Section four presents the results obtained. Finally, the last section reflects the conclusions derived from the work as a whole.

2 Literature Review

In recent years, numerous empirical studies have emerged that analyze FinTech firms from various perspectives. From the economic point of view, studies that analyze the role of FinTech firms among consumers, financial competitors and regulators stand out (Anifa et al., 2022; Milian et al., 2019; Sangwan et al., 2019). From the stock market point of view, studies that analyze the transmission of risks between these companies and other financial assets, as well as the stock market performance of these companies stand out (Chen et al., 2022; Franco et al., 2020; Gil-Corbacho et al., 2023; Le et al., 2021a, 2021b; Li et al., 2020; Najaf et al., 2021).

Apart from the empirical studies indicated above, a recent and promising line of research is that which analyzes the contribution of FinTech companies to sustainable development. The few existing studies are those of Atayah et al. (2023), Najaf et al. (2023) and Toumi et al. (2023). Specifically, Atayah et al. (2023) and Najaf et al. (2023), in two similar empirical studies, analyze the impact of ESG disclosure on the stock performance of a set of 193 FinTech and non-FinTech companies listed on the U.S. stock market over the period 2010–2019. Their findings reflect that the ESG disclosure of FinTech firms is lower, as well as their profitability measured through multifactor risk models. On the other hand, Toumi et al. (2023) first analyze the environmental disclosure practices of the 48 Fintech firms from the well-known Nasdaq Financial Technology Index (KFTX) during the period 2011–2021. Specifically, they compare these disclosure practices with those performed by a matched sample of non-FinTech firms. Finally, they analyze whether investors are sensitive to these disclosure practices. Their findings suggest that Fintech firms had better financial performance and environmental disclosure than non-Fintech firms during the pandemic period.

The three studies mentioned above focus on analyzing North American FinTech companies. However, as reflected in the introduction section, this is a global phenomenon. Therefore, we consider it essential to present international empirical evidence. Moreover, unlike previous empirical evidence, the objective of this research is to study the relevance of the value of not only disclosure practices but also verification of ESG information of a broad sample of FinTech companies.

From a theoretical point of view, there are two conflicting theories on whether sustainability practices have value relevance. These are the neoclassical theory advocated by Friedman (1962, 1970) and the stakeholder theory advocated by Freeman (1984). Specifically, Friedman (1970) argued that the responsibility of the firm is to use its resources in activities aimed at maximizing its profits, acting in accordance with the basic rules of society embodied in law and ethical mores. This theory holds, therefore, that company managers need only be concerned with generating maximum value for shareholders. Freeman (1984), on the other hand, argues that company managers should not only be concerned about their shareholders but also about everyone involved in the company. Therefore, managers should create value for all stakeholders of the firm, both internal (employees and shareholders) and external (consumers and local communities, among others).

Freeman's (1984) position has been supported by numerous authors in a more or less exalted form. On the one hand, Post et al. (2002) pointed out that companies should prioritize ESG issues regardless of whether this has value relevance or not. On the other hand, Porter and Kramer (2011) introduced the famous term of shared value creation for business and society. In addition, stakeholder theory has been supported by other theories such as legitimacy theory, information asymmetry theory and signaling theory. This is because companies use communication with their stakeholders as a means to manage their legitimacy (Deegan, 2019; Silva, 2021). In this sense, sustainability reports prepared and disclosed by companies are a one-way communication channel from the corporation to its stakeholders, providing a signal of its commitment to sustainable development and reducing information gaps between managers and shareholders. However, this information must be truthful, transparent and comparable and not simply limited to compliance with the law on non-financial disclosure.

As Miralles-Quirós et al. (2019, 2021) point out, the practice of disclosing sustainability information is especially common among listed companies due to the pressure exerted by financial stakeholders who demand reliable and comparable information to help them make their investment decisions or prepare their investment advice. Theoretically, by disclosing this information, listed companies can reduce information asymmetries between managers and financial stakeholders (Healy & Palepu, 2001), as well as their adverse selection costs (Diamond & Verrecchia, 1991; Kim & Verrecchia, 1994). In addition, financial analysts can reduce their forecasting errors because this information can mitigate stakeholder uncertainty about firms' future economic returns and their associated risks (Lang & Lundholm, 2000; Lang & Maffett, 2011). Therefore, all these aspects have direct positive effects, such as increasing the number of potential investors, increasing equity capital or attracting greater analyst coverage (Merton, 1987).

Finally, it should be noted that studying sustainability reporting assurance services responds to the demands of Kolk and Perego (2010) and Cohen and Simnett (2015), among others, to examine not only the adoption of sustainability reports but also the quality and reliability of the information provided. This is a prominent area of research, especially after the global financial crisis, which caused the loss of credibility and trust in sustainability information disclosed by companies themselves (Seguí-Mas et al., 2018). However, most studies on sustainability reporting assurance focus on analyzing the factors that influence the decision to start the process for external verification of this information (García-Sánchez, 2021).

3 Database

The initial sample of companies considered was formed by those companies that make up the Nasdaq Financial Technology Index (KFTX). This index is the one that previous empirical studies have mostly used to select this type of company (Franco et al., 2020; Gil-Corbacho et al., 2023; Le et al., 2021a, 2021b; Li et al., 2020; Toumi et al., 2023). However, the exclusive use of this index posed two initial problems: (i) the composition and access to the information of the component companies only allowed working with 43 companies; (ii) this index does not cover an international context but only considers companies listed on the Nasdaq, the North American market dedicated to the technology companies segment.

For these two reasons, we decided to extend the sample by including the composition of other international FinTech company indices. Specifically, we used the composition of the following indices:

- The Indxx Global Fintech Index, designed to track the performance of companies listed in developed markets that offer technology-driven financial services that are disrupting existing business models in the financial services and banking sectors.
- Two MSCI family indexes, namely Acwi Imi FinTech Inn Select ESG and FinTech ESG, which are characterized by screening FinTech companies associated with socially responsible investing.
- The regional indexes dedicated to FinTech companies prepared by Refinitiv Eikon, namely the Refinitiv Asia Pacific Finance Technology, Refinitiv Europe Finance Technology and Refinitiv North America Finance Technology indexes.

The final sample is made up of 202 companies. As can be seen in Table 1, although the objective was to provide an international sample of companies, a high percentage (66%) is made up of companies from the United States, and the rest is widely distributed among companies from countries in Europe, Asia and Oceania.

Another aspect to highlight is that the selected companies are very varied, belonging to various categories of digital financial services such as:

- Data analytics: providers of integrated financial information and solutions through data analytics.
- Digital banking: entities that offer banking services in the cloud.
- Wealth management: companies offering advice and/or management of financial assets for individuals.
- Means of payments: companies focused on facilitating real-time digital payments.
- Technology solutions: companies that provide software and information technology solutions for business processes.
- Trading: providers of market infrastructure and tradable products.

Table 1 Fintech firms grouped by region and country

America		Asia		Europe		Oceania	
Brazil	2	China	15	Denmark	1	Australia	8
Canada	8	Hong Kong	3	France	2	New Zealand	1
United States	124	India	3	Germany	2		
		Israel	2	Italy	1		
		Japan	8	Netherlands	4		
		Singapore	3	Sweden	1		
		South-Korea	2	Switzerland	2		
		Taiwan	3	United Kingdom	7		
Total	134		39		20		9

Source Own elaboration

However, it should also be noted that some of these companies have evolved into a conglomerate of digital financial services and, therefore, it is difficult to include them in a specific category of FinTech activity as they provide a wide variety of financial services.

Another important aspect to highlight is the fact that a significant percentage of these companies have gone public in recent years and more specifically in 2021. This is an important limitation of the sample under study, since, to avoid the results being conditioned by aspects related to the IPO of these companies, they have been eliminated from the final study sample.

On the other hand, the information related to the preparation and disclosure of sustainability reports, as well as the environmental, social and corporate governance performance of the FinTech companies in the sample was obtained from the Refinitiv Eikon database.

Specifically, in relation to the preparation and disclosure of sustainability reports by these FinTech companies, we have selected from the Refinitiv Eikon database the data corresponding to the issuance of sustainability reports, the use of GRI standards to report their contribution to sustainability and the fact that the information provided has been verified or assured by an auditor external to the company. As can be seen in Table 2, not all FinTech companies listed in the main developed markets prepare sustainability reports, and even fewer companies use GRI standards or verify the information provided to their stakeholders. On the other hand, it should also be noted that in all these aspects the trend has been increasing over the last few years.

On the other hand, in relation to the environmental, social and corporate governance performance measures used, it should be noted that the Refinitiv Eikon database calculates these scores as the weighted average of the scores achieved on more than 70 key performance indicators calculated from more than 400 observation points.

The environmental variable measures the company's contribution to issues such as minimizing the use of resources, reducing emissions and innovating environmental products and services. The social performance variable measures the actions taken by these companies in relation to providing customers with financial products and services that integrate ethical principles, their commitment to the community and employees that create a healthy and safe work environment, respecting diversity

Table 2 Fintech firms that prepare and disclose sustainability reports

	Fintech firms listed	Sustainability report	GRI standards	Assurance
2015	124	24	16	7
2016	128	32	22	10
2017	133	39	25	15
2018	145	51	33	20
2019	156	67	39	25
2020	174	89	50	30
2021	201	103	53	35

Source Own elaboration

Table 3 Environmental, social and corporate governance scores

	Fintech firms with score	ESG	Environmental	Social	Corporate Governance
2015	88	41.15	27.17	44.55	45.72
2016	106	41.86	24.58	45.07	47.83
2017	115	44.71	22.15	48.31	50.36
2018	128	44.36	23.48	48.02	49.03
2019	140	45.50	26.74	49.16	49.06
2020	160	47.93	28.95	51.83	51.24
2021	154	51.84	33.06	56.20	54.47

Source Own elaboration

and human rights, and providing equal opportunities. Meanwhile, corporate governance performance measures whether these companies include social responsibility values in their overall corporate vision and strategy, decision-making processes and communication practices in terms of sustainability reporting and disclosure.

As can be seen in Table 3, as with the previous sustainability variables considered, not all the companies in the initial sample have information on environmental, social and corporate governance performance. The positive aspect to note is that there is a slightly increasing trend in the scores achieved by the companies for which information is available. Finally, the environmental score is lower than the score obtained in the other two aspects considered, social and corporate governance.

4 Methodology

Once the sample of FinTech companies has been selected and the sustainability information has been obtained, this subsection indicates the methodology used to analyze whether these measures contribute to the creation of shareholder value. To this end, we use Tobin's Q ratio, which relates the market's assessment of the capacity to generate shareholder return (measured by the company's market capitalization or market value) to the replacement cost of assets (i.e., the cost of acquiring the company's productive capacity) approximated by the market value of equity and debt divided by the book value of the assets. Thus, if the Q ratio is greater than unity, the company is creating value, since the market is valuing the company higher than the replacement cost of its productive assets. Otherwise, the value is destroyed. As Jiao (2010) points out, this ratio is the most appropriate for this type of study since it is a variable based not only on historical data but also on future expectations.

Once the dependent variable was established, as explanatory variables we considered, in addition to the aforementioned sustainability variables, a series of moderating or control variables: a size variable to control for possible scale effects, a debt variable to control for the impact of the capital structure and a profitability variable to control for the economic moment the company is going through. Specifically, we used the number of employees as a proxy for the size of the firm, and long-term debt as a proxy for the level of indebtedness of the firm as a percentage of the firm's total capital. Finally, the profitability variable used is the return on assets, calculated as the ratio between the economic profit and the total assets of the company. Thus, the model to be estimated is as follows:

$$Q_{it} = \beta_0 + \beta_1 \cdot Q_{it-1} + \beta_2 \cdot Sustainablepractices_{it} + \beta_k \cdot \sum_{k=1}^{3} Control_{it}$$

It should be noted that the one-period lagged dependent variable has been included in the empirical modeling due to its persistent nature. In addition, the estimations have been carried out using the panel data technique. Unlike more simplistic techniques, panel data analysis allows us to model the unobservable heterogeneity that exists between companies. This is achieved by decomposing the error term into three components: a first component, representative of each firm, which includes those unobservable effects that affect it alone (unobservable heterogeneity); a second component, representative of the shocks that occurred in each year of the study and that affect all firms equally (macroeconomic effects); and a third component, which is a random variable. In addition, we thus managed to lengthen the sample, which means a substantial increase in the number of observations, the degrees of freedom of the model and, therefore, the consistency of the results. In addition, we use the System Generalized Method of Moments (System GMM) estimator that uses instrumental variables to estimate the model consistently (Arellano & Bond, 1991), thus avoiding the problem of endogeneity in the variables, i.e., correlation with the error term.

5 Empirical Results

As a preliminary step to the presentation and discussion of the empirical results obtained, the descriptive statistics of the variables used are shown below. Specifically, Table 4 presents the data relating to the mean, median, maximum, minimum and standard deviation of the variables.

The purpose of this table is to reflect the absence of outlier observations that could be conditioning the results. In this regard, we should point out several facts that motivated the final reduction in the number of companies considered in the empirical study. Firstly, it was decided to include in the final sample those companies that had at the beginning of the period a minimum duration on the stock exchange of 5 years, to avoid the results being conditioned by this fact. Secondly, all companies for which no sustainability information was available were eliminated from the sample. Finally, those companies for which there were outliers for the economic-financial variables considered were eliminated from the sample. In short, 72 FinTech companies were used in the empirical study.

Table 4 Descriptive statistics

	Media	Median	Maximum	Minimum	Standard deviation
Tobin's Q	6.07	4.02	53.71	0.40	6.36
Sustainability Info	0.48	0.00	1.00	0.00	0.50
GRI standards	0.31	0.00	1.00	0.00	0.46
Assurance	0.18	0.00	1.00	0.00	0.38
ESG performance	47.14	50.72	92.11	0.00	22.36
Environmental	28.85	20.72	94.35	0.00	28.35
Social	49.79	53.55	97.78	0.00	24.84
Government	52.09	55.71	96.12	0.00	25.05
Size	8.84	8.81	12.85	4.34	1.52
Leverage	0.34	0.35	0.91	0.00	0.22
Profitability	8.06	5.56	74.48	– 21.82	9.42

Source Own elaboration

Finally, we present the results obtained from the testing of the model presented above. The results are reflected in two tables. Table 5 presents the results related to sustainability disclosure activities. Table 6, on the other hand, reflects the results related to environmental, social and corporate governance performance. In both cases, the estimated coefficients of each of the explanatory variables are presented along with their *p*-value.

At the end of both tables, to reflect the correct specification of the models, we present the joint significance test of the coefficients of the explanatory variables,

Table 5 Impact of sustainability disclosures on value

	Model 1		Model 2		Model 3	
	Coef	*p*-value	Coef	*p*-value	Coef	*p*-value
Q (−1)	0.9700***	(0.00)	0.964***	(0.00)	0.955***	(0.00)
Sustainability Inf	0.2310	(0.47)				
GRI standards			0.624*	(0.06)		
Assurance					1.160***	(0.00)
Size	0.0084	(0.94)	– 0.042	(0.72)	– 0.099	(0.39)
Leverage	0.0116	(0.13)	0.012	(0.11)	0.014*	(0.06)
Profitability	0.0324*	(0.07)	0.033*	(0.07)	0.033*	(0.06)
Test F	459.01***		463.65***		471.44***	
Sargan test	23.539		17.196		23.670	
m_1	0.105*		0.105*		0.099*	
m_2	0.022		0.027		0.008	

Source Own elaboration
Note: *** significance at 1%, ** significance at 5% and * significance at 10%

Table 6 Impact of ESG performance on value

	Model 4		Model 5		Model 6		Model 7	
	Coef	*p*-value	Coef	*p*-value	Coef	*p*-value	Coef	*p*-value
Q (−1)	0.967***	(0.00)	0.964***	(0.00)	0.967***	(0.00)	0.969***	(0.00)
ESG	0.011	(0.15)						
Environmental			0.010*	(0.09)				
Social					0.008	(0.23)		
Government							0.007	(0.20)
Size	− 0.055	(0.66)	− 0.098	(0.46)	− 0.042	(0.74)	0.001	(0.99)
Leverage	0.012	(0.10)	0.013*	(0.07)	0.011	(0.11)	0.011	(0.12)
Profitability	0.032*	(0.07)	0.033*	(0.06)	0.033*	(0.06)	0.031*	(0.07)
Test F	461.37***		462.62***		460.41***		460.66***	
Sargan test	23.686		22.348		9.423		17.897	
m_1	0.107*		0.104*		0.106*		0.105*	
m_2	0.020		0.023		0.021		0.018	

Source Own elaboration
Note: *** significance at 1%, ** significance at 5% and * significance at 10%

the Sargan test of overidentification restrictions with which we test the absence of correlation between the instruments and the error term and the m_1 and m_2 statistics developed by Arellano and Bond (1991) to test the absence of first and second order serial correlation in the residuals in first differences.

As can be seen in Table 5, the coefficient that has the greatest explanatory power over Tobin's Q variable is this variable lagged by one period. As indicated in the methodological section, this is a very persistent variable. As for the sustainability information disclosure variables, it is observed that issuing social responsibility reports does not add value for the company, but the use of GRI standards or the verification of the information reported by an external auditor does. These results corroborate those obtained for other sectors of activity. As for the control variables, it is the profitability and indebtedness variables that have explanatory power over the company's value creation. On the other hand, the size variable, measured by the number of employees, is not significant in any case.

Similar results are shown in Table 6, which reflects the impact of environmental, social and corporate governance performance on the value creation variable. In this case, as can be seen, the only significant sustainability variable is the one associated with environmental performance. This result is not surprising because, as indicated in the theoretical background section, shareholders and other stakeholders are paying increasing attention to environmental performance. Although it may seem that FinTech companies cannot play any relevant role at the environmental level, they can do so mainly through environmental innovation in their financial products and services.

6 Conclusion

Finally, we should point out that this is an initial work that attempts to connect two aspects that are clearly relevant today: sustainability and the technological revolution applied to the financial sector. Although there is no doubt about the work that companies have to do to contribute to sustainable development, there is hardly any evidence of how technology-based companies that provide financial services through innovative solutions are doing it. We believe it is essential to analyze this very specific business field. Among the research options we could have carried out, we considered it appropriate to analyze whether sustainability-related activities provide shareholder value. The reason was to try to combine two objectives, an impact objective and an economic objective, and thus contribute to the creation of sustainable value.

The results obtained indicate that not all disclosure activities or sustainable performance generate value for the FinTech company. Only the use of GRI standards, the verification of reports and environmental performance generate value for these companies. These preliminary results may be interesting and useful for researchers, managers of these companies, public policymakers and investors. However, we believe that further research is needed to obtain conclusive results. It should be noted that the final database used was only 72 FinTech companies listed mainly in the U.S. stock markets, so it was not possible to differentiate between countries. This is an important limitation that should be overcome in future research. Finally, we believe that further studies should also consider analyzing the contributions of these companies to the Sustainable Development Goals set by the United Nations for 2030. This would help to provide information on the unique contribution of these companies to this global challenge.

References

Anifa, M., Ramakrishnan, S., Joghee, S., Kabiraj, S., & Bishnoi, M. M. (2022). Fintech innovations in the financial service industry. *Journal of Risk and Financial Management, 15*(7), 287. https://doi.org/10.3390/jrfm15070287

Arellano, M., & Bond, S. (1991). Some tests of specification for panel data: Monte Carlo evidence and application to employment equations. *Review of Economic Studies, 58*, 277–297. https://doi.org/10.2307/2297968

Arner, D. W., Buckley, R. P., Zetzsche, D. A., & Veidt, R. (2020). Sustainability, FinTech and financial inclusion. *European Business Organization Law Review, 21*, 7–35. https://doi.org/10.1007/s40804-020-00183-y

Atayah, O. F., Najaf, K., Ali, M. H., & Marashdeh, H. (2023). Sustainability, market performance and FinTech firms. *Meditari Accountancy Research, 61*, e12–e27. https://doi.org/10.1108/MEDAR-08-2021-1405

Baltgailis, J., & Simakhova, A. (2022). The technological innnovations of Fintech companies in ensuring the stability of financial system in pandemic times. *Marketing and Management of Innovations, 2*, 55–65.

Chen, Y., Chiu, J., & Chung, H. (2022). Givers or receivers? Return and volatility spillovers between Fintech and the traditional financial industry. *Finance Research Letters, 46*, 102458. https://doi.org/10.1016/j.frl.2021.102458

Cohen, J. R., & Simnett, R. (2015). CSR and assurance services: A research agenda. *Auditing: A Journal of Practice and Theory, 34*(1), 59–74.

Deegan, C. M. (2019). Legitimacy theory: Despite its enduring popularity and contribution, time is right for a necessary makeover. *Accounting, Auditing and Accountability Journal, 32*(8), 2307–2329. https://doi.org/10.1108/AAAJ-08-2018-3638

Diamond, D. W., & Verrecchia, R. E. (1991). Disclosure, liquidity, and the cost of capital. *Journal of Finance, 46*(4), 1325–1359. https://doi.org/10.1111/j.1540-6261.1991.tb04620.x

Financial Stability Board (2017). Financial stability implications from FinTech. https://www.fsb.org/2017/06/financial-stability-implications-from-fintech/

Franco, L., García, A. L., Husetovic, V., Lassiter, J. (2020). Does fintech contribute to systematic risk? Evidence from the US and Europe. Asian Development Bank Institute working paper 1132. https://www.adb.org/publications/does-fintech-contribute-systemic-risk-evidence-us-europe

Freeman, R. E. (1984). *Strategic management: A stakeholder approach.* Pitman.

Friedman, M. (1962). *Capitalism and freedom.* University of Chicago Press.

Friedman, M. (1970). Friedman doctrine: The social responsibility of business is to increase its profits. *New York times Magazine, 13*, 32–33.

García-Sánchez, I. M. (2021). Corporate social reporting and assurance: The state of the art. *Revista De Contabilidad, 24*(2), 241–269. https://doi.org/10.6018/rcsar.409441

Gil-Corbacho, A., Miralles-Quirós, M. M., & Miralles-Quirós, J. L. (2023). Stock market performance differences between fintech and traditional financial firms. *Estudios de Economía Aplicada/Studies of Applied Economics, 41*(1), 1–22.

Goldstein, I., Jiang, W., & Karolyi, G. A. (2019). To Fintech and beyond. *Review of Financial Studies, 32*(5), 1647–1661. https://doi.org/10.1093/rfs/hhz025

Gomber, P., Koch, J. A., & Siering, M. (2017). Digital finance and FinTech: Current research and future research directions. *Journal of Business Economics, 87*(5), 537–580. https://doi.org/10.1007/s11573-017-0852-x

Healy, P. M., & Palepu, K. G. (2001). Information asymmetry, corporate disclosure, and the capital markets: A review of the empirical disclosure literature. *Journal of Accounting and Economics, 31*(1–3), 405–440. https://doi.org/10.1016/S0165-4101(01)00018-0

Hoang, T. G., Nguyen, G. N. T., & Le, D. A. (2022). Developments in financial technologies for achieving the sustainable development goals (SDGs): Fintech and SDGs. In *Disruptive technologies and ecoinnovation for sustainable development* (pp 1–19). IGI Global

Jiao, Y. (2010). Stakeholder welfare and firm value. *Journal of Banking and Finance, 34*(10), 2549–2561. https://doi.org/10.1016/j.jbankfin.2010.04.013

Kim, O., & Verrecchia, R. E. (1994). Market liquidity and volume around earnings announcements. *Journal of Accounting and Economics, 17*(1/2), 41–67. https://doi.org/10.1016/0165-4101(94)90004-3

Kolk, A., & Perego, P. (2010). Determinants of the adoption of sustainability assurance statements: An international investigation. *Business Strategy and the Environment, 19*(3), 182–198. https://doi.org/10.1002/bse.643

KPMG (2022) Pulse of Fintech H2'21. https://kpmg.com/xx/en/home/insights/2022/01/pulse-of-fintech-h2-2021-global.html

Lang, M., & Lundholm, R. (2000). Voluntary disclosure and equity offerings: Reducing information asymmetry of hyping the stock. *Contemporary Accounting Research, 17*(4), 623–662. https://doi.org/10.1506/9N45-F0JX-AXVW-LBWJ

Lang, M., & Maffett, M. (2011). Economic effects of transparency in international equity markets: A review and suggestions for future research. *Foundations and Trends in Accounting, 5*(3), 175–241.

Le, T.-L., Abakah, E. J. A., & Tiwari, A. K. (2021a). Time and frequency domain connection and spill-over among fintech, green bonds and cryptocurrencies in the age of the fourth industrial revolution. *Technological Forecasting and Social Change, 162*, 120382. https://doi.org/10.1016/j.techfore.2020.120382

Le, T.-L., Yarovaya, L., & Nasir, M. A. (2021b). Did Covid-19 change spillover patterns between Fintech and other asset classes? *Research in International Business and Finance, 58*, 101441. https://doi.org/10.1016/j.ribaf.2021.101441

Li, J., Li, J., Zhu, X., Yao, Y., & Casu, B. (2020). Risk spillover between fintech and traditional financial institutions: Evidence from the US. *International Review of Financial Analysis, 71*, 101544. https://doi.org/10.1016/j.irfa.2020.101544

Merton, R. (1987). A simple model of capital market equilibrium with incomplete information. *Journal of Finance, 42*(3), 483–510. https://doi.org/10.1111/j.1540-6261.1987.tb04565.x

Milian, E. Z., de Spinola, M. M., & de Carvalho, M. M. (2019). Fintechs: A literature review and research agenda. *Electronic Commerce Research and Applications, 34*, 833. https://doi.org/10.1016/j.elerap.2019.100833

Miralles-Quirós, M. M., Miralles-Quirós, J. L., & Daza-Izquierdo, J. (2021). The assurance of sustainability reports and their impact on stock market prices. *Management Letters/cuadernos De Gestión, 21*(1), 47–60. https://doi.org/10.5295/cdg.201262mm

Miralles-Quirós, M. M., Miralles-Quirós, J. L., & Redondo-Hernández, J. (2019). ESG performance and shareholder value creation in the banking industry: International differences. *Sustainability, 11*(5), 1404.

Najaf, K., Haj Khalifa, A., Obaid, S. M., Rashidi, A. A., & Ataya, A. (2023). Does sustainability matter for Fintech firms? Evidence from United States firms. *Competitiveness Review, 33*(1), 161–180. https://doi.org/10.1108/CR-10-2021-0132

Najaf, K., Schinckus, C., & Chee Yoong, L. (2021). VaR and market value of fintech companies: An analysis and evidence from global data. *Managerial Finance, 47*(7), 915–936. https://doi.org/10.1108/MF-04-2020-0169

Ng, A. W., & Kwok, B. K. B. (2017). Emergence of Fintech and cybersecurity in a global financial centre: Strategic approach by a regulator. *Journal of Financial Regulation and Compliance, 25*(4), 422–434. https://doi.org/10.1108/JFRC-01-2017-0013

Porter, M. E., & Kramer, M. R. (2011). Creating shared value: How to reinvent capitalism—And unleash a wave of innovation and growth. *Harvard Business Review, 89*, 62–77.

Post, J., Preston, L., & Sachs, S. (2002). *Redefining the corporation: Stakeholder management and organizational wealth.* Stanford University Press.

Romanova, I., & Kudinska, M. (2016). Banking and fintech: A challenge or opportunity? *Contemporary Studies in Economic and Financial Analysis, 98*, 21–35. https://doi.org/10.1108/S1569-375920160000098002

Sánchez, M. A. (2022). A multi-level perspective on financial technology transitions. *Technological Forecasting and Social Change, 181*, 121766. https://doi.org/10.1016/j.techfore.2022.121766

Sangwan, V., Harsita, H., Prakash, P., & Singh, S. (2019). Financial technology: a review of extant literature. *Studies in Economics and Finance, 37*(1), 71–88. https://doi.org/10.1108/SEF-07-2019-0270

Seguí-Mas, E., Polo-Garrido, F., & Bollas-Araya, H. M. (2018). Sustainability assurance in socially-sensitive sectors: A worldwide analysis of the financial services industry. *Sustainability, 10*(8), 2777. https://doi.org/10.3390/su10082777

Silva, S. (2021). Corporate contributions to the sustainable development goals: An empirical analysis informed by legitimacy theory. *Journal of Cleaner Production, 292*, 125962. https://doi.org/10.1016/j.jclepro.2021.125962

Toumi, A., Najaf, K., Dhiaf, M. M., Li, N. S., & Kanagasabapathy, S. (2023). The role of Fintech firms' sustainability during the COVID-19 period. *Environmental Science and Pollution Research, 30*, 58855–58865. https://doi.org/10.1007/s11356-023-26530-3

Zveryakov, M., Kovalenko, V., Sheludko, S., & Sharah, E. (2019). FinTech sector and banking business: Competition or symbiosis? *Economic Annais-XXI, 175*(1–2), 53–57.

María Mar Miralles Quirós She is a Full Professor of Financial Economics and Accounting and the Director of the Research Group on Markets and Financial Assets at the University of Extremadura in Spain. Her research has focused on providing shareholders' investment strategies that would beat the market using sophisticated mathematical approaches, as well as the analysis of companies' sustainable activities and their connection to shareholders' value creation. Her research works have been published in prestigious journals such as Business Strategy and the Environment, Corporate Social Responsibility and Environmental Management, Energy Economics, International Review of Economics and Finance, Journal of Business Ethics, Journal of International Financial Markets, Institutions and Money, Mathematics, and North American Journal of Economics and Finance, among others.

José Luis Miralles Quirós He is an Associate Professor of Financial Economics and Accounting at the University of Extremadura in Spain. His research has primarily focused on providing shareholders' investment strategies that would beat the market in adverse situations using sophisticated mathematical approaches. Meanwhile, his research also includes the analysis of companies' environmental, social, and corporate governance activities and their connection to shareholders' value creation. These works have been published in prestigious journals such as Business Strategy and the Environment, Corporate Social Responsibility and Environmental Management, Energy Economics, International Review of Economics and Finance, Journal of Business Ethics, Journal of International Financial Markets, Institutions and Money, Mathematics, and North American Journal of Economics and Finance, among others.

Azahara Gil Corbacho She is an Assistant Professor of Economy at the University of Extremadura in Spain. She is writing her PhD under the supervision of María Mar Miralles-Quirós and José Luis Miralles-Quiros at the University of Extremadura. Her research focuses on analyzing the two most important challenges currently facing the financial sector, digitization and sustainable development. She is co-author of a paper published in Studies of Applied Economics and she is co-author of other studies submitted to other international journals.

Information Disclosure on the Integration of SDGs into Banking Management: The Mercosur Countries Case

María del Mar Gálvez-Rodríguez, Walter Daniel Ovelar-Fernández, and Carmen Caba-Pérez

1 Introduction

Banks play a leading role in achieving the 2030 agenda for sustainable development (Kumar & Prakash, 2020). The mediating role they play in investments and financing contributes to the economic welfare of a country (Kumar, 2022) and ensures environmental quality, social cohesion and prosperity for future generations. For example, banks, via their products, can condition economic support to their clients (companies, administrations, and families) with the compliance of socio-environmental criteria in addition to encouraging socially and environmentally responsible investments (Chatzitheodorou et al., 2021; Gunawan et al., 2022).

The actions related to the banking sector's commitment towards sustainable development goals (SDGs) should be transparent (Aguado-Correa et al., 2023) since being accountable to the public is the retrospective mirror of how far an organization truly aligns with the 2030 Agenda (Bexell & Jönsson, 2021). In this line, authors such as Chouaibi and Affes (2021) emphasize that organizations with a strong sense of social and ethical commitment usually engage more in disclosure regarding their sustainable practices. Moreover, the last assessments of global action for the SDGs show slow progress towards their achievement (United Nations, 2022). Consequently, there is a greater societal demand for information about specific actions that key actors,

M. M. Gálvez-Rodríguez (✉) · C. Caba-Pérez
University of Almería, Almería, Spain
e-mail: margalvez@ual.es

C. Caba-Pérez
e-mail: ccaba@ual.es

W. D. Ovelar-Fernández
Autonomous University of Asunción, Asunción, Paraguay
e-mail: wovelar@uaa.edu.py

M. d. C. Valls Martínez and J. M. Santos-Jaén (eds.), *Environmentally Sustainable Production*, https://doi.org/10.1007/978-3-031-52656-5_3

such as banking institutions, are implementing to achieve any of the SDGs (Germann et al., 2023; Zimmermann, 2019).

Despite the growing concern about the banking sector's disclosure practices around sustainability and whether their sustainability reporting practices provide detailed information about these actions, it is a field of study that remains under-researched, both within the industry and the literature (Baldissera, 2023; Khan et al., 2021). In this regard, few studies on corporate accountability related to SDGs within the banking sector exist. Indeed, the literature principally focuses on European countries (Aguado-Correa et al., 2023; Avrampou et al., 2019; Cosma et al., 2020; Sardianou et al., 2021). In the context of emerging economies, studies related to the SDGs reported by the banking sector are even more scarce and mainly focus on countries such as Pakistan (Ijaz et al., 2020) or Brazil (Chagas et al., 2022). Hence, more attention is needed to the bank's dissemination related to SDGs from less developed countries such as those from the Latin American region. Moreover, there is a need for a greater understanding of the organizational characteristics influencing banks' policies on sustainability reporting aligned with SDGs (Cosma et al., 2020).

Focused on the banking sector of Latin American countries that form part of the Mercosur economic community composed of Argentina, Brazil, Paraguay, and Uruguay, this chapter is twofold. Firstly, the authors analyze the level of detailed information that Mercosur banks disclose about their sustainability practices aligned to SDGs (hereafter, detailed SDG disclosure). Secondly, grounded on the stakeholder and legitimacy theory, an analysis is conducted of organizational factors that characterize the levels of detailed SDG disclosure achieved.

In general terms, the findings of this study may help to compare the SDG reporting practices of developed countries (e.g., European countries) with those of less developed countries (e.g., Latin American countries). In addition, the findings will address the silence that the literature remains on the academic and practical understanding of factors related to the decision to provide detailed sustainability disclosure (Khan et al., 2021). This article can help bank practitioners in Mercosur countries become aware of divergences regarding their level of detailed information on their commitment to SDGs in each of the Mercosur countries.

To fulfill the objectives of this chapter, the following structure is outlined. The second section briefly summarizes the banking reporting practices aligned to the 2030 Agenda. The third section reviews the determining factors of sustainable disclosure and develops research hypotheses. The fourth section explains the methodology used to reach the objectives of this study, and the following section presents the results of the analyses. The future implications of this study are shown in the sixth section. Finally, the seventh section states the most relevant conclusions.

2 The SDGs, Corporate Sustainability Reporting and Banking Sector

Since 2015, all business organizations have faced new challenges related to the corporate sustainability issue as a result of the introduction of the 2030 Agenda by the United Nations (UN), which includes a set of 17 SDGs and 169 targets (Biswas et al., 2021; Van der Waal & Thijssens, 2020). These global goals aim to define and implement actions to end poverty and hunger, combat inequalities, build peaceful, just and inclusive societies, protect human rights, promote gender equality and ensure the lasting protection of the planet and its natural resources with a determination to create conditions for sustainable, inclusive and sustained economic growth, shared prosperity and decent work for all (see Fig. 1) (United Nations General Assembly, 2015).

Sustainable development refers to actions that aim to improve the quality of life and well-being of both present and future generations without compromising and exhausting the available natural resources and without affecting the environment. The SDGs are grouped into the three dimensions of sustainable development: Economic growth, social inclusion, and environmental protection. SDGs 8, 9, 10, 12 and 17 are related to the economic dimension, SDGs 1, 2, 3, 4, 5, 7, 11 and 16 are associated with the social dimension, and finally, SDGs 6, 13, 14 and 15 are linked to the environmental dimension. The interlinkages and integrated nature of the SDGs are crucially important for ensuring that the purpose of the new Agenda is fulfilled (United Nations General Assembly, 2015).

The call for greater transparency in sustainability reporting is increasing because banking corporations are considered key players in helping achieve every SDG (UN Global Compact, 2015). Sustainability reporting is the practice of measuring,

Fig. 1 The 17 sustainable development goals. *Source* https://www.un.org/sustainabledevelopment/news/communicationsmaterial/. © United Nations

disclosing and being accountable to internal and external stakeholders for organizational performance towards the goal of sustainable development. It involves reporting on how an organization considers sustainability issues in its operations and its environmental, social and economic impacts (Scott & McGill, 2018; European Commission, 2019; Maama, 2021). A sustainability report also presents the organization's values and governance model and demonstrates the link between its strategy and its commitment to a sustainable global economy (Higgins & Coffey, 2016). Sustainability reports should be connected to how organizations like banks address the SDGs. In this regard, rather than making generic statements on their commitment to the SDGs, they should give a comprehensive view of the actions taken to realize such commitments (Avrampou et al., 2019; Cosma et al., 2020).

SDG reporting in the banking sector is still at an early stage (Al Lawati & Hussainey, 2022). Focusing on the papers that carried out a content analysis of banking sustainability reports, it seems that the contributions to SDGs remain scarce and unanimously focus on SDG 8 (Decent work and economic growth). Despite there being more heterogeneity in the less reported SDGs, SDG 15 (Life On Land) and SDG14 (Life Below Water) seem to be the least considered in the banking sector's commitment to the 2030 Agenda (see Table 1).

The majority of the research addresses the European context. Avrampou et al. (2019) find a low and incomplete disclosure across the reporting of five leading European banks. Nevertheless, the authors conclude that goals 4, 8, and 16 are the best-reported SDGs. Focusing on 262 European banks from 22 European Union countries, Cosma et al. (2020) state that the "scope" of contribution to SDGs from the European banks is narrow. Furthermore, the authors highlight an increasing interest from emerging countries toward sustainable development goals, particularly from Estonia, Croatia, and Poland. In addition, that study also observed that the goals prioritized goals by the banks, and thus more referred to in their reports, are SDGs 8, 13 and 4. Sardianou et al. (2021) use a sample of 37 European Banks to show that the majority of the European banks promote SDGs more connected with their core business. In the previous research analyses of SDG disclosure under the GRI reporting model, Hassan et al. (2022) investigated the disclosure of the SDGs in integrated reports of the European financial services sector, indicating that overall,

Table 1 Studies SDGs reporting in the banking sector

Publication year	Author(s)	Country	Sample	High disclosure	Low disclosure
2023	Aguado-Correa et al.	Spain	12	8, 4 & 13	2 & 14
2022	Chagas et al.	Brazil	17	8	11
2021	Sardianou et al.	Europe	37	8, 9 & 16	6, 14 & 15
2021	Hassan et al.	Europe	9	13 & 7	2, 14 & 15
2020	Ijaz et al.	Pakistan	6	4, 3 & 8	16, 15 & 14
2020	Cosma et al.	Europe	262	8, 13 & 4	14, 15 & 6
2019	Avrampou et al.	Europe	5	4, 8 & 16	6, 17 & 15

the highest goals are SDG 13 and 7. At a country level, Aguado-Correa et al. (2023) analyzed 12 Spanish banks, highlighting that banks studied released information on the SDG where their activity had a greater capacity to address it, specifically SDGs 8, 13, and 4.

In the context of less developed countries, there are many fewer studies. For example, Ijaz et al. (2020) state that Pakistani banks still lag behind in SDGs implementation and proper reporting in annual sustainability reports. In this regard, Pakistan's banks seem to engage more with the SDGs, 8, 3, 4. Moreover, Chagas et al. (2022) examination of 11 reports of the largest banks operating in Brazil noted the low quality of these reports. The authors show the direct and positive relationship between the number of material topics reported, foreseen in the GRI standard, and the level of SDG reporting. Thus, the overall knowledge regarding SDGs reporting in this sector remains low, with a great need to delve further into the context of less developed countries such as Latin American countries.

3 Determining Factors of Sustainable Reporting in the Banking Sector and Hypotheses Development

Legitimacy and Stakeholder Theory are among the most used approaches in corporate disclosure on sustainability, as corporations engage in sustainability reporting to seek legitimacy and for strategic reasons such as improving their relationships with stakeholders (Islam & Deegan, 2008 et al.). In particular, according to Legitimacy Theory, a company's sustainability reporting aims to claim legitimacy to external stakeholders by demonstrating the company's adherence to social norms and expectations (Dowling & Pfeffer, 1975; Lodhia et al., 2022). To maintain, reinforce and repair their image as agents that serve society, corporations such as banks need to voluntarily disclose information that explains how their activities impact economic growth, social inclusion, and environmental protection, as well as the measures instituted to mitigate the negative impacts of their activities (Pizzi et al., 2021). In this regard, there is no doubt that sustainable behavior nowadays should be aligned with the SDGs. Hence, SDGs disclosure is crucial for businesses, such as banks, since it allows them to demonstrate their sustainability successes to stakeholders and, in doing so, gain an excellent reputation and legitimacy (Garcia-Sanchez et al., 2020a, 2020b; Silva, 2021).

Like Legitimacy Theory, Stakeholder Theory asserts that the external environment must be influenced. Freeman (2010) argued that companies able to manage their stakeholders successfully could proactively anticipate stakeholder concerns and attempt to impact their environment. This involves addressing stakeholder needs and concerns, engaging in communication processes, negotiating with them and seeking voluntary agreements concerning any issues. Hence, banks carry out reporting practices as a means of cultivating long-term relationships with stakeholders, including investors, creditors and shareholders (Githaiga & Kosgei, 2023). In this regard, the SDG initiative is backed by a broad range of stakeholders (García-Meca and Martínez-Ferrero, 2021), in which banks are considered key players. The disclosure practices related to the integration of SDG in banks are a way of responding to the actual information needs on stakeholders' demands and expectations regarding the roles of banks vis-à-vis achieving the 2030 Agenda (Ferrero-Ferrero et al., 2023).

Based on legitimacy theory and stakeholder theory, this study aims to investigate the association between the characteristics of the banks (their age, size, ownership structure and listing status) and the SDGs disclosure practices of banks from Mercosur countries.

Bank Age

Current studies have evidenced a significant association between a corporation's age and sustainable reporting of corporations from the industrial services sector in Jordan (Alkayed & Omar, 2023) or various business sectors from Latin American countries (Correa-García et al., 2020). In the banking sector, Chakroun et al. (2017) and Terán and Cortés (2023) confirm the influence of bank age in the sustainable disclosure of Tunisian and Ecuadorian banks, respectively. The study by Bidari and Djajadikerta (2020) observed that Napalese banks' age influences their transparency practices, specifically, the social and environmental impact of their activities. In general, this positive relation can be explained since older banks have received more benefits from society over time, resulting in a mature relationship where businesses undertake a more significant leadership role and develop an increased sense of the importance of sustainable development. In addition, more than others, older banks realize the value of high sustainability disclosure to improve relationships with key stakeholders as their sustainable behavior boosts their corporate image. In contrast, previous studies also found that an organization's age is negatively associated with banks' disclosure of sustainability information (Bose et al., 2016). The majority of the aforementioned studies are not specific to SDGs reporting; thus, more research on this issue is needed. Based on previous literature, the following hypothesis is formulated:

H1. The age of the bank positively influences detailed SDG disclosure.

Bank Size

Van der Waal and Thijssens (2020) found evidence of a positive relationship between corporate size and corporate involvement in the SDGs. Specific to SDGs reporting, recent studies have evidenced a positive relation between firm size and the level of disclosure of companies from Gulf Cooperation Council countries (Al-Qudah & Houcine, 2023) and Malaysia (Buniamin et al., 2021) as well as analyzing organizations worldwide (Rosati & Faria, 2019). The above authors argue that large companies would perceive SDG disclosure differently from small companies due to their different impacts on society. Large firms have a more significant impact than smaller ones. They are more visible and, thus, subject to increased stakeholder scrutiny and pressure. In addition, these organizations possess more resources to respond to the information needs of their stakeholders. Nevertheless, in some recent studies, such as the study by Krasodomska et al. (2022) in large public interest entities in the European Union, the size of the firms does not seem to be a relevant factor influencing SDG reporting practices. There is scant literature focused on the banking sector's SDG disclosure practices relative to the size of the bank. However, previous research has found a positive relationship between banks' accountability on their sustainable conduct and the organization's size (Bidari & Djajadikerta, 2020; Buallay & Al-Ajmi, 2020; Terán & Cortés, 2023). Thus, the third hypothesis is as follows:

H2. The size of the bank positively influences detailed SDG disclosure.

Type of Ownership

Foreign banks are usually defined as those with more than 50% of the shares owned by non-domestic residents. This implies that a bank may be a domestic bank in one country but a foreign bank everywhere else (Lensink et al., 2008: 835). Previous literature states that foreign investment makes banks more open to the stakeholders' demands (Bonifácio Neto & Branco, 2019). Foreign investors tend to have knowledge and values from their international market engagement; hence, firms with foreign ownership are more incentivized to provide higher sustainability reporting (Bose et al., 2016). This situation is most evident in the context of emerging countries, as Haladu and Salim (2016) noted that foreign ownership used to come from developed nations and entail multinational corporations with very high compliance with sustainability standards. In the banking sector, this relationship has been confirmed at a country level, such as in the case of Indonesia (Sumarta et al., 2023) and across different countries (Fang et al., 2019). Nevertheless, the results of other studies are not as conclusive as they find no significant relation between bank foreign ownership and sustainable reporting practices (Bose et al., 2018; Khan et al., 2021; Terán & Cortés, 2023). The aforementioned cases do not focus on SDGs reporting; thus, more research is needed. Considering that overall literature, it would be logical to expect Mercosur banks with foreign ownership to be more "incentivized" to follow the global reporting trend of greater disclosure. Thus, the following hypothesis is formulated:

H3. Foreign ownership of the bank positively influences detailed SDG disclosure.

Listing Status

Listed firms are more likely than unlisted firms to disclose information about their contribution to the SDGs. The reason for this could be their desire to use reporting practices to ensure a positive relationship with key stakeholders, affecting a firm's overall market value (Pizzi et al., 2021). In addition, some stock markets have regulatory guidelines that pressure and "pull" companies into providing sustainability disclosures by incorporating them as listing requirements (Dong et al., 2022) or recommending sustainability reporting (Hamad et al., 2023). In emerging countries, listed firms are more motivated to perform and better report their sustainability practices than unlisted companies to enhance their international comparability and accountability on sustainability issues (Al-Hajaya, 2023). In the banking sector, Terán and Cortés (2023) evidence that Ecuador's banks listed in their country's stock market are more transparent than those not listed. However, the analysis of European banks by Cosma et al. (2020) shows that the SDG reporting preferences in listed and non-listed banks are similar. The abovementioned cases do not focus on SDGs reporting; thus, more research is needed. Considering the overall extant literature, it would be logical to expect that Mercosur banks listed in their "native" country's stock market would be more willing to follow global reporting trends such as disseminating the bank's actions linked with SDGs. Thus, the following hypothesis is formulated:

H4. The listing status of a bank positively influences detailed SDG disclosure.

4 Methodology

4.1 Sample

Initially, the overall sample was composed of 269 banks. These banks are recognized by the respective central banks of the four Mercosur signatory countries (Argentina, Brazil, Paraguay, and Uruguay) as of December 31, 2022. The final sample of banks was selected based on those whose reports are available on their official website and content information regarding SDGs. The final sample consisted of 38 banks, 14% of the original sample (Table 2). Likewise, we can observe that the type of report used to disseminate information regarding their commitment to the 2030 Agenda mainly provides a sustainability report based on GRI standards. To a lesser extent, the accountability of the integration of SDGs in banking management is published in the integrated report format. Few of the banks analyzed provide it in the sustainability section in their annual report or through their Environmental, Social and Governance Report (see Table 3).

Table 2 Sample

	Mercosur	Central Bank of Argentina	Central Bank of Brazil	Central Bank of Paraguay	Central Bank of Uruguay
Banks	269	80	161	18[a]	10
Banks with SDG reporting	38	10	20	4	4
Coverage (%)	14	13	12	22	40

[a] Includes the Development Financial Agency (DFA), a second-tier state bank in Paraguay
Source Authors' own compilation

Table 3 Types of reports

Names	Mercosur	Argentina	Brazil	Paraguay	Uruguay
Sustainability report	22	5	9	4	4
Integrated report	9	5	4	0	0
Annual report	4	0	4	0	0
ESG report	3	0	3	0	0
Total	38	10	20	4	4

Source Authors' own compilation

4.2 Disclosure Index

To analyze the first objective of this study, the level of detailed SDG disclosure from Mercosur banks, a content analysis is performed in line with similar studies (Aguado-Correa et al., 2023; Gunawan et al., 2022). This content analysis examines quantitative and qualitative content specific to 17 SDGs within the reports by the banks from Mercosur countries, mentioned in Sect. 4.1.

The analysis included a dichotomous scoring system and a series of assessments. The first assessment observes whether the banks express their commitment to any of the 17 goals in a generic/narrative manner that offers few details. The second assessment examines whether the bank provides detailed information regarding the actions taken to address such commitment. These two forms of information are based on the work of Beck et al. (2010) and Sardianou et al. (2021). Manual coding was chosen instead of content analysis software, as these tools cannot always process much of the information associated with the 17 SDGs (Cosma et al., 2020). In line with Krippendorff (2018), a single researcher performed the analysis following a preliminary assessment conducted by the entire research team, which avoided the risks of inter-coder reliability.

In accordance with previous studies, a global disclosure index is used to evaluate the information provided regarding the banks' commitment to the SDGs. The index is a synthetic indicator approximating the overall disclosure degree on SDG engagement (Cosma et al., 2020; Githaiga & Kosgei, 2023; Gunawan et al., 2022). In line with the abovementioned studies, we developed two indexes. The first, "Generic SDG disclosure index" (GENSDG), shows the extent of the banks' generic statements regarding their commitment to SDGs. The second, "Detailed SDG disclosures index" (DETSDG), indicates to what extent Mercosur banks provide information about the actions related to their sustainability commitment to SDGs. Hence, in both indexes, the generic/detailed disclosure of each bank analyzed, 17 items were used. Each item (SDG_i) was scored as 1 if the information was available in the sustainability report and 0 if not. The calculation was determined by the ratio of the sum of the scores for all the items ($\sum SDG_i$) and the total number of items observed. The results are expressed as percentages in line with the following formula:

$$\text{GENSDG/DETSDG index}: \frac{\sum_{j=1}^{17,38} SDGi}{38} * 100.$$

Delving further into the detailed SDG disclosure, three more indexes are calculated, grouping the information into the three dimensions of sustainable development: Economic growth, social inclusion, and environmental protection. SDGs 8, 9, 10, 12 and 17 are related to economic detailed disclosure (EcDD), SDGs 1, 2, 3, 4, 5, 7, 11 and 16 are associated with social detailed disclosure (SDD), and, finally, SDGs 6, 13, 14 and 15 are linked to environmental detailed disclosure (EnDD). Hence, in these instances, the formula will be the following:

$$\text{EcDD} = \frac{\sum_{j=1}^{5,38} SDGi}{38} * 100; \ \text{SDD} = \frac{\sum_{j=1}^{8,38} SDGi}{38} * 100;$$
$$\text{EnDD} = \frac{\sum_{j=1}^{4,38} SDGi}{38} * 100.$$

The results regarding the level of Generic SDG disclosures, Detailed SDG disclosures and the three sustainable development dimensions aim to explain to readers the extent of information disclosed in the banking sustainability report, which, in turn, provides a picture of the integration of the SDGs into the management practices of Mercosur banks.

4.3 Regression Analysis

Regarding the second objective of this study, to analyze organizational factors that characterize the levels of detailed SDG disclosure achieved, a regression analysis is performed in line with previous studies related to reporting practices (e.g., Hamad et al., 2023). The dependent variable is the index of the Detailed SDG disclosures (DETSDG). Following the hypothesis development outlined in Sect. 3, the independent variables are bank size, bank age, the bank's ownership structure and its listing status, are summarized in Table 4 and expressed in the following formula:

$$DETSDG := \alpha + \beta_1 AGEBAN + \beta_2 SIZEBAN + \beta_3 OWNBAN + \beta_4 LISTBAN + \mu$$

where α is the constant term, X_{ij} represents the variables that influence the information disclosure on the Web, β_j is a coefficient vector to be calculated, and μ is the random error term, presumably with identical and independent distribution, with an average of 0. To assess the linear relation between independent variables, an analysis of Pearson's Correlation Coefficient was carried out according to Githaiga and Kosgei (2023).

Table 4 Independent variables definitions and measurements

Independent variables			Description	Authors	Expected results
H1	Bank age	**AGEBAN**	Years old of the bank	Terán and Cortés (2023), Al Qudah & Houcine (2023), Hamad et al. (2023), Correa-García et al. (2020), Khan et al. (2021)	(+)
H2	Bank size	**SIZEBAN**	Natural logarithm of total Assets as of 12/31/2022	Terán and Cortés (2023), Al Lawati and Hussainey (2022), Buniamin et al. (2021), Hamad et al. (2023), Pizzi et al. (2021), Rosati and Faria (2019)	(+)
H3	Bank ownership	**OWNBAN**	Dichotomous: 1 for foreign property and 0 for local property	Terán and Cortés (2023), Correa-García et al. (2020), Khan et al. (2021)	(+)
H4	Listing status of the bank	**LISBAN**	Dichotomous: 1 if listed on its country's stock exchange and 0 if unlisted	Terán and Cortés (2023), Cosma et al. (2020), Pizzi et al. (2021), Khan et al. (2021)	(+)

AGEBAN: Bank age, the number of years since the bank was created; SIZEBAN: Bank size in terms of total assets; OWNBAN: Bank ownership, the owners of a bank; LISBAN: Listing status of a bank which indicates if the bank is listed in the stock exchange of its country

5 Results and Discussion

5.1 SDGs Disclosure

As indicated in Sect. 4.1, *14% of the Mercosur banks (38 of 269) report their commitment towards the 2030 Agenda, indicating the low level of contribution these entities demonstrate towards achieving the SDGs. Among the banks that report on them, we can observe that, overall, the Mercosur banks provide detailed information on less than half of the SDGs (see Table 5). This finding, therefore, supports previous international studies in the context of European countries that call for the need to increase the contribution of banks towards the SDGs (Avrampou et al., 2019).

If we disaggregate the above general outcomes into the levels of SDG disclosure for each country, the banks of each of the four countries provide very low levels of complete information regarding their actions taken to contribute to any of the SDGs.

Table 5 Detailed SDG disclosures of Mercosur banks

	Mercosur (%)	Argentina (%)	Brazil (%)	Paraguay (%)	Uruguay (%)
Detailed SDG disclosure	48	17	17	7	8
Economic dimension (SDGs 8, 9, 10, 12, 17)	65	24	23	8	10
Social dimension (SDGs 1, 2, 3, 4, 5, 7, 11, 16)	48	17	16	7	8
Environmental dimension (SDGs 6, 13, 14, 15)	25	6	9	5	5

In this regard, banks operating in Argentina and Brazil are the most transparent, followed by those in Paraguay and Uruguay. These findings align with those of Chagas et al. (2022), who found that Brazilian banks' reports were of low quality.

Regarding the three dimensions of sustainable development, data show that the detailed information on integrating SDGs into banking management prioritizes those SDGs directly related to progress in the country's economic growth. The second priority is accountability regarding actions taken to support social inclusion. Finally, the actions taken for environmental protection are the least disclosed. This outcome also supports previous research both at a generic level on reporting of sustainability practices (Maama, 2021) and reports explicitly related to SDGs, albeit in the European context (Sardianou et al., 2021). The authors state that the low level of disclosure and banks' commitment to the environment could be because it is not considered guilty of significant pollution (unlike other industries, such as manufacturing). Instead, the sector focuses on reducing the impact of its daily energy consumption inside and outside the organization.

Delving further into the disclosure level of each SDG, Table 6 shows a slightly greater level of generic disclosure than detailed disclosure of the bank's actions. Moreover, the best-reported goals are SDG 8 (Decent Work and Economic Growth), followed by SDG 5 (Gender Equality). The third position is shared by SDG 9 (Industry, Innovation, and Infrastructure) and SDG 12 (Responsible Consumption and Production). Thus, Mercosur banks appear to be more engaged with SDGs that are more directly related to their core business, perhaps because they have a greater capacity to address them, according to previous studies (Avrampou et al., 2019; Cosma et al., 2020; Ijaz et al., 2020; Brazil Chagas et al., 2022; Aguado-Correa et al., 2023). The least generic and detailed information given is related to SDG 14 (Life Below Water), SDG 15 (Life On Land), and SDG 6 (Clean Water and Sanitation). Despite not being in the same order, these findings coincide with the research on European banks (Cosma et al., 2020; Sardianou et al., 2021) and partially agree with the research conducted on Pakistan (Ijaz et al., 2020).

Table 6 Generic and detailed SDG disclosures

SDGs	Generic SDG disclosure (%)	Detailed SDG disclosure (%)
SDG 1 = No poverty	58	55
SDG 2 = Zero hunger	32	29
SDG 3 = Good health and well-being	34	32
SDG 4 = Quality education	66	58
SDG 5 = Gender equality	82	71
SDG 6 = Clean water and sanitation	26	26
SDG 7 = Affordable and clean energy	53	50
SDG 8 = Decent work and economic growth	89	76
SDG 9 = Industry, innovation, and infrastructure	79	68
SDG 10 = Reduce inequalities	76	68
SDG 11 = Sustainable cities and communities	37	37
SDG 12 = Responsible consumption and production	79	68
SDG 13 = Climate action	76	63
SDG 14 = Life below water	5	3
SDG 15 = Life on land	24	21
SDG 16 = Peace, justice and strong institutions	66	53
SDG 17 = Partnerships for the goals	55	45

Table 7 shows the top three most transparent banks. The Development Bank details activities that contribute to all 17 goals. In addition, there is no clear "country reporting trend" since the most remarkable sustainable conduct belongs to banks operating in different countries.

Table 7 Mercosur banks with the most detailed SDG disclosures

Banks	Country	Detailed SDG disclosure
Development Bank (BNDES)	Brazil	17
Scotia Bank	Uruguay	16
Development Financial Agency (AFD)	Paraguay	15

5.2 Factors Influencing SDG Reporting Practices

Pearson's Correlation Coefficient was performed for all the independent variables included in the model. According to Taylor (1990), values above 0.7 show a high linear relationship between two variables. The results indicate a low correlation between the independent variables bank age, bank size, and ownership structure of the banks. The highest correlation is less than 0.57. Moreover, the test quantified the variance inflation factor (VIF), which assesses the extent to which the variance of an estimated regression coefficient increases if the explanatory variables are correlated. In VIF, values above ten indicate the absence of independence among the variables (Neter et al., 1996). After examining both (see Tables 8 and 9), we concluded that the models do not present multicollinearity problems.

The independent variable AGEBANK presents no significant effect on DETSDG; therefore, we reject hypothesis H1. This result does not support previous results regarding the positive effect of the age of the organization and the sustainability reporting practices of companies from different sectors in Latin American countries (Correa-García et al., 2020) and in Ecuador banks (Terán & Cortés, 2023) as well as in other contexts (Alkayed & Omar, 2023; Bidari & Djajadikerta, 2020). As there are no conclusive results related to the significant influence of SIZEBANK in DETSDG, the H2 cannot be confirmed. This outcome aligns with previous research that found no significant relation between the organization size and the SDG reporting practices of public entities in the European Union (Krasodomska et al., 2022). However, it does not agree with previous studies concerning sustainable reporting practices in the

Table 8 Correlations analysis

	AGEBAN	SIZEBAN	LISBAN	OWNBAN
AGEBAN	1			
SIZEBAN	0.559**	1		
OWNBAN	– 0.278*	0.092	1	
LISBAN	0.159	0.014	– 0.161	1

* The correlation is significant at the 0.1 level; ** the correlation is significant at the 0.01 level

Table 9 Regression analysis

Model	Coefficients	Significance	VIF
(Constant)	0.858		
AGEBAN	0.118	0.544	1.759
SIZEBAN	0.125	0.502	1.615
OWNBAN	0.418	0.013*	1.212
LISBAN	– 0.283	0.064*	1.045

R: 0.556; R2: 0.309; R Fitted square: 0.226
* Sig. at 10%

banking sector, albeit not specific to SDGs (Bidari & Djajadikerta, 2020; Buallay & Al-Ajmi, 2020; Terán & Cortés, 2023).

Our findings show that OWNBANK has a significant and positive effect on DETSDG. Thus, hypothesis H3 is accepted. This result is consistent with previous studies that evidence that foreign ownership is a factor that positively influences information disclosure on sustainability practices in the banking sector both at a country level, such as in the case of Indonesia (Sumarta et al., 2023) and across different countries (Fang et al., 2019). The H4 is not supported as the data indicates that LISBAN is a factor that negatively affects banks' detailed disclosure regarding their commitment to SDGs. This result does not support previous findings concerning the positive effect of listing status on access to detailed information regarding the sustainability conduct of Ecuador's banks (Terán & Cortés, 2023). Indeed, this finding does not coincide with previous research that stated that European banks' SDG reporting preferences are similar in listed and non-listed banks (Cosma et al., 2020).

6 Conclusion

The reports in which Mercosur banks indicate their involvement with the 2030 Agenda are very scarce; when they do, the reports usually follow the GRI model. The reports do not address the bank's commitment to all the goals and instead focus most of the attention on actions concerning its role in the country's economic growth. In addition, we observe certain opacity in their sustainable behavior as banks generally include a "commitment statement" to SDGs in their reports rather than a "detailed disclosure" of the actions taken. The low level of disclosure is a convergent conduct among the four countries. However, banks operating in Argentina and Brazil seem more engaged with SDGs than those in Paraguay and Uruguay. Where previous studies emphasized the low contribution of banks to SDGs in European countries (Aguado-Correa et al., 2023; Avrampou et al., 2019; Cosma et al., 2020) and in Asian countries (Ijaz et al., 2020), this chapter contributes to this trend by analyzing the context of Latin American countries, specifically, those belonging to the Mercosur trading bloc.

Moreover, the results of this study contribute to previous knowledge as to the extent that foreign investment makes banks more open to stakeholders' demands (Bonifácio Neto and Branco, 2019). The Mercosur banks that provide more comprehensive information on their commitment to SDGs are those with a foreign ownership structure since, in emerging countries, foreign investors mainly come from developed countries with higher sustainability practice concerns (Haladu & Salim, 2016). Hence, such banks are more engaged in sustainable development and accountability practices in order to accomplish the high sustainability reporting expected from their key stakeholders and the society overall.

Surprisingly, banks not listed are more transparent than those listed on the stock market where they operate. This could be because Capital markets in Latin America remain relatively small compared to other regions and could, therefore, be less subject to stakeholder scrutiny. In addition, while most of the stock exchanges in Argentina, Brazil, Paraguay and Uruguay have rules that state that corporations must provide access to relevant information on their operations, information regarding sustainability practices is voluntary (Global Reporting Initiative & AG Sustentable, 2020). In this regard, it seems that banks from Mercosur countries not listed on the stock market consider achieving the 2030 Agenda to be more important and value more comprehensive sustainability disclosure to attract investors and boost corporate legitimacy.

Bank managers from Mercosur countries should be more aware of the considerable challenge they face to increase their bank's involvement with the 2030 Agenda. Banks need to realize that their managerial practice impacts economic growth. They also have a significant indirect effect at a social and environmental level as they play a crucial role in encouraging social and environmental behavior among their clients (companies, public administrations, and families). This engagement must manifest itself in specific actions that, in turn, must be made visible to society at large through the publication of comprehensive reports. By doing so, banks and society will win.

This chapter contributes to the lack of SDG reporting practices in the banking sector from less developed countries, such as those operating in Mercosur countries. However, further research is needed into the Latin American region and other emerging economies. In addition, more studies that address external and internal factors that affect the sustainable conduct towards SDG reporting are required. Likewise, the perceptions of bank managers regarding SDG reporting practices and the role of banks vis-à-vis the 2030 Agenda remain unexplored.

References

Aguado-Correa, F., de la Vega-Jiménez, J. J., López-Jiménez, J. M., Padilla-Garrido, N., & Rabadán-Martín, I. (2023). Evaluation of non-financial information and its contribution to advancing the sustainable development goals within the Spanish banking sector. *European Research on Management and Business EcoNomics, 29*(1), 100211. https://doi.org/10.1016/j.iedeen.2022.100211

Al Lawati, H., & Hussainey, K. (2022). Does sustainable development goals disclosure affect corporate financial performance? *Sustainability, 14*(13), 7815. https://doi.org/10.3390/su14137815

Alkayed, H., & Omar, B. F. (2023). Determinants of the extent and quality of corporate social responsibility disclosure in the industrial and services sectors: the case of Jordan. *Journal of Financial Reporting and Accounting, 21*(5), 1206–1245

Al-Hajaya, K. (2023). Adherence level to materiality in stand-alone GRI sustainability reporting in Arab nations: Does listing status matter? *Afro-Asian Journal of Finance and Accounting, 13*(3), 369–398. https://doi.org/10.1504/AAJFA.2023.132205

Al-Qudah, A. A., & Houcine, A. (2023). Firms' characteristics, corporate governance, and the adoption of sustainability reporting: evidence from Gulf Cooperation Council countries. Journal of Financial Reporting and Accounting. Vol. aheadof-print No. ahead-of-print. https://doi.org/10.1108/JFRA-02-2023-0066

Avrampou, A., Skouloudis, A., Iliopoulos, G., & Khan, N. (2019). Advancing the sustainable development goals: Evidence from leading European banks. *Sustainable Development, 27*(4), 743–757. https://doi.org/10.1002/sd.1938

Baldissera, A. (2023). Sustainability reporting in banks: History of studies and a conceptual framework for thinking about the future by learning from the past. *Corporate Social Responsibility and Environmental Management*. https://doi.org/10.1002/csr.2491

Beck, A. C., Campbell, D., & Shrives, P. J. (2010). Content analysis in environmental reporting research: Enrichment and rehearsal of the method in a British–German context. *The British Accounting Review, 42*(3), 207–222.

Bexell, M., & Jönsson, K. (2021). *The politics of the sustainable development goals: Legitimacy, responsibility, and accountability* (p. 185). Taylor & Francis. https://doi.org/10.4324/9781003043614

Bidari, G., & Djajadikerta, G. (2020). Factors influencing corporate social responsibility disclosures in Nepalese banks. *Asian Journal of Accounting Research, 5*(2), 209–224. https://doi.org/10.1108/AJAR-03-2020-0013

Biswas, S. S., Ahad, M. A., Nafis, M. T., Alam, M. A., & Biswas, R. (2021). Introducing "α-sustainable development" for transforming our world: A proposal for the 2030 agenda. *Journal of Cleaner Production, 321*, 129030. https://doi.org/10.1016/j.jclepro.2021.129030

Bonifácio Neto, J., & Branco, M. C. (2019). Controversial sectors in banks' sustainability reporting. *International Journal of Sustainable Development & World Ecology, 26*(6), 495–505. https://doi.org/10.1080/13504509.2019.1605546

Bose, S., Bhattacharyya, A., & Islam, S. (2016). Dynamics of firm-level financial inclusion: Empirical evidence from an emerging economy. *Journal of Banking and Finance Law and Practice, 27*(1), 47–68.

Bose, S., Khan, H. Z., Rashid, A., & Islam, S. (2018). What drives green banking disclosure? An institutional and corporate governance perspective. *Asia Pacific Journal of Management, 35*, 501–527. https://doi.org/10.1007/s10490-017-9528-x

Buallay, A., & Al-Ajmi, J. (2020). The role of audit committee attributes in corporate sustainability reporting: Evidence from banks in the Gulf Cooperation Council. *Journal of Applied Accounting Research, 21*(2), 249–264. https://doi.org/10.1108/JAAR-06-2018-0085

Buniamin, S., Jaffar, R., Ahmad, N., & Johari, N. H. (2021). Exploring SDGs disclosure among public listed companies in Malaysia: A case of energy-related SDGs. *Global Business & Management Research, 13.*

Chagas, E. J. M. D., Albuquerque, J. D. L., Maia Filho, L. F. A., & Ceolin, A. C. (2022). Sustainable development, disclosure to stakeholders and the Sustainable Development Goals: Evidence from Brazilian banks' non-financial reports. *Sustainable Development, 30*(6), 1975–1986. https://doi.org/10.1002/sd.2363

Chakroun, R., Matoussi, H., & Mbirki, S. (2017). Determinants of CSR disclosure of Tunisian listed banks: A multi support analysis. *Social Responsibility Journal, 13*(3), 552–558.

Chatzitheodorou, K., Tsalis, T. A., Tsagarakis, K. P., Evangelos, G., & Ioannis, N. (2021). A new practical methodology for the banking sector to assess corporate sustainability risks with an application in the energy sector. *Sustainable Production and Consumption, 27*, 1473–1487. https://doi.org/10.1016/j.spc.2021.03.005

Chouaibi, S., & Affes, H. (2021). The effect of social and ethical practices on environmental disclosure: Evidence from an international ESG data. *Corporate Governance: THe International Journal of Business in Society, 21*(7), 1293–1317. https://doi.org/10.1108/CG-03-2020-0087

Correa-Garcia, J. A., Garcia-Benau, M. A., & Garcia-Meca, E. (2020). Corporate governance and its implications for sustainability reporting quality in Latin American business groups. *Journal of Cleaner Production, 260*, 121142. https://doi.org/10.1016/j.jclepro.2020.121142

Cosma, S., Venturelli, A., Schwizer, P., & Boscia, V. (2020). Sustainable development and European banks: A non-financial disclosure analysis. *Sustainability, 12*, 6146. https://doi.org/10.3390/su12156146

Dong, S., Xu, L., & McIver, R. P. (2022). Sustainability reporting quality and the financial sector: Evidence from China. *Meditari Accountancy Research*. https://doi.org/10.1108/MEDAR-05-2020-0899

Dowling, J., & Pfeffer, J. (1975). Organizational legitimacy: Social values and organizational behavior. *Pacific Sociological Review, 18*(1), 122–136. https://doi.org/10.2307/1388226

García-Meca, E., & Martínez-Ferrero J. (2021). Is SDG reporting substantial or symbolic? An examination of controversial and environmentally sensitive industries. *Journal of Cleaner Production, 298*, 126781. https://doi.org/10.1016/j.jclepro.2021.126781.

European Commission. (2019). *JRC Technical Report, Interlinkages and policy coherence for the sustainable development goals implementation, 2019*, p. 4: http://publications.jrc.ec.europa.eu/repository/bitstream/JRC115163/sdg_interlinkages_jrc115163_final_on_line.pdf

Fang, Y., Hasan, I., Leung, W. S., & Wang, Q. (2019). Foreign ownership, bank information environments, and the international mobility of corporate governance. *Journal of International Business Studies, 50*, 1566–1593. https://doi.org/10.1057/s41267-019-00240-w

Ferrero-Ferrero, I., Muñoz-Torres, M. J., Rivera-Lirio, J. M., Escrig-Olmedo, E., & Fernández-Izquierdo, M. Á. (2023). SDG reporting: An analysis of corporate sustainability leaders. *Marketing Intelligence & Planning, 41*(4), 457–472. https://doi.org/10.1108/MIP-07-2022-0332

Freeman, R. E. (2010). *Strategic management: A stakeholder approach.* Cambridge University Press.

Garcia-Sanchez, I. M., Aibar-Guzman, B., Aibar-Guzman, C., & Rodriguez-Ariza, L. (2020a). "Sell" recommendations by analysts in response to business communication strategies concerning the sustainable development goals and the SDG compass. *Journal of Cleaner Production, 255*, 120194. https://doi.org/10.1016/j.jclepro.2020.120194

Garcia-Sanchez, I.-M., Rodriguez-Ariza, L., Aibar-Guzman, B. A., & Aibar-Guzman, C. (2020b). Do institutional investors drive corporate transparency regarding business contribution to the sustainable development goals? *Business, Strategy and the Environment*. https://doi.org/10.1002/bse.2485

Germann, V., Borgwardt, F., Fischer, J., Fuchs-Hanusch, D., Regelsberger, M., Schubert, G., ... & Langergraber, G. (2023). Development and evaluation of options for action to progress on the SDG 6 targets in Austria. *Journal of Environmental Management, 325*, 116487. https://doi.org/10.1016/j.jenvman.2022.116487

Githaiga, P. N., & Kosgei, J. K. (2023). Board characteristics and sustainability reporting: A case of listed firms in East Africa. *Corporate Governance: THe International Journal of Business in Society, 23*(1), 3–17. https://doi.org/10.1108/CG-12-2021-0449

Gunawan, J., Permatasari, P., & Sharma, U. (2022). Exploring sustainability and green banking disclosures: A study of banking sector. *Environment, Development and Sustainability, 24*(9), 11153–11194. https://doi.org/10.1007/s10668-021-01901-3

Haladu, A., & Salim, B. B. (2016). Corporate ownership and sustainability reporting: Environmental agencies' moderating effects. *International Journal of Economics and Financial Issues, 6*(4), 1784–1790.

Hamad, S., Lai, F. W., Shad, M. K., Khatib, S. F., & Ali, S. E. A. (2023). Assessing the implementation of sustainable development goals: Does integrated reporting matter? *Sustainability Accounting, Management and Policy Journal, 14*(1), 49–74. https://doi.org/10.1108/SAMPJ-01-2022-0029

Hassan, A. N. O., Győri, Z., & Abdelrahman, R. E. K. (2022). Sustainable development goals disclosure practices through integrated reporting: An empirical analysis on European Financial Institutions. *Edukacja Ekonomistów I Menedżerów, 62*(4). https://doi.org/10.33119/EEIM.2021.62.1

Higgins, C., & Coffey, B. (2016). Improving how sustainability reports drive change: A critical discourse analysis. *Journal of Cleaner Production, 136*, 18–29. https://doi.org/10.1016/j.jclepro.2016.01.101

Global Reporting Initiative & AG Sustentable (2020). Divulgación ASG y de sostenibilidad en mercados de capitales. Una mirada a América Latina. Retrieved from: https://www.globalreporting.org/media/gnelv4de/informe_divulgacion_gri.pdf

Ijaz, A., Siddiqui, S. H., & Rasheed, R. (2020). SDGs Implementation and Future Recommendations: A Case of Banking Sector in Pakistan. *Pakistan Journal of Social Sciences, 40*(4), 1687–1699. Retrieved from https://pjss.bzu.edu.pk/index.php/pjss/article/view/969

Islam, A. M., & Deegan, C. (2008). Motivations for an organisation within a developing country to report social responsibility information: Evidence from Bangladesh. *Accounting, Auditing & Accountability Journal, 21*(6), 850–874. https://doi.org/10.1108/09513570810893272

Khan, H. Z., Bose, S., Mollik, A. T., & Harun, H. (2021). "Green washing" or "authentic effort"? An empirical investigation of the quality of sustainability reporting by banks. *Accounting, Auditing & Accountability Journal, 34*(2), 338. https://doi.org/10.1108/AAAJ-01-2018-3330

Krasodomska, J., Zieniuk, P., & Kostrzewska, J. (2022). Reporting on sustainable development goals in the European Union: What drives companies' decisions? *Competitiveness Review: An International Business Journal, 33*(1), 120–146. https://doi.org/10.1108/CR-12-2021-0179

Krippendorff, K. (2018). *Content analysis: An introduction to its methodology*. Sage publications.

Kumar, K. (2022). Emerging phenomenon of corporate sustainability reporting: Evidence from top 100 NSE listed companies in India. *Journal of public affairs, 22*(1). https://doi.org/10.1002/pa.2368

Kumar, K., & Prakash, A. (2020). Managing sustainability in banking: Extent of sustainable banking adaptations of banking sector in India. *Environment, Development and Sustainability, 22*, 5199–5217. https://doi.org/10.1007/s10668-019-00421-5

Lensink, R., Meesters, A., & Naaborg, I. (2008). Bank efficiency and foreign ownership: Do good institutions matter? *Journal of Banking & Finance, 32*(5), 834–844. https://doi.org/10.1016/j.jbankfin.2007.06.001

Lodhia, S., Kaur, A., & Kuruppu, S. C. (2022). The disclosure of sustainable development goals (SDGs) by the top 50 Australian companies: Substantive or symbolic legitimation? *Meditari Accountancy Research*. https://doi.org/10.1108/MEDAR-05-2021-1297

Maama, H. (2021). Achieving financial sustainability in Ghana's banking sector: Is environmental, social and governance reporting contributive? *Global Business Review*. https://doi.org/10.1177/09721509211044300

Neter, J., Kutner, M. H., Nachtsheim, C. J., & Wasserman, W. (1996). Applied linear statistical models.

Pizzi, S., Rosati, F., & Venturelli, A. (2021). The determinants of business contribution to the 2030 Agenda: Introducing the SDG Reporting Score. *Business Strategy and the Environment, 30*(1), 404–421. https://doi.org/10.1002/bse.2628

Rosati, F., & Faria, L. G. D. (2019). Business contribution to the sustainable development agenda: Organizational factors related to early adoption of SDG reporting. *Corporate Social Responsibility and Environmental Management, 26*(3), 588–597. https://doi.org/10.1002/csr.1705

Sardianou, E., Stauropoulou, A., Evangelinos, K., & Nikolaou, I. (2021). A materiality analysis framework to assess sustainable development goals of banking sector through sustainability reports. *Sustainable Production and Consumption, 27*, 1775–1793. https://doi.org/10.1016/j.spc.2021.04.020

Scott, L., & McGill, A. (2018). *From promise to reality: Does business really care about the SDGs*. PriceWaterhouseCoopers.

Silva, S. (2021). Corporate contributions to the sustainable development goals: An empirical analysis informed by legitimacy theory. *Journal of Cleaner Production, 292*, 125962. https://doi.org/10.1007/s10668-021-01371-7

Sumarta, N. H., Rahardjo, M., Satriya, K. K. T., Supriyono, E., & Amidjaya, P. G. (2023). Bank ownership structure and reputation through sustainability reporting in Indonesia. *Social Responsibility Journal, 19*(6), 989–1002.

Taylor, R. (1990). Interpretation of the correlation coefficient: A basic review. *Journal of Diagnostic Medical Sonography, 6*(1), 35–39. https://doi.org/10.1177/875647939000600106

Terán, R. S., & Cortés, D. L. (2023). Divulgación sobre sostenibilidad en el sector bancario de Ecuador: medios utilizados, factores determinantes y tipo de información revelada. *Entramado, 20*(2). https://doi.org/10.18041/1900-3803/entramado.2.10186

UN Global Compact (2015). SDG industry matrix: Financial Services. United Nations. Retrieved from: https://d306pr3pise04h.cloudfront.net/docs/issues_doc%2Fdevelopment%2FSDGMatrix_FinancialSvcs.pdf.

United Nations (2022). The Sustainable Development Goals: Report 2022. Department of Economic and Social Affairs, United Nations. Retrieved from https://home.kpmg/xx/en/home/insights/2020/11/the-time-hascome-survey-of-sustainability-reporting.html.

United Nations General Assembly (2015). Transforming our world: The 2030 Agenda for Sustainable Development. United Nations. Retrieved from https://www.un.org/en/development/desa/population/migration/generalassembly/docs/globalcompact/A_RES_70_1_E.pdf.

Van der Waal, J. W., & Thijssens, T. (2020). Corporate involvement in sustainable development goals: Exploring the territory. *Journal of Cleaner Production, 252*, 119625. https://doi.org/10.1016/j.jclepro.2019.119625

Zimmermann, S. (2019). Same but different: How and why banks approach sustainability. *Sustainability, 11*(8), 2267–2286. https://doi.org/10.3390/su11082267

Dr. Maria del Mar Gálvez-Rodríguez is Associate Professor of Management in the Department of Economics and Business at the University of Almería (Spain). Her research interests are focused on good governance and the strategic management of Information Technologies. She is author of numerous articles in national and international journals, including PLOS ONE, Tourisms Management Perspective, Online Information Review, Internet Research, Voluntas, Public Relations Review, Information Development, Information, Communication and Society etc. Also, she has written some book chapters (IGI Global and Springer).

Walter Daniel Ovelar Fernandez is Bachelor Degree in Accounting & Management by the Universidad Nacional de Asunción (UNA), Paraguay; Master Degree in Accounting & Information Systems by the University of Kansas (KU), USA. He is currently pursuing his Ph.D. regarding information disclosure practices in the banking sector. He teaches undergraduates, masters and doctoral courses in auditing, internal controls, management information systems and strategic management in public and private universities in Paraguay. He is a researcher of the Paraguayan National Council of Science and Technology (Consejo Nacional de Ciencia y Tecnología, CONACYT). He has published five books in his area of expertise and has a number of international paper publications to his credit. He has worked as a management consultant for many international and Paraguayan public and private organizations in the past.

Dr. Carmen Caba-Perez is a Professor in the Public Management Sector at the University of Almeria. Her research is concerning to transparency of public administration; nonprofit organizations and Venture Capital sectors. She is author of several articles published in JCR publications including The American Review of Public Administration, International Review of Administrative Science, Online Information Review, Government Information Quarterly, Voluntas, Latin American Research Review or Public Administration and Development, Public Relations Review, Information Development, etc. Also, she has written more than 15 book chapters (Kluwer Academic Publishers; IGI Global, Springer, Cappelen Akademisk Forlag).

Environmentally Sustainable Agricultural Production

Role of Sustainability and Circular Economy in Europe's Common Agricultural Policy

Francisco José Castillo-Díaz, Luis J. Belmonte-Ureña, Juan Fernando Álvarez-Rodríguez, and Francisco Camacho-Ferre

Acronyms

CAP	Common Agricultural Policy
GDP	Gross Domestic Product
GVA	Gross Value Added
EAGGF	European Agricultural Guidance and Guarantee Fund
CMO	Common Market Organization
EEC	European Economic Community
CEEC	Central and Eastern European Countries
BE-WA	Belgium-Walloon
BE-FL	Belgium-Flanders
BG	Bulgaria
CZ	Czech Republic
DK	Denmark
DE	Germany
EE	Estonia
IE	Ireland
EL	Greece
ES	Spain
FR	France
HR	Croatia

F. J. Castillo-Díaz (✉) · F. Camacho-Ferre
Department of Agronomy, University of Almeria, Almería, Spain
e-mail: fcastillo@ual.es; fcamacho@ual.es

L. J. Belmonte-Ureña
Department of Economy and Business, University of Almeria, Almería, Spain

J. F. Álvarez-Rodríguez
Faculty of Environmental and Rural Studies, Pontificia Universidad Javeriana, Bogota, Colombia
e-mail: juanfernandoalvarez@uniandes.edu.co

M. d. C. Valls Martínez and J. M. Santos-Jaén (eds.), *Environmentally Sustainable Production*, https://doi.org/10.1007/978-3-031-52656-5_4

IT	Italy
CY	Cyprus
LV	Latvia
LT	Lithuania
LU	Luxembourg
HU	Hungary
MT	Malta
NL	Netherlands
AT	Austria
PL	Poland
PT	Portugal
RO	Germany
SI	Slovenia
SK	Slovakia
FI	Finland
SE	Sweden

1 Introduction

Increasing productivity in the agri-food sector is an effective tool to reduce the poverty rate in some countries. China reduced its poverty rate by 51% in just four years thanks to this strategy (The World Bank, 2008), thus helping to safeguard food sovereignty and security of nations (European Commission, 2022d; Maudos & Salamanca, 2022a).

Food production generates significant wealth, mainly in rural areas (Maudos & Salamanca, 2022a, 2022b). The agri-food sector in the EU-27 accounted for 3.8% of GDP and 6.5% of total employment in 2021. However, the regional influence may be even more significant. Actually, the agri-food system of some Spanish regions accounted for 18.8% of Gross Value Added (GVA) and 21.4% of employment (Maudos & Salamanca, 2022a, 2022b). This situation can lead to the emergence of local productive systems that can benefit from the employment and wealth indirectly produced through their auxiliary industry (Honoré et al., 2019; Valera-Martínez et al., 2017).

However, food production causes negative environmental impacts: loss of genetic diversity, erosion, desertification and soil degradation, decrease and losses in water quality, high emissions of greenhouse gases, or substantial alteration of endogenous fauna and flora. Poor management of agricultural inputs or residues is usually the causal agent (Castillo-Díaz et al., 2021a, 2021b, 2022; Dhaoui et al., 2022; Duque-Acevedo et al., 2020, 2022; European Union, 2019; Gómez-Tenorio et al., 2021; IPCC, 2019; Jacobs et al., 2022; López-Serrano et al., 2021, 2022; Pedraza et al., 2015; Región de Murcia, 2019).

1.1 The Role of Agricultural Policy in Food Security

The specific characteristics of the agricultural sector make it an activity in which governments intervene to balance agricultural income with that of the rest of the economy (European Commission, 2022a; Khafagy & Vigani, 2022; Pe & Lakner, 2020). The main reason for this intervention is the inelastic nature of the demand and supply curve for food products. From an economic point of view, the solution is to reduce the number of farms and concentrate production units (Pañeda-Fernández, 1999). However, the social consequences of such a transformation could be devastating for the equilibrium of many territories since agriculture is the main source of income in many rural environments (Honoré et al., 2019; Maudos & Salamanca, 2022b, 2022c). For this reason, agricultural policies, such as the European Union's CAP, which has been periodically updated to meet changing needs, play an essential role in maintaining the sustainability of agricultural systems and avoiding the influence of external agents (Khafagy & Vigani, 2022; Massot Martí, 2000; Pañeda-Fernández, 1999; Pe & Lakner, 2020; Sáen-Lorite & Cejudo-García, 2013).

Recently, sustainable development has become an essential criterion for the economic expansion of many states. Since 2015, these states have increased their commitments towards sustainability (UN, 2015) due to their adherence to the 2030 Agenda. Sustainability is composed of three subcomponents: social, economic, and environmental. Many economic activities, including the primary sector, have neglected the environmental component (Duque-Acevedo et al., 2020; López-Serrano et al., 2023; Prados-peña et al., 2022; Tsalis et al., 2022). The European Union has founded its economic progress on a system based on the circular economy and it is seeking to neutralize its environmental footprint by 2050. This economic system aims to reduce inputs, reuse by-products, recycle waste, and repair damage. To meet this objective, it has developed a set of strategies containing various goals, which are currently aimed at 2030 but will be systematically reviewed until 2050. In the current period, these goals pursue a 50% reduction in the use pesticides, a 20% cutback of fertilizers, and 50% drop in antimicrobials (Cifuentes-Faura, 2022; European Commission, 2018, 2019, 2020a, 2020b, 2020c; European Union, 2015; Kirchherr et al., 2017; Mazur-Wierzbicka, 2021).

In recent years, the reformulation of the CAP for the period 2023–2027 has been negotiated. The influence of the new environmental policies has been a key factor in this modification. The CAP has incorporated new mechanisms to ensure compliance with sustainability, based, in some cases, on its previous historical periods (European Council, 2021a; Pañeda-Fernández, 1999). Therefore, the objectives of this chapter were (i) to identify the main historical milestones that have marked the course of the CAP up to the 2023–2027 period; (ii) to characterize the new features of the 2023–2027 CAP; (iii) to identify the perceptions of European society regarding the CAP; (iv) to identify the main changes in the CAP in the 2023–2027 period; and (v) to identify the main changes in the CAP in the 2023–2027 period. For this purpose, a systematic review of specialized literature and official reports of the European Union was carried out using the snowball method (Batlles-delaFuente et al., 2022).

2 History of the European Common Agricultural Policy

After World War II, the foundations were laid for an economic and political alliance that gave rise to the European Economic Community (EEC). In 1955, Messina Conference started work on the 1956 Spaak Report, which laid down the principles of the Treaty of Rome. The treaty identified the need to establish a common agricultural policy to address the problems arising from the instability of agricultural prices, the European farms being managed by families, and the application of an unequal agricultural policy. The initial CAP sought to increase the productivity of the European countryside through the modernization of agricultural holdings so that they would offer a higher level of income and employment to those dependent on the sector while ensuring the sovereignty and food security of the EEC. These objectives were intended to expand social and economic sustainability, but the environmental aspect was neglected (European Community, 2002).

The Treaty of Rome created the possibility of Common Market Organizations (CMOs) and one or more sources of funding for their financing (European Community, 2002). A conference to lay the final foundations of the CAP was also recommended. At The Stresa Conference in 1958, it was decided that the CAP should be based on the principles of family farms, the maintenance of trade relations with third world countries, the policy of pricing, markets, and agricultural infrastructure, and the establishment of mechanisms for action against overproduction (Conférence de Stresa, 1958).

In 1962, the first CMOs and the European Agricultural Guidance and Guarantee Fund (EAGGF) were created. The EAGGF constituted the first pillar of the CAP. The first CAP was based on the principles of market unity, a single price throughout the Community, Community preference, protectionism against the external market, and financial solidarity (Pañeda-Fernández, 1999).

2.1 *Actual Effects of the CAP*

The CAP market unity brought about strong disputes between member nations (e.g., Germany and France), which resulted in the price of cereals being set almost 50% higher than the stable world price. The solution hurt consumers and livestock farmers in the foreign market and financial solidarity (Borrel & Hubbard, 2000; Pañeda-Fernández, 1999).

The CAP market unity brought about strong disputes between member nations (e.g., Germany and France), which resulted in the price of cereals being set almost 50% higher than the stable world price. The solution hurt consumers, livestock farmers in the foreign market, and financial solidarity (Borrel & Hubbard, 2000; Pañeda-Fernández, 1999).

2.2 *Mansholt Plan*

At the end of the 1970s, the production of certain food products exceeded demand. The solution proposed by the EEC was to reduce production units and the number of farmers, given the social impossibility of eliminating the price policy. As a result, the Mansholt Plan was born, contrary to some of the initial principles of the CAP. The objectives of this plan were not achieved, due to the limited resources allocated and the periods of economic crisis suffered by the cities, which did not provide sufficient employment for the rural population and found the primary sector as a refuge (Pañeda-Fernández, 1999; Stead & Jahre, 2007).

2.3 *A Decade in Search of Financial Sustainability: Early CAP Reforms*

In the early 1980s, EAGGF Guarantee expenditure accounted for 95% of total EAGGF expenditure, with the dairy sector accounting for nearly 40%. EAGGF Guarantee disbursement rose by 32% from 1978 to 1980, which led to the unsustainability of the CAP. In 1977, the co-responsibility levy was established in the dairy sector. In 1981, the guarantee thresholds were created, which reduced the intervention price if production exceeded the pre-set threshold for the sector as a whole. Finally, in 1984, the first production limits per farm were set (e.g., milk quotas). In 1986, the co-responsibility levy was established for cereals, quotas in the dairy sector were reduced, and the abandonment of farmland was encouraged through compensation payments. In 1988, agricultural expenditure was again modulated with the introduction of the figure of the stabilizers, which set maximum guaranteed amounts based on the average for each sector in 1987. If the amounts were exceeded, the intervention price was reduced (Pañeda-Fernández, 1999).

After establishing production quotas, the Community suffered serious budgetary problems because expenses could grow at a faster rate than resources. The metamorphosis of the EEC from an international organization with an importing vocation to an exporting one also played a fundamental role. At this time, sustainability was based on economic equilibrium, which was achieved by reducing expenses. In addition, the need was identified for the EEC taxpayer to finance part of the CAP instead of the responsibility falling entirely on the consumer (Pañeda-Fernández, 1999).

2.4 *The 1992 McSharry Reform*

The main difference between this reform and the reforms carried out in the 1980s is that expenditure control becomes a "secondary" issue after limiting the price of interventions. It also emphasized two points of interest: a better internal distribution of farm income to modulate it towards smaller farmers and the reinterpretation of the principle of financial solidarity (Pañeda-Fernández, 1999).

The McSharry reform came into effect in the 1993–1994 marketing year and affected mainly arable crops, tobacco, beef and dairy cattle, meat and dairy sheep/goats, and dairy products. After the implementation of the reform, the internal price of cereals fell by 29% from 1991/1992 to 1995/1996, which resulted in a drop in the production costs of livestock farming. The price decrease was compensated by direct payments per hectare or, in the case of livestock, per head, which transferred the cost from consumers to taxpayers, and by the set-aside of the crop area. Regarding cattle, the intervention system was preserved, with a 15% price reduction from 1993/1994 to 1995/1996. However, the level of intervention purchases was restricted. To compensate for these initiatives, the allowance per non-breeding cow and male bovine was increased. Payments were subject to a density factor. In the dairy sector, quotas were maintained until the 2000/2001 marketing year. The co-responsibility levy was eliminated in 1993, but the levy was maintained in case of exceeding each farm's quota. However, not all sub-sectors received compensation for the fall in prices. The CMOs for fruit and vegetables and wine and oil were reformed in accordance with the principle of financial equilibrium and therefore suffered losses (Camacho-García et al., 2011).

The 1992 reform laid the foundations for reducing trade distortions, although it did not affect all subsectors and the decoupling of aid was partial. In addition, the CAP was allowed to accommodate the commitments acquired in the World Trade Organization's Agreement on Agriculture. Likewise, consumers were the main beneficiaries of the decrease in food costs (Camacho-García et al., 2011).

On the other hand, the McSharry reform included environmental initiatives through the so-called accompanying measures. Council Regulation (EEC) 2078/92 of June 30, 1992, encouraged the reduction of agrochemicals, the expansion of organic farming, or the reduction of the number of animals per farm or per hectare to allow the extensification of the productions included in Council Regulation (EEC) 2328/91 of July 15, 1991. Also, Council Regulation (EEC) 2080/92 of June 30, 1992, established a framework for subsidizing reforestation and, therefore, expanding the Community's forestry resources (European Council, 1991, 1992a, 1992b).

2.5 *The Challenges Posed by Agenda 2000 in the 1999 CAP Reform*

A review of the European Union's policies was carried out with the Agenda 2000. Regarding agriculture, it was established that this sector had to be sustainable and adapt to the changes brought about by political and commercial exchanges (European Commission, 1997). The European agricultural model had to be built on the principle of multifunctionality and based on the weight of productive factors (water and soil resources, employment, wealth, etc.), their positive externalities (conservation of the environment and biodiversity, food security and sovereignty, rural development, etc.), and the need to regulate agricultural markets to increase their commercial openness to the outside world (Massot Martí, 2000). For this reason, the reform focused on four main points according to Camacho-García et al. (2011):

- A new adjustment of EU prices in relation to world prices was made, which was compensated with direct aid to producers. In addition, the budget was limited to 74% of the European Union's GDP.
- A second pillar of the CAP was created for rural development, as opposed to the first pillar, which dealt with price and market measures. The measures of the second pillar were grouped into a horizontal regulation with socio-structural initiatives, modernization, first installation or early cessation; measures of agro-environmental and environmental nature or reforestation programs; and measures for the improvement and diversification of the rural environment. The second pillar was combined with the LEADER program created in 1991.
- First pillar aid was modulated, allowing a 20% reduction in the amount received, which was to be used for agri-environmental aid, and facilitating the transfer of funds between the first and second pillars. However, it was voluntary, so it was only implemented in France and the United Kingdom.
- Eco-conditionality was implemented. For the first time, Member States could make the receipt of direct payments conditional on the achievement of environmental targets and reduce or withdraw the allocation granted.

2.6 2003 Mid-Term Review

This revision became an ambitious and substantial renewal of the European agricultural policy. One of its goals was to control the overall budget of the European Union. Therefore, in 2002, the European Council implemented the financial discipline mechanism that froze the CAP spending. Under this mechanism, agricultural expenditure contemplated in the first pillar (prices and markets) could not exceed the value set for 2006 for the following period (2007–2013) (Dixon & Matthews, 2006).

On the other hand, the adhesion of the Central and Eastern European Countries (CEECs) fueled and hardened the budgetary debate because the EU-15 regarded the CEECs as a threat. First, because of the competition that their agri-food products could cause in EU markets. Secondly, the EU-15 feared that the CEECs would monopolize most of the funds of the first and second pillars of the CAP, which would negatively affect the sustainability of many of their agri-food systems dependent on CAP funds and may make them lose their privileged position.

The solution found by the European Union was to transform direct payments into a single payment per farm not linked to production. The single payment was fixed based on the amounts received by producers in previous historical periods. In addition, a special regime was implemented for the CEECs, which maintained the privileged position of the EU-15 Member States until they converged in the 2013 reform.

The single payment solution led to two issues that could reduce the sustainability of European primary production. First, the decision to cultivate crops was made based on market considerations. Second, the single payment would lead to a progressive abandonment of farmland and, therefore, to rural depopulation. The solution was to establish flexibility clauses that allowed the partial maintenance of aid per hectare or head of livestock. Furthermore, an important element of the 2003 reform was that, although the granting of aid under the first pillar was not linked to production, it was related to the application of appropriate agronomic or environmental practices, which were defined by each Member State. In addition, from 2005 onwards, a voluntary external farm audit procedure was included.

Finally, the 2003 revision intensified the transfer of economic funds between the first and second pillars due to modulation. This modulation became compulsory and resulted in aid reductions of 3–5%. In 2007, a single CMO was created to simplify regulations and unify similar CMOs.

2.7 2008 Health Check-Up

The 2008 reform made it possible to consolidate and legitimize the modifications proposed in the 2003 review. The European Court of Auditors (2006) highlighted certain shortcomings in the 2003 CAP aid. This institution indicated that some Member States had not correctly applied certain elements of the single payment, allocating payment entitlements to owners of rented land, tenant farmers, or landowners who did not carry out primary activities. Regarding rural development and agri-environmental measures, a large number of errors were identified, one of them being primary producers not complying with the eligibility conditions. Moreover, in many cases these conditions were complex to apply, which resulted in a deficit payment control not verified in the field. Sáen-Lorite and Cejudo-García (2013) indicated that rural areas and primary producers did not fully benefit from the CAP as a result of the general goals of the policy.

With the 2008 health check, the direct payment was replaced by the single payment (except for sheep, goat, and suckler cow premiums) and controls increased. In addition, the funds of the first pillar were oriented in favor of rural development. Modulation was strengthened by cutting the amount subsidized to larger farms and eliminating small direct payments. Funds were transferred to the second pillar to address the so-called "new challenges." The "new challenges" planned by the CAP were those related to climate change, ecosystem biodiversity, water management, renewable energies, agri-food innovation, and accompanying measures to structure the dairy sector (European Council, 2009).

2.8 2013 Reform

In 2013, the agricultural policy was reformed, which was the last step to put the CAP on the path of sustainable development. The two funding pillars had to respond to the economic, environmental, and territorial challenges (European Council, 2013).

The figure of multifunctional support for specific objectives, which was granted to active farmers, was also introduced. The single payment was replaced by a quota subdivided into seven levels that were granted based on the applicant's profile, namely: (a) basic payment; (b) green payment; (c) payment to young farmers; (d) a redistributive payment; (e) payment for natural constraints; (f) production-linked aid; (g) payment to small farmers. Direct payments were adjusted based on the extreme convergence mechanism so that Member States could access a minimum payment per hectare.

The green payment was granted to producers who applied environmental practices such as crop diversification, maintenance of permanent pastures, or had areas of ecological interest. Farmers who were automatically entitled to the payment were those who carried out organic farming, were under the small farmer regime—as they were not obliged to do so—, had permanent crops (vineyards, olive groves, wheat, fruit trees), or remained on the land for at least five years and did not enter into the crop rotation.

In addition, the modulation figure to transfer funds from the first to the second pillar was eliminated and replaced by a mandatory reduction for basic payments exceeding 150,000 euros. However, from 2015 moving forward, the transfer of funds between both pillars and in either direction is allowed, with a limit of 15% for transfers from the first pillar to the second pillar and 25% in the opposite direction. This way, rural development measures implemented in the second pillar acquire a more integrated, territorial, and specific character. The set of initiatives identified in this fund has been simplified. These initiatives focus on improving competitiveness, agri-food innovation, knowledge, promotion of young farmers, sustainable environmental management, and territorial articulation. The tools of the single CMO created in previous stages were consolidated, creating a reserve to face crises. In addition, the measures that controlled the supply of agri-food products (i.e., the dairy, grapevine, and sugar sectors) were eliminated.

3 A New Policy Reformulation: CAP 2023–2027

Figure 1 shows the evolution, projection, and composition of the CAP budget over the period of 1980–2027. On average, 9% of Europeans have some detailed information about the CAP, 64% of Europeans have heard of the CAP, and 27% of Europeans are unaware of its existence (European Commission, 2020). The majority of Europeans think that the CAP budget allocation is justified because it makes it possible to

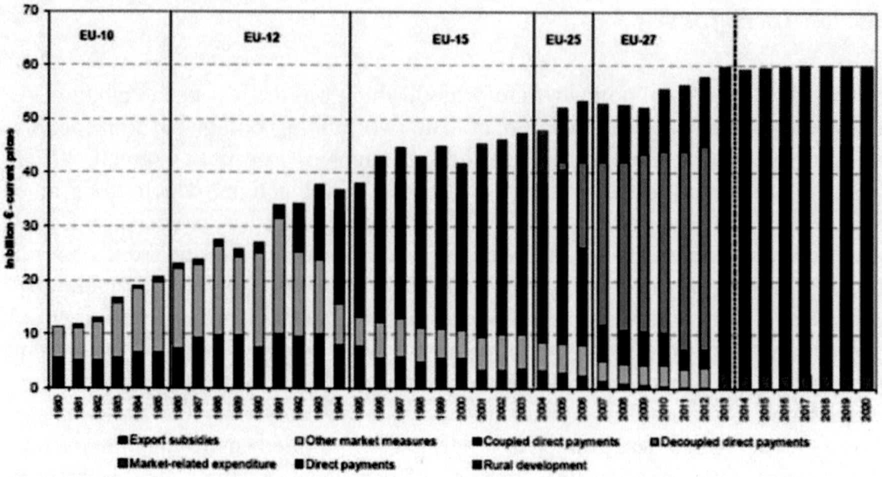

Fig. 1 Evolution and projection of the CAP budget for the period 1980–2027. *Source* EC-DG AGRI (2022)

guarantee food security in the EU, help achieve sustainable agriculture, and deal with the additional costs of European primary production (Fig. 2).

The 2023–2027 CAP places greater emphasis on climate change mitigation, equity for producers to achieve income redistribution, and improving conditions for small and medium-sized farms. It is also based on the performance and results of agricultural holdings, aimed at financing those farmers and stockbreeders who add value to the agri-food chain. The new objectives of the CAP have been shaped by the demands of European society, as shown in Fig. 3, where food safety and sustainability have taken on a fundamental role (Fig. 4).

The European Union has 28 strategic plans for the CAP 2023–2027 period, one per Member State, except for Belgium, which has two: Wallonia and Flanders. Europe has also allowed its Member States to adapt their policies to local conditions. The European Council plans to review these plans in 2025 and 2027, reserving the right to include new initiatives to catalyze the achievement of CAP objectives. The main content of the strategic plans prepared by the Member States is as follows (European Commission, 2022c; European Council, 2021b):

3.1 Socioeconomic Aspects

3.1.1 Active Farmer

Legislation changes have led to the introduction of a mandatory definition of active farmer in order to grant various types of subsidies (e.g. direct aid), leaving the Member States freedom to set these criteria. However, they must comply with the

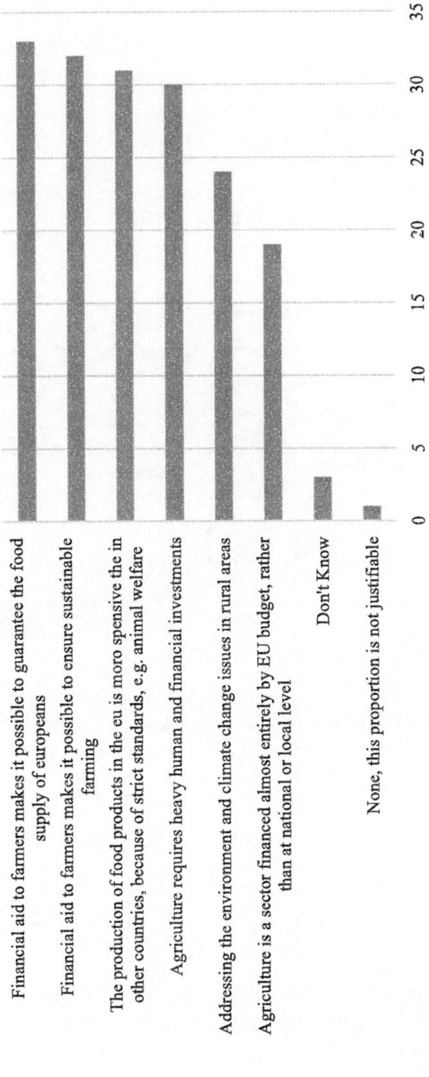

Fig. 2 Reasons selected by Europeans to justify their share of the European Union budget (n = 27,237). *Source* Own elaboration based on European Commission (2020)

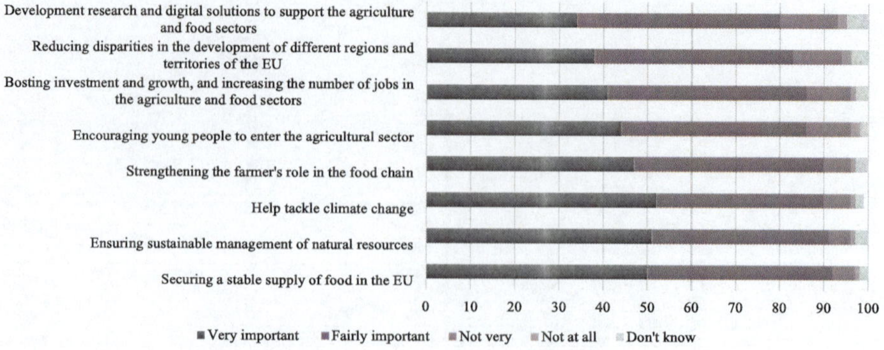

Fig. 3 Main demands of European society regarding agricultural policy objectives (n = 27,237). *Source* Own elaboration based on European Commission (2020)

Fig. 4 CAP objectives 2023–2027. *Source* European Commission (2022b)

minimum principles of agricultural activity, must not exclude multi-active and part-time farmers, and must include a list of activities that are not eligible for subsidies (Table 1).

Table 1 Criteria used by EU Member States to define the concept of active farmer

Concept	Specifications
Active farmer	• ES, BE-WA, FR, HR, HU, LT, MT, PL, RO, and SI opt for a negative, but a not complementary, list to identify non-active farmers • AT, BE-WA, BG, CY, DE, EL, ES, FI, HR, HU, IT, LT, LV, NL, PL, PT, RO, SE, SI, and SK set a maximum exemption threshold from the previous year's direct payments (usually EUR 5,000) • BG, CY, CZ, DK, EE, HU, IE, LU, PT1, and SE propose a minimum area or specific requirements • Direct payment entitlement has been based on an area threshold (0.3–0.4 ha) or financial thresholds (100–500 €). BE-FL, EL, ES, FI, FR, IE, IT, LU, and NL opt for the latter

Source Own elaboration based on European Commission (2022c)

Table 2 Criteria for the redistribution of aid proposed by EU Member States (i.e., decoupled aid)

Concept	Specifications
Redistributive support Complementary to Income for Sustainability (CRISS)	• In the strategic plans submitted, more than 10% of direct payments are allocated to CRISS • AT, DE, ES, HU, LT, LU, LV, and SK foresee more than one range of areas for the granting of support, and ES, EL, and AT have fixed unit amounts for groups of territories
Basic Income in Support of Sustainability (BISS)	• It accounts for more than 50% of the allocation of direct payments at the European level, varying from 31 to 75% • AT, EL, ES, FI, LV, and FR have territorialized this income
Payment to small farmers	• Allocation varies from 0.3 to 9.0% • Only MT, LV, CZ, PT, and BG have identified the simplified payment figure

Source Own elaboration based on European Commission (2022c)

3.1.2 Aid Redistribution

The new CAP aims to be fairer concerning income support, mainly for those producers with small or medium-sized farms. Thus, a mandatory redistribution aid mechanism has been incorporated, which must reach at least one-tenth of direct payments. However, Member States can derogate from this mechanism in other areas, like payment reduction, internal convergence, or territorial distribution of the basic payment. In addition, an optional mechanism to allow Member States to limit the granting of aid to larger farms has been established. The European Administration has agreed to replace direct payments with a single payment in the case of small farmers. Table 2 summarizes the main developments in this regard.

3.1.3 Reinforce the Position of Producers in the Agri-Food Chain

The new CAP has strengthened the position of farmers in the agrifood chain by boosting competitiveness. To this end, the following initiatives have been included:

- Sector-specific interventions. The new strategic plans extend the support of Producer Organizations, as in the fruit and vegetable subsector, but not in wine growing and beekeeping (Table 3). The specific subsectoral plans under the CMO become part of the strategic plans and are governed by the principle of results.
- Exceptions to the competition law. To encourage collaboration between producers and/or within other links in the chain, the exceptions to the competition law have been extended.

Table 3 Specific subsectoral measures of the new CAP (i.e., coupled support)

Concept	Specifications
Sectoral interventions and farmers positioning	• 24 Member States have implemented interventions only in the fruit and vegetable sector. Only EE, LU, and MT will not implement interventions as they do not have recognized producer organizations in this sector • All Member States are planning interventions in beekeeping • ES, FR, and IT propose an intervention in olive groves (oil and table olives) • 16 Member States propose intervention in the vine sector • Specific interventions: • **BG**: milk and dairy products • **CZ**: eggs, potatoes, and ornamental plants • **LV**: multiple sectors under one intervention • **IT**: potatoes • **SK**: potatoes, milk and dairy products, pork, sheep, and goat meat • **ES**: traditional olive groves
Coupled income support (CIS)	• Only NL does not plan to apply CIS • MT, BE-WA, PT, FI, HU, PL, HR, CZ, LV, LT, SK, SI, BG, RO, ES, FR, SE, EE, EL, and LU allocate more than 10% to the CIS • The CIS is mainly dedicated to the livestock sector, with 70% of the budget allocation. Although it also supports pulses/protein crops (BE-WA, BG, HR, CZ, FR, EL, HU, IE, IT, LV, LT, LU, PL, PT, RO, SK, SI, ES, and FI), fruits and vegetables (BG, HR, CY, CZ, EE, FR, EL, HU, IT, LV, LT, LU, PL, PT, RO, SK, SI, ES, and FI), fruits and vegetables (BG, HR, CY, CZ, EE, FR, EL, HU, IT, LV, LT, LU, MT, PL, PT, RO, SK, and ES), sugar beet (HR, CZ, EL, HU, IT, LT, PL, RO, SK, and ES), rice (FR, EL, HU, IT, PT, RO, and ES) and other cereals (EE, FR, EL, IT, LV, and PT)

Source Own elaboration based on European Commission (2022c)

- Supply side of PDO and PGI schemes. The supply side regulation scheme for PDO and PGI products, which was limited to ham and cheese producers, has been expanded to all subsectors.

A relevant fact is that Member States can allocate a limited fraction of their direct payment budget to support some strategic agricultural subsectors.

3.2 Market Orientation

The European Commission identifies the need for the CAP to be market-oriented to align agri-food exports with internal and external demand. However, regarding imports from third world countries, and in line with the review of EU trade policy and the framework of the circumstances defined by the CMO, Member States must

Table 4 Measures identified by the strategic plans to manage risks derived from food production

Concept	Specifications
Plans to manage risks	• DE, EE, FR, HR, HU, IT, LT, LV, NL, PL, PT, RO, and SK have identified risk management tools, such as insurance premium schemes (15 measures) and investment fund support schemes (7 schemes) • DE, EE, HR, LV, NL, PT, and SK have proposed to support insurance premiums exclusively, while BG, FR, HU, IT, LT, PL, and RO have set to combine insurance premiums with mutual funds or other management tools • BG (1.5%), IT (3.0%), and RO (3.0%) have allocated funds from direct payments to risk management

Source Own elaboration based on European Commission (2022c)

require compliance with internal production standards (i.e., sanitary, animal welfare, environmental) and the objectives of the European Green Pact and the Farm to Market Strategy. This criterion coincides with the demand of the European Union itself because 56% of Europeans think that agri-food imports should only enter EU territory if they comply with EU quality standards (European Commission, 2020).

It also identifies the need to create an emergency reserve of at least 450 million euros in case of emergency purchases and private storage aid (Table 4).

3.2.1 Internal and External Convergence

Member States that continue to make direct payments based on historical series must ensure that there is internal convergence of payments and that these payment entitlements are at 85% of the average value in 2026. In addition, DE, AT, DK, EL, FI, LU, MT, NL, PT, SE, and SI have decided to abolish payment entitlements, while the rest of the countries have opted for the 2026 convergence rule.

As for external convergence, countries that do not reach 90% of the average value of direct payments given by the European Union will increase their budget to match the level of support. Payments are expected to reach €200/ha in 2022 and €215/ha in 2027.

3.2.2 Social Dimension

The granting of aid is linked to compliance with various European Union labor laws. The state authority must inform the agricultural payment agencies of the procedures. In this regard, a mechanism to reduce the amount received by the producer in the event of non-compliance with these regulations has been included. In order for farmers and stockbreeders not to see their payment entitlement reduced, they must offer transparent and predictable employment conditions, they must inform their workers of the employment conditions (i.e., beginning and end of the contractual relationship, trial period, working hours, work location, vacations, social security), and ensure

Table 5 Main social initiatives of the new CAP

Concept	Specifications
Laboral social conditionality	• FR, IT, and AT will apply labor conditionality from 2023. ES and PT will do it from 2024. The other Member States will apply it from 2025
Generational renewal and support for young farmers	• Only DK and PT have not set up income support for young farmers • Only IE has not set up installation aid for young farmers • CY, CZ, FR, NL, PT, and SK have proposed to expand the granting of investment support to young farmers • FR, IT, HU, NL, IE, and ES have identified cooperation for generational renewal
Gender equality	• AT, DE, ES, IE, and IT have established specific initiatives to support rural women, although only ES and IE specifically address the goal of expanding women's participation in agriculture

Source Own elaboration based on European Commission (2022c)

the safety and health of their workers concerning the use of machinery and work equipment (i.e., clothing, protective equipment, handling of hazardous substances).

The new CAP also promotes the generational replacement of agricultural holdings, with a minimum of 3% of direct payments allocated for initiatives related to young farmers in the form of income, investment, or installation aid. In addition, the role of women in agriculture is expanded. The initiatives mentioned above are listed in Table 5.

3.3 *Environmental Criteria*

The new reformulation of the CAP is committed to transitioning toward sustainability (Table 6). To this end, a new green architecture that favors the integration of the two pillars of the CAP into a strategic plan has been designed. Green architecture goals are aligned with the CAP's environmental and climate objectives. The Member States' strategic plans should be more environmentally ambitious regarding their CAP 2014–2022 objectives, based on the principle of "no backsliding".

The stricter requirements result in reinforced cross-compliance to access basic aid. These requirements comprise statutory management conditions and Good Agricultural and Environmental Conditions (GAEC; Table 6). Farmers must comply with these to receive the basic payment entitlement (i.e., reinforced cross-compliance). These practices allow the principles of the circular economy to be applied.

Table 6 Main management conditions and Good Agricultural and Environmental Conditions

Concept	Specifications
Management conditions and other agri-environmental measures	
Biodiversity	• Member States have set the goal of preserving habitats and species, eleven of them have set as a requirement to cover from 0 to 20% of the Utilised Agricultural Area (UAA). In 5 the stratum has been set from 21 to 40% of the UAA, and in 7 the coefficient exceeds 41% of the UAA • The maintenance of landscape characteristics has led 15 Member States to propose a requirement to cover from 0 to 10% UAA, in 5 the stratum has been set at 11 to 40% UAA and in 7 the coefficient exceeds 41% UAA Ecological plans: • Biodiversity: AT, RO, and SE are the only Member States that do not include them • Pest/pesticide management: BE-FL, BE-WA, BG, CY, DE, DE, EE, HU, LU, MT, NL, PL, and PT include it • Nutrient management: BG, CY, DK, FI, HR, HU, IE, LU, LV, PL, SI, and SK include it • Extensive farming: BE-FL, BE-WA, BG, DE, ES, HR, IE, LT, LV, NL, PL, SI, and SK are included • Organic farming: BE-FL, BG, DK, EE, EL, FR, LT, LV, NL, PL, PT, and SE are included
Climate	• BE-FL, BE-WA, BG, CZ, DE, DK, EE, IE, EL, ES, FR, HR, LV, LU, HU, NL, AT, PT, RO, SI, SK, and FI foresee for carbon sequestration in soil and biomass, with target figures ranging from 2 to 86% • BE-FL, IE, EL, LV, LU, AT, PT, SK, and FI have set targets ranging from 1 to 60% of livestock units to meet ammonia or GHG commitments • CY, IT, LT, and SE are the only countries that have not set targets for soil protection • BE-FL, BE-WA, BG, CZ, DK, DE, EL, ES, FR, HR, HU, MT, AT, PL, PT, RO, SI, SK, and FI have set targets to cover UAA under climate adaptation • BE-FL, EE, IE, EL, ES, HR, HU, MT, PT, SI, SK, and FI foresee solar thermal power plants
Support for organic farming	• The 27 Member States support the growth of organic areas • Twenty-five Member States aim to increase organic farming areas through CAP support • Regarding the organic schemes and/or agri-environmental commitments, a mechanism has been set up to catalyze the conversion and maintenance of organic areas
Supporting multifunctional and sustainable forest management	• Only FI, IE, LU, NL and SE do not foresee any forestry aid • Less than half of the Member States have not established sustainable forest management objectives

(continued)

Table 6 (continued)

Concept	Specifications
Animal Welfare and Antimicrobial Resistance (AMR)	• Promotion of organic farming, and initiatives outside the CAP • BG, CZ, LV, EE, EL, FI, PT, SK, BE-FL y BE-WA have included the antimicrobial resistance indicator in their strategic plans • AT, CY, IT, LT, PL, PT, RO, and SK have planned interventions related to animal welfare in the ecological plans • AT, BE-FL, BE-WA, BG, EE, FI, FR, DE, EL, HU, IE, IT, LV, LT, LU, NL, PT, SK, SL, and ES are going to invest in improving their farms under animal welfare • AT, BE-FL, BE-WA, BG, HR, CY, CZ, EE, FI, FR, DE, EL, HU, IE, IT, LV, LU, PT, RO, SK, SL, ES, and SE have included animal welfare measures in their agri-environmental commitments
Good Agricultural and Environmental Conditions	
GAEC 2-Peatlands and moisture	• Nine Member States will apply the standard in 2023 (AT, BE-FL, BE-WA, DE, DK, LU, NL, FI, SE, and RO), and sixteen Member States have requested an extension (IE, FR, LT, SI, BG, CZ, EE, EL, ES, HR, IT, LV, HU, PL, PT, and SK)
GAEC 4-Establishment of buffer strips along watercourses	• Twenty-four Member States have set a minimum width of at least 3 m (AT, BE-FL, BE-WA, BG, CZ, CY, DE, DK, IE, ES, EL, FR, HR, IT, LT, LV, LU, HU, MT, PL, RO, SI, SK, FI and PT), two Member States have set a minimum width of more than 10 m (CY and LV), and three Member States have set the width of less than 3 m (EE, NL, and SE)
GAEC 7- Crop rotation	• Eleven Member States have not modified the crop rotation requirements (CZ, DE, HR, IT, CY, LT, MT, NL, RO, SI, and SK) • Eight Member States have identified exceptions in crop diversification (AT, BG, FR, LV, LU, PL, FI, and SE) • Eight Member States have proposed a combination of crop rotation and crop diversification (BE-FL, BE-WA, DK, EE, EL, ES, HU, IE, and PT)
GAEC 8- Surfaces and non-productive characteristics	• CZ and HU are the only countries that do not offer the basic option of 4% of non-arable land • Fourteen strategic plans have identified the supplement to the organic scheme (BE-WA, BG, EE, EL, ES, HR, LT, NL, PL, PT, SI, DK, LU, and RO) and sixteen Member States have included catch crops or nitrogen-fixing crops (BE-FL, BE-WA, BG, CZ, EE, EL, ES, HR, HU, LT, NL, PL, PT, SI, FR, CY, and SK) • Eleven Member States offer all three options (4% rule, organic, and catch crops or nitrogen-fixing crops (BE-WA, BG, EE, EL, ES, HR, LT, NL, PL, PT, and SI) • Twenty-five strategic plans apply exceptions to this rule (AT, BE-FL, BE-WA, BG, CZ, DE, EE, EL, ES, FR, HR, IT, CY, LV, LT, LU, HU, MT, PL, PT, RO, SI, SK, FI, and SE)

Source Own elaboration based on European Commission (2022c)

In the CAP 2014–2022 they were determined by practices such as crop diversification, maintenance of permanent pastures, and dedication of land to areas of ecological interest (Table 6).

3.4 Eco-Schemes

Eco-schemes are one of the main novelties of the new CAP. Thus, 25% of the direct aid budget is earmarked for eco-schemes, which consist of implementing practices of ecological interest that make it possible to improve the sustainability of agro-systems. These eco-schemes are annual or multiannual commitments, as established in the different strategic plans of the European Union Member States. The amount is set as a compensatory payment for losing profitability or as an incentive.

In this sense, the eco-schemes are compulsory for the Member States but voluntary for agricultural producers. Member States have a phase of adaptation during the 2023 and 2024 fiscal years and may allocate less than 25% to these architectures, provided that they are compensated in subsequent years. Member countries have identified reduction coefficients (i.e., capping or degressivity) to control spending (Table 7).

Table 7 Eco-schemes characteristics

Concept	Specifications
Eco-schemes	• BE-WA, CZ, EE, EL, IT, LV, NL, PT, RO, and SK have allocated more than the minimum 25% coefficient • Eight Member States have proposed the rebate mechanism (AT, CY, DE, DK, ES, FI, SE, SI) • One hundred and eighty-four different eco-schemes have been proposed, including 256 practices. Of these, 30% are additional payments • FR, HU, IE, and NL have proposed a single eco-scheme composed of several practices, while the rest of the States have defined several eco-schemes • CZ, FR, HU, IE, NL, and SK have defined an eco-scheme, which applies to the whole farm, with several commitments • The preferred practices are those based on soil conservation, maintenance of landscape and non-productive areas, preservation and expansion of biodiversity, carbon farming, permanent pasture, integrated pest management, agrochemical reduction, and animal welfare
Limitation and degressivity	• Twelve strategic plans foresee degressivity (IE, ES, SK, BE-FL, BE-WA, LV, LT, AT, BG, PT, SI and SE) • Six strategic plans aim to reduce labor costs before applying limitations and/or degressivity (LT, ES, SK, LV, BG)

Source: own elaboration based on European Commission (2022c)

Table 8 Support for areas with constraints and handicaps

Concept	Specifications
Constraints and handicaps	• Seventeen-percent of ERDF funds are destined for areas with natural or specific constraints, as identified by twenty-three Member States (IE, IT, HR, EL, CY, PT, AT, ES, MT, SI, PL, SE, DK, FR, FI, LT, LU, CZ, DE, SK, BG, RO, BE) • Zero point eight-percent of ERDF funds are earmarked for payments for the Natura 2000 Network and/or the Water Framework Directive

Source Own elaboration based on European Commission (2022c)

3.5 Rural Development

The rural development measures of the second pillar of the CAP mostly include actions associated with the climate, the environment, animal welfare, and areas with natural limitations. The CAP 2014–2022 allocation will increase by 5% (reaching 35%), destined for agri-environmental management, the Natura 2000 Network, the Water Framework Directive, and animal welfare. The endorsement will be reduced to 50% of payments to areas with natural limitations (Table 8).

3.5.1 Climate Monitoring

The commission will propose a methodology associated with a set of indicators to identify the contribution of the CAP to the climate and the environment. It will be implemented through a delegated regulation in 2025 moving forward.

4 Conclusion

Given the contents of this chapter, it can be observed how sustainability has played a role in the CAP. It is true that, in the initial periods, sustainability was based on expanding the economic and social subcomponents of European food production. Over time, forestry and agri-environmental measures have gained weight in the CAP. In addition, these measures were initially included as accompanying measures during the McSharry reform. The 1999 reform included the eco-conditionality mechanism, which allowed Member States to restrict aid under the first pillar of the CAP on environmental grounds. Agri-environmental measures were placed under the second pillar of the CAP. In successive periods, the environmental conditionality of aid has been profiled and some figures, such as the green payment, have been incorporated.

In the current period of the CAP (2023–2027), the Commission has strengthened the social and environmental role of the Common Agricultural Policy and has directed the granting of aid towards a policy of results while modernizing agricultural holdings through the implementation of new technologies resulting from digitalization. The latter is intended to meet the goals of the 2030 Agenda and the objectives set out in the European Green Pact and Circular Economy Strategy. To this end, the European Union has approved 28 Strategic Plans, one per Member State, with the exception of Belgium, which has submitted two.

Regarding the first pillar of the CAP, which has received the largest number of changes, the definition of an active farmer stands out. This definition aims to channel aid to primary producers who carry out agricultural activities and prevent CAP funds from being used for activities other than food production. This figure ensures that the aid mechanisms designed to improve the conditions of small and young farmers, for example, depend mainly on them. In addition, labor conditionality measures have been included to ensure the safety of workers in the primary sector. The role of the Producers' Organizations has been strengthened to defend the subsectors, and specific subsectoral measures have been established. In the area of sustainability, the figure of reinforced conditionality is highlighted, and producers must certify the application of various practices classified as sustainable to be granted basic aid. A revolutionary aspect of the CAP is the eco-schemes. For the first time in the history of the CAP, agri-environmental measures will be financed with funds from the first pillar of the CAP. They are mandatory for the Member States and voluntary for primary producers. In addition, the 28 Strategic Plans define 184 organic schemes. It should be noted that to receive aid under the eco-schemes it is not necessary to have basic payment entitlements. The measures under the second pillar of the CAP are mainly aimed at the agri-environmental improvement of agricultural systems, climate change mitigation, animal welfare, and improving conditions on sites with natural constraints.

In short, the reformulation of the CAP for the period 2023–2027 seeks to expand the triple aspect of the sustainability of agricultural systems and rural environments based on accumulated knowledge and society's demands.

References

Batlles-delaFuente, A., Abad-segura, E., González, M., & Cortés, F. (2022). An evolutionary approach on the framework of circular economy applied to agriculture. *Agronomy, 12*, 620. https://doi.org/10.3390/agronomy12030620

Borrel, B., & Hubbard, L. (2000). Global economic effects of the EU Common Agricultural Policy. *Economic Affairs, 20*(1), 16–26. https://doi.org/10.1111/1468-0270.00218

Camacho-García, A., Abellán-Gómez, J., García-Fernández, C., & López-Hernández, M. (2011). *Aproyo público a la agricultura española 2003–2010*. Ministerio de Medio Ambiente y Medio Rural y Marino.

Castillo Díaz, F. J., Marín-Guirao, J. I., Belmonte-Ureña, L. J., & Tello-marquina, J. C. (2021). Effect of repeated plant debris reutilization as organic amendment on greenhouse soil fertility. *International Journal of Environmental Research and Public Health, 18*(21), 11544. https://doi.org/10.3390/ijerph182111544

Castillo-Díaz, F. J., Belmonte-Ureña, L. J., Batlles-delaFuente, A., & Camacho-Ferre, F. (2022). Impact of the new measures related to the circular economy on the management of agrochemical packaging in Spanish agriculture and the use of biodegradable plastics. *Environmental Sciences Europe, 34*(1), 94. https://doi.org/10.1186/s12302-022-00671-7

Cifuentes-Faura, J. (2022). European Union policies and their role in combating climate change over the years. *Air Quality, Atmosphere and Health, 15*(8), 1333–1340. https://doi.org/10.1007/s11869-022-01156-5

Conférence de Stresa. (1958). Recueil des documents de la Conferénce Agricole des Éstats membres de la Communauté Économique Européenne. *A Stresa Du 3 Au 12 Juillet 1958*, 251.

Council, E. (1991). Council Regulation (EEC) No. 2328/91 of July 15, 1991 on improving the efficiency of agricultural structures. *Official Journal of the European Communities, L, 218*, 1–21.

Council, E. (1992a). Council Regulation (EEC) No. 2078/92 of 30 June 1992 on agricultural production methods compatible with the requirements of the protection of the environment and the maintenance of the countryside. *Official Journal of the European Communities, L, 215*, 85–92.

Council, E. (1992b). Council Regulation (EEC) No. 2080/92 of 30 June 1992 establishing a Community support scheme for forestry measures in agriculture. *Official Journal of the European Communities, L, 215*, 96–99.

Council, E. (2009). Council Regulation (EC) No 73/2009 of 19 January 2009 establishing common rules for direct support schemes for farmers under the common agricultural policy and establishing certain support schemes for farmers, amending Regulations (EC) No 1290/2005, (EC). *Official Journal of the European Union, L, 30*, 16–99.

Council, E. (2013). Regulation (EU) No 1307/2013 of the European Parliament and of the Council of 17 December 2013 laying down rules applicable to direct payments to farmers under support schemes included in the framework of the Common Agricultural Policy and repealing Counc. *Official Journal of the European Union, L, 347*, 608–670.

Council, E. (2021a). Regulation (EU) 2021/2115 of the European Parliament and of the Council of 2 December 2021 establishing rules on support for strategic plans to be drawn up by Member States under the common agricultural policy (CAP Strategic Plans) and financed by the Eur. *Official Journal of the European Union, L, 435*(2), 1–186.

Council, E. (2021b). Regulation (EU) 2021/2115 of the European Parliament and of the European Council of 2 December 2021 laying down rules concerning support for the strategic plans to be drawn up by the Member States under the common agricultural policy (CAP strategic plans). *Official Journal of the European Union, 2021*(2), 1–186.

Dhaoui, O., Antunes, I., Agoubi, B., & Kharroubi, A. (2022). Integration of water contamination indicators and vulnerability indices on groundwater management in Menzel Habib area, south-eastern Tunisia. *Environmental Research, 205*(December 2021), 112491. https://doi.org/10.1016/j.envres.2021.112491

Dixon, J., & Matthews, A. (2006). Impact of the 2003 mid-term review of the common agricultural. *Economic and Social Research Institute*, 1–17.

Duque-Acevedo, M., Belmonte-Ureña, L. J., Terán-Yépez, E., & Camacho-Ferre, F. (2022). Sustainability and circularity in fruit and vegetable production. Perceptions and practices of reduction and valorization of agricultural waste biomass in south-eastern Spain. *Journal of Environmental Management, 316*, 115270. https://doi.org/10.1016/j.jenvman.2022.115270

Duque-Acevedo, M., Belmonte-Ureña, L. J., Plaza-Úbeda, J. A., & Camacho-Ferre, F. (2020). The management of agricultural waste biomass in the framework of circular economy and bioeconomy: An opportunity for greenhouse agriculture in Southeast Spain. *Agronomy, 10*(4), 489. https://doi.org/10.3390/agronomy10040489

EC-DG AGRI. (2022). *Cap budget in perspective 80* (Issue July 2020). https://www.google.com/url?sa=t&rct=j&q=&esrc=s&source=web&cd=&cad=rja&uact=8&ved=2ahUKEwjF_ZHF2a38AhXZUqQEHcCwDyAQFnoECBcQAQ&url=https%3A%2F%2Ffood.ec.europa.eu%2Fsystem%2Ffiles%2F2021-10%2Ff2f_conf_20211015_pres-09.pdf&usg=AOvVaw3edwVULpJrUfw7k3MCb2h9

European Commission. (1997). *The European Commission presents "Agenda 2000": for a stronger and wider Europe*. Office of the European Union.

European Community. (2002). Consolidated version of the treaty establishing the european community. *Official Journal of the European Communities.*

European Union. (2015). *Communication from the Commission to the European Parliament, the Council, the European Economic and Social Committee and the Committee of the Regions. Closing the loop—An EU Action Plan for the Circular Economy.* Office of the European Union

European Commission. (2018). *A European strategy for plastics in a circular economy.* Office of the European Union.

European Comission. (2019). *The European Green Deal.* Office of the European Union.

European Union. (2019). *Audit preview. Biodiversity in farming. May.* Office of the European Union.

European Commission. (2020d). *Europeans, Agriculture and the CAP. Special Eurobarometer 504.* Office of the European Union.

European Comission. (2020a). *A farm to fork strategy for a fair, healthy and environmentally-friendly food system.* Office of the European Union.

European Comission. (2020b). A *new circular economy Action Plan. In For a cleaner and more competitive Europe.* Office of the European Union.

European Comission. (2020c). *EU Biodiversity strategy for 2030. Bringing Nature back into our lives.* Office of the European Union.

European Commission. (2022b). *Main strategic objectives of the new CAP.* https://agriculture.ec.europa.eu/common-agricultural-policy/cap-overview/new-cap-2023-27/key-policy-objectives-new-cap_es

European Commission. (2022a). *Common agricultural policy indicators.* https://agriculture.ec.europa.eu/cap-my-country/performance-agricultural-policy/cap-indicators_es

European Commission. (2022c). *Proposed CAP Strategic Plans (CSPs). Key facts and figures.* Official Journal of the European Union.

European Commission. (2022d). *Safeguarding food security and reinforcing the resilience of food systems.* Official Journal of the European Union.

Gómez-Tenorio, M. Á., Magdaleno-González, J., & Tello-Marquina, J. C. (2021). *Evaluación e implementación de técnicas regenerativas para la mejora de la fertilidad en el cultivo del almendro en las provincias de Almería y Granada.* Editorial Tecniagrícola de España.

Honoré, M. N., Belmonte-Ureña, L. J., Navarro-Velasco, A., & Camacho-Ferre, F. (2019). Profit analysis of papaya crops under greenhouses as an alternative to traditional intensive horticulture in Southeast Spain. *International Journal of Environmental Research and Public Health, 16*(16), 2908. https://doi.org/10.3390/ijerph16162908

IPCC. (2019). An IPCC Special Report on climate change, desertification, land degradation, sustainable land management, food security, and greenhouse gas fluxes in terrestrial ecosystems. *Food and Agriculture Organization of the United Nations and UNED.*

Jacobs, A. A., Evans, R. S., Allison, J. K., Garner, E. R., Kingery, W. L., Mcculley, R. L., Major, G., Resource, L., Soil, A., Office, S., & Resources, N. (2022). Soil & Tillage Research Cover crops and no-tillage reduce crop production costs and soil loss, compensating for lack of short-term soil quality improvement in a maize and soybean production system. *Soil & Tillage Research, 218*, 105310. https://doi.org/10.1016/j.still.2021.105310

Khafagy, A., & Vigani, M. (2022). Technical change and the Common Agricultural Policy. *Food Policy, 109*, 102267. https://doi.org/10.1016/j.foodpol.2022.102267

Kirchherr, J., Reike, D., & Hekkert, M. (2017). Resources, Conservation & Recycling Conceptualizing the circular economy: An analysis of 114 de fi nitions. *Resources, Conservation and Recycling, 127*, 221–232. https://doi.org/10.1016/j.resconrec.2017.09.005

López-Serrano, M. J., Lakho, F. H., Van Hulle, S. W., & Batlles-delafuente, A. (2023). Life cycle cost assessment and economic analysis of a decentralized wastewater treatment to achieve water sustainability within the framework of circular economy. *Oeconomia Copernicana, 14*(1). https://doi.org/10.24136/oc.2023.003

López-Serrano, M. A., Velasco-Muñoz, J. F., Aznar-Sánchez, J. A., & Román-Sánchez, I. M. (2021). Financial evaluation of the use of reclaimed water in agriculture in Southeastern Spain, A Mediterranean Region. *Agronomy, 11*, 2218. https://doi.org/10.3390/agronomy11112218

López-Serrano, M. J., Velasco-Muñoz, J. F., Aznar-Sánchez, J. A., & Román-Sánchez, I. M. (2022). Farmers' attitudes towards irrigating crops with reclaimed water in the framework of a circular economy. *Agronomy, 12*, 435. https://doi.org/10.3390/agronomy12020435

Massot Martí, A. (2000). La PAC, entre la Agenda 2000 y la Ronda del Milenio: ¿A la búsqueda de una política en defensa de la multifuncionalidad agraria? *Estudios Agrosociales y Pesquero, 188*, 9–66.

Maudos, J., & Salamanca, J. (2022a). *Observatorio sobre el sector agroalimentario de las regiones españolas. Informe 2021*. Cajamar Caja Rural.

Maudos, J., & Salamanca, J. (2022b). *Observatorio sobre el sector agroalimentario español en el contexto europeo Informe 2021*. Cajamar Caja Rural.

Mazur-Wierzbicka, E. (2021). Circular economy: advancement of European Union countries. *Environmental Sciences Europe, 33*(1). https://doi.org/10.1186/s12302-021-00549-0

Pañeda-Fernández, C. (1999). Teoría e Historia de la PAC. *Revista Asturiana De Economía, 14*, 7–25.

Pe, G., & Lakner, S. (2020). Preview The EU ' s Common agricultural policy could be spent much more efficiently to address challenges for farmers, climate, and biodiversity. *One Earth*, 173–175. https://doi.org/10.1016/j.oneear.2020.08.004

Pedraza, A. C., Díaz, A. R., & Soto, I. E. (2015). Cambios paisajísticos y efectos medioambientales debidos a la agricultura intensiva en la Comarca de Campo de Cartagena-Mar Menor (Murcia). *Estudios Geograficos, 76*(279), 473–498. https://doi.org/10.3989/estgeogr.201517

Prados-peña, M. B., Gálvez-sánchez, F. J., & García-lópez, A. (2022). Sustainable crafts: Describing conceptual evolution through a bibliometric analysis and systematic literature review. *Frontiers in Environmental Science, 10*(June), 1–16. https://doi.org/10.3389/fenvs.2022.949681

Región de Murcia, C. de G. (2019). Decreto-Ley 2/2019 de 26 de Diciembre, de Protección Integral del Mar Menor. *Boletín Oficial de La Región de Murcia, 298, 27 de diciembre*, 36008–36089.

Sáen-Lorite, M., & Cejudo-García, E. (2013). La Política de Desarrollo Rural de la Unión Europea para 2007–2013. *Investigaciones Geográficas, 46*(2008), 5–30.

Stead, D. R., & Jahre, D. M. (2007). The Mansholt Plan forty years on. *EuroChoices, 6*(3), 40–45.

The World Bank. (2008). *World development report 2008. Agriculture for Development*. The World Bank. https://doi.org/10.1596/978-0-8213-7233-3

Tsalis, T. A., Terzaki, M., Koulouriotis, D., Tsagarakis, K. P., & Nikolaou, I. E. (2022). The nexus of United Nations' 2030 agenda and corporate sustainability reports. *Sustainable Development, July 2021*, 1–13. https://doi.org/10.1002/sd.2419

UN. (2015). Draft outcome document of the United Nations summit for the adoption of the post-2015 development agenda. In *The World Bank Group A to Z 2016* (Vol. 13689). https://doi.org/10.1596/978-1-4648-0484-7_post_2015_development_agenda

Valera-Martínez, D. L., Belmonte-Ureña, L. J., Molina Aiz, F. D., & Camacho-Ferre, F. (2017). The greenhouses of Almería, Spain: Technological analysis and profitability. *Acta Horticulturae, 1170*, 2019–2226. https://doi.org/10.17660/ActaHortic.2017.1170.25

Francisco José Castillo Díaz is Ph.D. in Protected Agriculture from the University of Almería and acts as a researcher in the SEJ-668 Research Group "Circular Economy, Agri-Food and Socioeconomic Development". His main lines of research are waste management, circular economy and agriculture. He accumulates more than 25 publications between scientific articles and book chapters. He has also participated in 8 national and international congresses.

Luis Jesús Belmonte-Ureña is a professor in the Area of Applied Economics at the University of Almeria, with more than 20 years of teaching experience, where he has taught several courses of master and degree. His main line of research is related to the circular economy in agriculture. Currently, Professor Belmonte-Ureña has more than 70 publications, including articles in scientific journals, books and book chapters. He has participated in 19 national and international congresses. He has also participated in 11 competitive research projects, acting in two of them as principal investigator. According to the Scopus database, Professor Belmonte-Ureña has an h-index of 22 and more than 1400 citations.

Juan Fernando Álvarez-Rodríguez is professor from Rural and ambiental studies from Pontificia Universidad Javeriana de Colombia. Economist from Los Andes University in Venezuela, master's in social economy from Barcelona University and Doctor in Social Science from Lisboa University. President from scientific committee from CIRIEC Colombia and member from the observatory iberoamerican of employment and social and cooperative economy in Latin America. His publications cover topics related to satelital accounts, social balance, impact evaluation and public politics.

Francisco Camacho-Ferre From 1977 to 1999, he managed protected crop protected crops. He holds a Master's degree in Agricultural Business Management and a Ph. Agricultural Engineer from the University of Almeria. Professor of intensive horticulture at the Agronomy Department of the Department of Agronomy at the University of Almeria. He has participated in more than more than 200 congresses. He has supervised 175 undergraduate theses and 17 Ph.D. theses. He is author or co-author of 23 books, 37 chapters and 128 national and international scientific articles. international scientific articles. He was an International Expert Consultant (United Nations) in the methyl bromide alternatives project between 2003 and 2015. He was Technical Director of the UAL-ANECOOP Foundation in the period 2003–2008, Director of the Cajamar Chair of Economics and Agri-Food in the period 2011–2015. He has carried out technical consultancies in Argentina, Chile, Uruguay, Bolivia, Peru, Venezuela, Cuba, Honduras, Guatemala, Dominican Republic and Mexico. Mexico.

The Role of Stakeholders on the Intention to Implement Sustainable Practices: An Exploratory Research in the Agri-Business Spanish Sector

Ana Isabel García-Agüero, Eduardo Terán-Yépez, Luis J. Belmonte-Ureña, and Francisco Camacho-Ferre

1 Introduction

Sustainability has emerged as one of the paramount challenges of our era, commanding significant attention as a pressing societal concern (Riegler et al., 2023). Mili and Arovuori (2023) underscore the pivotal hurdles we confront, foremost among them being the degradation of natural resources and the specter of climate change. Spain, owing to its entrenched agricultural heritage, stands as a key global producer, increasingly subject to mounting pressures to embrace sustainable practices (Duque-Acevedo et al., 2022). Within this intricate business landscape, stakeholders assume a pivotal role in shaping the trajectory of the agro-industrial sector. For instance, the findings by Ordonez-Ponce et al. (2021), stemming from their examination of 71 companies across four major intersectoral associations, elucidate the nexus between corporate sustainability at local and global scales, reaffirming the potential for these entities to substantively contribute to their objectives.

A. I. García-Agüero (✉) · L. J. Belmonte-Ureña
Department of Economics and Business, University of Almeria (ceiA3), Ctra. Sacramento, S/N, 04120 La Cañada, Almeria, Spain
e-mail: aga202@ual.es

L. J. Belmonte-Ureña
e-mail: lbelmont@ual.es

E. Terán-Yépez
Department of Economics and Business, CIMEDES Research Center, University of Almeria (ceiA3), Ctra. Sacramento, S/N, 04120 La Cañada, Almeria, Spain
e-mail: eduardo.teran@ual.es

F. Camacho-Ferre
Department of Agronomy and Sustainable Protected Agriculture Research Network, University of Almeria (ceiA3), Ctra. Sacramento, S/N, 04120 La Cañada, Almeria, Spain
e-mail: fcamacho@ual.es

M. d. C. Valls Martínez and J. M. Santos-Jaén (eds.), *Environmentally Sustainable Production*, https://doi.org/10.1007/978-3-031-52656-5_5

This study endeavors to deepen our comprehension of the influence wielded by stakeholders on the strategic orientation of Spanish agricultural enterprises as they embark on sustainable initiatives. The Spanish agricultural domain grapples with multifarious challenges encompassing extreme weather events, water scarcity, biodiversity loss, and waste management (Du Plessis, 2019; Vargas-Amelin & Pindado, 2013). Amidst this backdrop of environmental exigency, stakeholders are poised to exert a foundational impact on the direction and velocity of sustainable practice adoption (Castillo-Díaz et al., 2023).

While research on the influence of stakeholders in the corporate realm abounds (e.g., De Falco et al., 2021; Greenwood, 2007; Hillis et al., 2018), scant attention has been devoted to their interplay within the sphere of agricultural production. Thus, this study's objective lies in analyzing the impact of stakeholders on the intention to implement sustainable practices within the agricultural companies in Spain and to investigate the multifaceted dynamics that influence their roles in either advancing or impeding sustainable practices. By illuminating the intensity of commercial interactions among stakeholders in the agribusiness sector, this research aspires to furnish a roadmap for the more effective engagement of interested parties in organizational decisions that align with a more sustainable future for Spanish agriculture. Hillis et al. (2018) elucidate in their study of sustainability associations among agricultural production companies that such associations yield superior management outcomes for these enterprises. De Falco et al. (2021) conducted an analysis of 225 Italian SMEs, evaluating the pressures exerted by diverse stakeholders (networks, investors, employees, government and regulators, support institutions, and competitors), and their findings affirm that innovative companies demonstrate "a discernible sensitivity and attentiveness to external pressures from stakeholders related to environmental, social, and governance criteria."

Loh and Tan (2020) contend that stakeholder awareness and engagement in shaping corporate policies, strategies, and long-term objectives confer a competitive advantage. Given the diversity of the agricultural firms under scrutiny, sustainable production opportunities frequently evolve within highly uncertain environments, characterized by divergent perspectives that are relatively challenging to forecast and control (Céspedes-Lorente & Galdeano-Gómez, 2004). Consequently, this study draws conclusions about the impact of stakeholders and the extent of influence they exert on each of the interviewed companies, by answering the following research question: How do a diverse range of stakeholders, including top leadership and management, other staff, suppliers, intermediaries, marketplace, wholesalers, customers, end users, governments, laws, and activist groups, influence the intentions of agri-businesses in Spain to adopt and implement sustainable practices, and what are the underlying dynamics that shape their roles in promoting or obstructing sustainability initiatives? This chapter seeks to advance our comprehension of this dynamic by scrutinizing the role played by the implementation of sustainable practices in agricultural companies and the manner in which managerial and sustainability-focused personnel shape these efforts.

2 Theoretical Background and Literature Review

2.1 *Business Sustainability Practices in the Agri-Business Sector*

Business sustainability encompasses a company's commitment to extend its focus beyond mere profitability, incorporating strategies to effectively manage its environmental, social, and economic effects on both the marketplace and society at large (Andersson et al., 2022). The exploration of corporate sustainability had its origins in the 1970s, marked by seminal studies conducted by Bowman and Haire (1975) and Carroll (1979). These initial investigations primarily delved into the "what" and "why" of sustainability, leaving the "how" aspect of sustainability within companies relatively uncharted. Consequently, attention gradually shifted toward the examination of the driving forces propelling organizations to embrace sustainability strategies (Engert & Baumgartner, 2016), along with the identification of suitable practices for achieving sustainability goals. As time advanced, it became increasingly apparent that businesses needed to undergo transformative changes in their business models to move beyond the traditional profit-centric approach. These changes were seen as essential for effectively addressing the ecological, social, and economic challenges confronting companies (Alonso-Martinez et al., 2021; Lüdeke-Freund et al., 2018). According to Avila et al. (2023), "sustainable practices manifest through behavioral shifts towards more rational and efficient resource management, resulting in reduced environmental pressure and impact."

Similar to developments in the wider business landscape, the agricultural sector has recognized the imperative of incorporating sustainable practices into its production chain. Many companies within this sector have initiated efforts to create more sustainable products and services, pivoting away from a solely consumer-centric business orientation (Damke et al., 2021). The Food and Agriculture Organization (FAO, 2023) underscores that sustainable practices within the agricultural sector are indispensable for promoting the responsible and efficient utilization of natural resources, safeguarding their long-term viability for future generations, and ensuring food security. Moreover, both legislative mandates and societal expectations compel fruit and vegetable companies to couple innovations in products, services, processes, and business models with responsible sustainable development to mitigate the impacts of their production processes (Kneipp et al., 2018).

Hence, companies are implementing a diverse array of sustainable practices, reflective of the adaptable nature and diversity of productive activities within these organizations (Céspedes-Lorente & Galdeano-Gómez, 2004; Duque-Acevedo et al., 2022). These sustainable improvements are aimed at aligning with the standards established by the Global Reporting Initiative's (GRI) sustainability reports for the agricultural sector (GRI, 2022), which advocate for reducing carbon footprints, managing emissions generated during production activities, optimizing energy consumption, fostering recycling of products at the end of their useful life while minimizing waste generation, reducing the water footprint, and contributing to environmental conservation, among other imperatives (Spanish Ministry of Agriculture, 2023).

Nevertheless, there remains a degree of ambiguity concerning what farmers and producers define as sustainable practices. A recent study by Jaworski et al. (2023) sheds light on this issue, revealing that farmers employ various practices in diverse combinations, and not all of these combinations adhere strictly to the comprehensive principles of sustainable agriculture. These principles encompass productivity, stability, resilience, and sustainability, emphasizing the capacity to maintain consistent levels of production over time (Fereres & Villalobos, 2016). It's important to note that stakeholders also wield influence over the management of natural resources within the agricultural sector. However, it's crucial to acknowledge that social interests may not always align harmoniously with the interests of the private sector (Jaworski et al., 2023).

2.2 *Stakeholder Theory and Categorization*

For several decades, management theorists have observed a growing alignment between companies' interests and their broader environment, encompassing both internal and external factors. This alignment has led to a shift in organizational planning, with a heightened emphasis on stakeholders (Fernández & Bajo, 2012). As we entered the twenty-first century, Peter White noted that companies were increasingly recognizing their responsibility for the impact on the physical environment (White, 2009). Freeman (1984) highlighted that stakeholder theory, from a strategic standpoint, posits that companies should formulate their business strategies and policies while considering the interests of various stakeholders that have an influence on the company. Building on this, Donaldson and Preston (1995) argued that this theory serves to identify the connections, or lack thereof, between stakeholder management and the achievement of traditional corporate objectives.

Subsequently, the relationship between a company and its stakeholders began to undergo comprehensive analysis, giving rise to various perspectives. According to Evans and Sawyer (2010), stakeholders encompass all facets of the company that contribute to its overall health and vitality. Numerous studies have consistently emphasized the enduring significance of this concept, highlighting that the theoretical underpinning of contemporary organizational strategic vision centers on the idea that companies capable of adapting to a dynamic environment and leveraging both internal and external resources are best positioned to generate value for the organization. Moreover, such adaptability fosters sustainable competitive advantages for the company (Liang et al., 2022; Teece, 2007).

Given the multifaceted nature of stakeholder theory, companies must navigate a diverse array of stakeholders, extending beyond their markets and business networks. This theory originally emerged with the intent of systematically categorizing the various stakeholders associated with an organization and gaining insights into their impact and influence within the company's operational context (Mitchell et al., 1997).

Stakeholders can be categorized in various ways (Svensson et al., 2018). According to Dansky and Gamm (2004), there are three primary categorizations of stakeholders that form the basis for understanding a company's relationships. The first category comprises internal stakeholders, including management and employees. The second category encompasses external stakeholders, who possess a certain degree of involvement or engagement with the company. Lastly, the third category comprises boundary stakeholders who operate at the interface between the company and its broader stakeholder network. In contrast, Svensson et al. (2016) present a categorization scheme that distinguishes between internal and external stakeholders:

- Upstream Stakeholders such as suppliers, suppliers' suppliers, manufacturers and raw material producers;
- The focal company refers to one's own organization, top leadership and management, the executive board, chief executive officer, managers and other staff;
- Downstream stakeholders such as wholesalers, retailers, sales outlets and intermediaries;
- Market stakeholders such as customers, end users, the marketplace; and
- Societal stakeholders such as government, laws, activist groups, interest groups and general public.

While numerous categorizations exist within the scientific literature, the fundamental concept remains consistent: all stakeholders are interconnected and must engage with one another during the course of their commercial endeavors. We contend that this principle holds true, especially when implementing sustainable business practices.

2.3 Business Sustainability Practices in the Agri-Business Sector and Stakeholder

Stakeholders in organizations are interconnected, exerting a direct influence on strategic decision-making. This interdependence fosters stability within companies, facilitating a collective pursuit of common objectives (Valentinov & Chia, 2022). Notably, trust among stakeholders plays a pivotal role in encouraging companies to incorporate sustainable practices across all facets of their operations (Gupta & Gupta, 2020). This trust can transform into a driving force, compelling all stakeholders to collectively embrace sustainability through collaboration. Consequently, it represents the path towards enhancing companies' sustainability performance, heralding a transition from the traditional command-and-control model to a more progressive approach characterized by support and collaboration (Svensson et al., 2018).

The agricultural sector faces substantial pressure from a diverse array of stakeholders, including members of society, regulatory bodies, consumers, suppliers, and more. As demonstrated by Massoud et al. (2019) in their research on water resource reuse within agriculture, these stakeholders exert influence through various barriers that encompass ethical, religious, social, and economic dimensions. Van Gorp and Van Der Goot (2012) acknowledge the multifaceted nature of stakeholders in sustainable agriculture. They recognize that these stakeholders possess a wide spectrum of interests, ranging from profit-driven objectives to a mission centered on promoting healthy food or safeguarding the well-being of farmers.

Companies are compelled to adopt various sustainability measures, including the elimination of plastic in organic farming and the stringent regulation of pesticides and fertilizers, as stipulated by the European Commission (2023). Furthermore, growing societal consciousness towards sustainability, as highlighted by Knaggård et al. (2019), adds to this imperative. In response to these demands, companies must implement robust risk management practices, often in the form of standards and certifications. These measures serve the dual purpose of monitoring environmental and social impacts while averting potential penalties and disapproval from stakeholders, as emphasized by Seuring and Mü Ller (2008). Quality certifications assume a pivotal role in creating strategic value for companies while bolstering their reputation and commitment to societal well-being, as underscored by Mora-Córdova et al. (2020).

Analyzing the existing literature, we can find various examples of scientific research work on the impact of stakeholders in different areas. For example, in the stakeholders when formulating conservation policies for a natural space (van den Broek, 2019). Rawlinson et al. (2022) present a study on stakeholders affecting barriers in healthcare from various linguistic regions. Moermond and De Rooy (2022) analyzes stakeholders from health sectors on wastewater treatment with pharmaceutical waste. Or the management approach for stakeholders and corporate environmental sustainability (Góes et al., 2023). The main contribution of Walker and Laplume (2014) is the nexus between stakeholder influence strategies and the collective goal of sustainability.

On the flip side, prior research exploring the impact of stakeholders has predominantly been examined within the context of our study. Mohammadi-Nasrabadi et al. (2020), for instance, centered their investigation on the viewpoints and recommendations of stakeholders concerning the political challenges associated with food advertising. Their study concluded that an ongoing necessity exists for the implementation of new regulations governing food advertising.

MacLeod et al. (2022) directed their efforts towards amplifying the voices of stakeholders engaged in the management of New Zealand's agricultural landscape. Their goal was to establish biodiversity priorities on agricultural land that resonate with stakeholders and communities, emphasizing the most relevant management practices required to achieve these objectives. Their conclusions underscored the importance of integrating broader considerations encompassing environmental, social, economic, and political dimensions. In the work of Van Gorp and Van Der Goot (2012), the primary aim was to comprehend how key stakeholders strategically employ frameworks in their public discourse pertaining to sustainable food and agriculture. Embracing a stakeholder model, they acknowledged the diverse interests of various stakeholders in sustainable agriculture and the food industry. Some stakeholders prioritize profit objectives, while others are driven by the mission of promoting healthy food and caring for the well-being of farmers.

The study conducted by Reimer et al. (2023) delved into the challenges associated with conservation agriculture, recognizing the imperative for fundamental shifts in the prevailing commodity production model. This transformation includes the adoption of practices such as reduced tillage, expanded coverage, and crop rotations aimed at minimizing soil disturbance while enhancing biological diversity. Their analysis unveiled insights into the perspectives of key stakeholders, the barriers they encounter, and emerging approaches that impact these agricultural challenges. Their findings highlighted a significant gap between research within the field and the practical application of its outcomes.

3 Methodology

This study is meticulously crafted to provide valuable insights into the intricate ways in which stakeholders exert influence on a company's intentions to adopt sustainable business practices. As mentioned earlier, while previous research has examined stakeholder roles in promoting sustainability in various sectors, the agri-business domain remains relatively uncharted territory. To shed light on this crucial aspect, we employ the case study method, chosen for its manifold advantages over alternative approaches. By focusing on specific instances that exemplify how stakeholders shape the intentions of agri-businesses to embrace sustainability, our research acquires depth and specificity within the context of this sector (Rashid et al., 2019). Attempting to collect detailed information on stakeholder impact through methods other than case studies would be fraught with considerable challenges. The incorporation of open-ended questions in our interviews with agri-business managers offers a unique opportunity for these individuals to explore and articulate nuanced aspects related to the multifaceted influence of stakeholders (De Massis & Kotlar, 2014). Furthermore, the case study methodology aligns seamlessly with our research goal of addressing the "how and why" questions, which are pivotal in unraveling the complex dynamics at play (Punch, 2005). Consequently, this approach proves highly suitable for addressing the overarching research question: How do a diverse range of stakeholders, including top leadership and management, other company staff, suppliers, intermediaries, wholesalers, customers, end users, governmental bodies, legal frameworks, and activist groups, influence the intentions of agri-businesses in Spain to adopt and implement sustainable practices, and what are the underlying dynamics that shape their roles in promoting or obstructing sustainability initiatives?

Following the precedents set by previous studies (cf. Andersson et al., 2022; Svensson et al., 2018), this research focused on key informants with in-depth knowledge of their respective companies' sustainable business practices within the agricultural sector. Specifically, individuals holding roles such as Production Managers or Food Safety and Quality Managers were chosen as the primary subjects of inquiry. From a broader pool of Spanish agri-businesses (and their executives), a subset of cases was carefully selected for examination. This larger sample encompassed managerial representatives from 11 agri-business companies, intentionally drawn from diverse sources, including media coverage and databases of agricultural companies from Almeria (a province located in the southeastern region of Spain). Almeria served as an ideal backdrop for this study due to its distinctive status as a central hub of intensive agriculture in Spain, a fact underscored by prior research (e.g., Castillo-Díaz et al., 2023; Duque-Acevedo et al., 2022). The province's remarkable agricultural landscape, characterized by vast greenhouse crop cultivation and associated environmental intricacies, provided a rich and contextually relevant setting for scrutinizing the multifaceted dynamics of stakeholder influence on sustainability intentions within the agribusiness sector. Out of the initial 11 companies and their managerial representatives, four met specific selection criteria for inclusion in the study. These criteria encompassed, firstly, two screening questions to assess the competency of respondents. The questions which were taken from the study by Svensson et al., 2018 addressed (1) the knowledge of the respondent about his/her company's sustainable business practices and (2) the knowledge of the respondent about his/her company's sustainable business practices in the whole business network. Responses were measured on a 5-point Likert scale (1 = I do not have any knowledge, 5 = I have a lot of knowledge), with only those managers scoring 4 or 5 in both responses considered as appropriate. Secondly, the selection of firms and managers was contingent on the availability of comprehensive information regarding their activities, a pivotal factor for conducting a comprehensive analysis of stakeholder impact.

Each of the selected managers participated in personal interviews conducted during the period from July to September 2023. These interviews were comprehensive, lasting between 60 to 75 min. A portion of the interview time was dedicated to inquiries about the company's overall operations, while the primary focus of the interviews was on exploring the intricate interplay of stakeholders and their impact on the intention to implement sustainable practices within their respective businesses. In the second part of the interview, we asked the key informants to answer the following question in relation to each stakeholder shown in Table 1: "How do/does [name of a stakeholder (e.g., your suppliers)] affect your intentions to implement sustainable business practices in your company?". To ensure a thorough record of the discussions, all interviews were recorded and subsequently transcribed for analysis. Additionally, supplementary information regarding the selected firms was collected through diligent internet research. Table 1 and the upcoming section of this study provide concise yet informative descriptions of each case study. In order to preserve the confidentiality of the firms under examination, their actual names have been anonymized.

To answer the research question, the interview material was analyzed with reference to the influence of each stakeholder on the intention of agri-businesses to implement business sustainable practices. Stakeholders were defined as any collective or individual entity with the capacity to exert influence on or be influenced by the attainment of a company's goals (Freeman, 1984). Business sustainability is defined as a company's commitment to extend its focus beyond mere profitability, incorporating strategies to effectively manage its environmental, social, and economic

Table 1 Selected stakeholders

Upstream stakeholders Suppliers	
Focal company Top leadership/management Other staff	
Downstream stakeholders Wholesalers Intermediaries	
Market stakeholders End users Marketplace Surrounding society	
Societal stakeholders Government Laws Activist groups	

Source Own elaboration based onAndersson et al. (2022)

effects on both the marketplace and society at large (Andersson et al., 2022). Based on this definition, for this study we conceptualize business sustainable practices as the strategies, actions, and initiatives adopted by a company to ensure its long-term viability and success while minimizing negative impacts on the environment, society, and the economy.

4 The Cases

Company 1

Founded around 1950, this family company initially started with the vision of producing and marketing fruit and vegetable products. In 2005, it officially transformed into an Agrarian Transformation company. Presently, the company boasts a diverse portfolio of 35 products, which are cultivated across various farms throughout Spain, taking into account different crop cycles and locations. This company is committed to integrated and ecological production, making use of active materials derived from natural synthesis. Their agronomic management revolves around two primary strategies. Firstly, they emphasize reducing their environmental footprint by striving for zero waste, thereby progressively enhancing the sustainability of their products. Secondly, they are dedicated to minimizing the usage of single-use plastics, opting instead for recycled and biodegradable materials.

Company 2

Established in 1998 as a family-owned fruit and vegetable marketing venture, this company has evolved significantly. Today, it stands as a recognized national leader in Spain renowned for its sustainable cultivation of fruits and vegetables. This organization maintains a steadfast commitment to environmental enhancement and the preservation of indigenous flora and fauna. They employ cutting-edge machinery that not only optimizes resource utilization, including water and energy but also adeptly caters to the specific requirements of their crops. Concurrently, the company remains dedicated to developing infrastructure that facilitates the expedient and eco-friendly transportation of their merchandise.

Company 3

Established in 2005, this agricultural cooperative resulted from the amalgamation of two akin cooperative agricultural entities. Notably, their cooperative framework remains open to the inclusion of new members. This collaborative initiative marked the inception of a pioneering and ambitious venture within the fruit and vegetable sector, driven by a vision to deliver unparalleled quality, quantity, and variety in their products and services. With a clientele that spans across diverse countries, including the United States, Canada, and numerous European Union member states, their reach is truly international. Additionally, they function as producers of natural predators, which are strategically employed in their crop management practices. Moreover, this company is resolutely committed to achieving 100% organic production in the near future, underscoring their dedication to various social initiatives.

Company 4

Established in 2004 as a family-owned enterprise, this business has embraced a resolute dedication to environmental sustainability through the cultivation of organic fruits and vegetables in the most natural and eco-conscious manner possible. It specializes in the production of organic vegetables and stands as a paragon of innovation, sustainability, and social responsibility. While adhering to traditional agricultural production methods, the company champions an agricultural model that prioritizes the welfare of the planet and its precious resources. Their cultivation practices are inherently non-invasive, guaranteeing both maximum food safety and unwavering environmental sustainability. In pursuit of these ideals, the company actively engages in various environmental initiatives. These include the establishment of small reserves for fauna and flora on their farms, enriching their crops and fostering biodiversity. Furthermore, they have instituted a biodynamic composting plant, aimed at recycling remnants from prior crops into high-quality natural fertilizers, thereby nurturing the soil and preserving its vitality.

5 Cross-Case Analysis

Focal Company-Own Organization

All selected interviewers agree that *top leadership and management* play a relevant role in the adoption of sustainable practices of agribusiness firms. However, when we delve deeper into the contributions of *other staff* within the company, we observe disparities in the responses provided by the interviewees.

In most instances, the workforce demonstrates a willingness to embrace and adhere to the established sustainability guidelines and regulations. Nevertheless, there are instances where certain employees exhibit reluctance to adopt these sustainable measures mandated by the organization. This resistance may stem from various factors, such as long-standing employees who find it challenging to alter their accustomed work methods or individuals who resist conforming to new patterns that may require additional effort, such as segregating waste for recycling.

Company 4: The commitment to sustainability is deeply ingrained in the company's DNA, dating back to its inception. As evidenced by our eco-friendly facility design, with recyclable materials comprising the entire company façade, this commitment has been an integral part of our company ethos from the very beginning. Whether it's the sustainable tableware we use when hosting clients or our long-standing practice of eliminating plastic materials, this approach to sustainability has been seamlessly integrated into our operational methods. This commitment is evident to both long-serving employees and newcomers, making it a fundamental aspect of our company culture.

Company 2: Some areas within our organization exhibit resistance to change, driven by the mentality of 'if it's worked this way for so long, why change it?' It's essential to recognize that measurement is the key to improvement. In my role as the production manager, I've witnessed the transition from a 'measure-less' approach to one that emphasizes precision. Previously, everything was done by estimation and without measurements. For instance, no company was measuring the plant's evapotranspiration to determine its water requirements.

Company 3: Within our workforce, there exists a spectrum of attitudes towards sustainability. While some employees are highly proactive, others may remain steadfast in their own ways despite efforts to instill sustainable practices. However, it's encouraging to note that the majority make genuine attempts to align with our sustainability initiatives.

6 Upstream Stakeholders

All managers interviewed acknowledge the pivotal role that *suppliers* play in shaping their businesses' performance and charting the course for the implementation of sustainable practices. They concur that championing sustainability is a shared responsibility, with everyone actively supporting efforts to reduce carbon emissions, regulate water consumption, and more. To achieve these goals, various strategies are employed, such as optimizing transportation load capacities and deploying larger machinery to minimize emissions, among other initiatives.

Company 1: The sustainability of our suppliers directly impacts our own sustainability. They significantly contribute to our value chain, enhancing the sustainability of our final products. When our suppliers, particularly those providing packaging materials, implement sustainability measures, it inherently enhances the sustainability of our end products that incorporate their materials.

Company 4: Our primary suppliers include organic farmers who provide seedlings to our farms. While our warehouse division works with other suppliers, our auxiliary material providers, particularly those supplying cardboard, are equally committed to sustainability. These suppliers actively seek alternatives to plastic, utilizing materials like cellulose meshes and natural threads such as jute raffia, which we've been using for an extended period. Sustainability efforts are ingrained not only within our farm suppliers but also among our warehouse suppliers.

7 Downstream, Market and Societal Stakeholders

All the companies participating in this study emphasize their direct sales approach, with no *intermediaries* involved. They exclusively sell their products to *wholesalers* (or supermarkets), rather than catering to *end consumers*. In this supply chain, they recognize the substantial influence wielded by their *customers* (*wholesalers*) when it comes to sustainable practices, as these customers (*wholesalers*) demand products meeting specific environmental criteria, demands that ultimately originate from end users.

Company 4: We operate without intermediaries, serving wholesalers who distribute to European retail outlets. These wholesalers are environmentally conscious and prioritize factors like water and carbon footprints. Meeting these demands is essential for our business. Consequently, we have proactively undertaken projects to reduce plastic usage, not only in production but also in our warehousing operations. We have long been exploring alternative materials like biodegradable cellulose and compostable options, emphasizing the use of paper over plastic. The only remaining plastic use is in returnable packaging, which is part of a closed-loop system.

It's worth noting that while *end users* can influence certain consumer behaviors, they do not exert direct control over the environmental practices adopted by agricultural production companies.

Company 2: Our ultimate customers are wholesalers, meaning our agricultural practices and choices—such as seed selection or resource consumption—do not directly impact the end consumer. For example, when we purchase seeds for watermelons, lettuce, or cauliflower, the environmental practices of the seed supplier, like reducing inputs or improving energy efficiency, do not directly reach the end consumer. This underscores the importance of consumer awareness and labeling to bridge this gap.

Company 3: Two years ago, I received an email from a Norwegian customer (wholesaler) who frequently purchased tricolor peppers. They inquired about the possibility of using a newly developed compostable film for packaging. While I found this idea intriguing, it underscored the need for commercial viability in adopting such sustainable practices.

Marketplace trends and *surrounding society* awareness exert significant pressure on the sustainability decisions made by the interviewed companies. While three of the companies acknowledge this influence, they also recognize the practical constraints that may limit full sustainability implementation. Notably, one of the companies emphasizes that all market factors are currently aligned in their favor.

Company 3: The marketplace often necessitates a cautious approach, as not every sustainability initiative is financially viable. Our commitment to sustainability is contingent upon whether customers are willing to purchase sustainable products.

Company 4: Our experience in the marketplace is notably favorable. Customers actively demand sustainable practices, such as packaging in recycled cellulose mesh rather than plastic bags. In some cases, customer demands are even ahead of conventional market expectations. The topic of sustainability can feel repetitive at times, but it underscores the critical importance of adopting sustainable practices. Climate change is a reality we all face, and embracing sustainability is no longer a choice but a necessity.

Company 1: Similarly, we've found that the presence of an environmentally conscious public, who prefer products without plastic and with recyclable packaging, does not directly impact our operations.

Another pivotal factor to consider in our analysis is the profound influence of *government regulations* and *laws*, a point unanimously acknowledged by all interviewees. However, two of the interviewed companies express a sense of demotivation, primarily stemming from the lack of government assistance in implementing these regulations through direct financing or research support.

Company 3: The impact of government measures is contingent on various factors, often characterized by a discrepancy in pace. There are instances where government regulations lag behind the innovations we are implementing, or they demand actions that are operationally challenging for our specific context. This misalignment between the political sphere and the business world can be frustrating at times.

Company 2: The government's regulations do affect us significantly, as they are imposed upon us. However, what can be disheartening is the absence of substantial assistance in their implementation. For instance, as we approach 2030, there are impending restrictions on herbicides and various active materials in agriculture. The question arises, 'How are we to produce? and what alternatives are available?' Without government support or research initiatives, companies like ours are compelled to bear the full burden of compliance. Fortunately, we've initiated multiple research projects in collaboration with universities, such as the University of Córdoba and the University of Alicante, aimed at innovation and increased efficiency. These projects encompass areas like plant waste treatment and the development of more efficient machinery for our operations. While the regulations are imposed upon us, we find ourselves navigating these challenges largely unsupported.

When it comes to the influence exerted by *activist groups*, the responses from those interviewed demonstrate a range of perspectives. On one hand, two of the companies assert that activist groups do not impact their operations in any significant manner. In contrast, company 2 underscores their efforts to seek mutually beneficial solutions, while company 4 perceives the influence of activist groups as a potentially indirect yet noteworthy factor.

Company 2: We haven't encountered any major issues with activist groups here. We do face occasional challenges with the neighboring town's mayor due to factors like noise, odors, and agricultural treatments, given our proximity to an urban center. Nonetheless, we strive for harmonious coexistence, recognizing that we operate within a rural environment. While we don't have any organized associations causing trouble for us, occasional individual complaints are part of the norm. In such cases, we engage in dialogue, offering insights into our operations and sharing information about the substances we use. We also seek their input on preferred hours for noise-producing activities.

Company 4: Directly speaking, activist groups don't significantly affect our operations, but their influence may manifest indirectly. These activist groups can sway political decisions, and in turn, these political decisions can impact companies, either positively or negatively.

8 Discussion and Conclusion

8.1 Discussion

The objective of this study was to analyze the impact of stakeholders on the intention to implement sustainable practices within the agricultural companies in Spain and to investigate the multifaceted dynamics that influence their roles in either advancing or impeding sustainable practices. The case studies of four distinct agri-businesses in Spain provide a rich and diverse set of perspectives to address this objective, revealing the various approaches and challenges encountered in adopting sustainable practices. This approach not only adds depth to the research but also allows for the identification of common patterns and unique characteristics within the sector.

The discussion of the influence of stakeholders in the agri-business sector is illuminating. It highlights the significance of top leadership and management in driving sustainability initiatives. However, it also acknowledges the potential resistance from some employees, emphasizing the importance of change management strategies within these organizations (Lozano et al., 2016). The impact of upstream stakeholders, particularly suppliers, is recognized as pivotal in enhancing the sustainability of agri-businesses. This aligns with the broader concept of supply chain sustainability, where the actions of suppliers have a direct bearing on a company's environmental and social footprint (Andersson et al., 2022; Svensson et al., 2018). The discussion emphasizes the interconnectedness of the supply chain in achieving sustainability goals (Svensson et al., 2016).

The exploration of downstream stakeholders, including wholesalers, and their role in influencing sustainability practices is noteworthy. The study acknowledges that while end users may not directly impact agricultural practices, their demands for sustainable products can drive change within the supply chain. This underscores the importance of consumer awareness and labeling as mechanisms to bridge the gap between production and consumption (Zander et al., 2018). The study also underscores the influence of government regulations and laws, a force that all interviewed companies acknowledge. However, the challenges posed by regulatory compliance, including the need for research and innovation, are highlighted (Svensson et al., 2018). This points to a potential area where government support and collaboration with the private sector can lead to more effective sustainability initiatives. In contrast, the influence of activist groups and interest groups varies among the companies, highlighting the diverse landscape of stakeholder engagement. While some companies do not perceive significant impacts from these groups, others recognize the potential indirect influence they can exert through political decisions (Andersson et al., 2022).

As a summary of the findings detected when analyzing the impact of stakeholders on the intention to implement sustainable practices in Spanish agricultural companies, we present in Fig. 1 a comparison of the levels of influence that we have been able to detect that each stakeholder possesses. Thus, for example, we propose that top leadership and management, customers, governments, and laws are the four stakeholders that exert the greatest pressure on the intention to implement sustainable practices in Spanish agricultural companies. On the other hand, intermediaries and other staff are the ones that exert the least pressure.

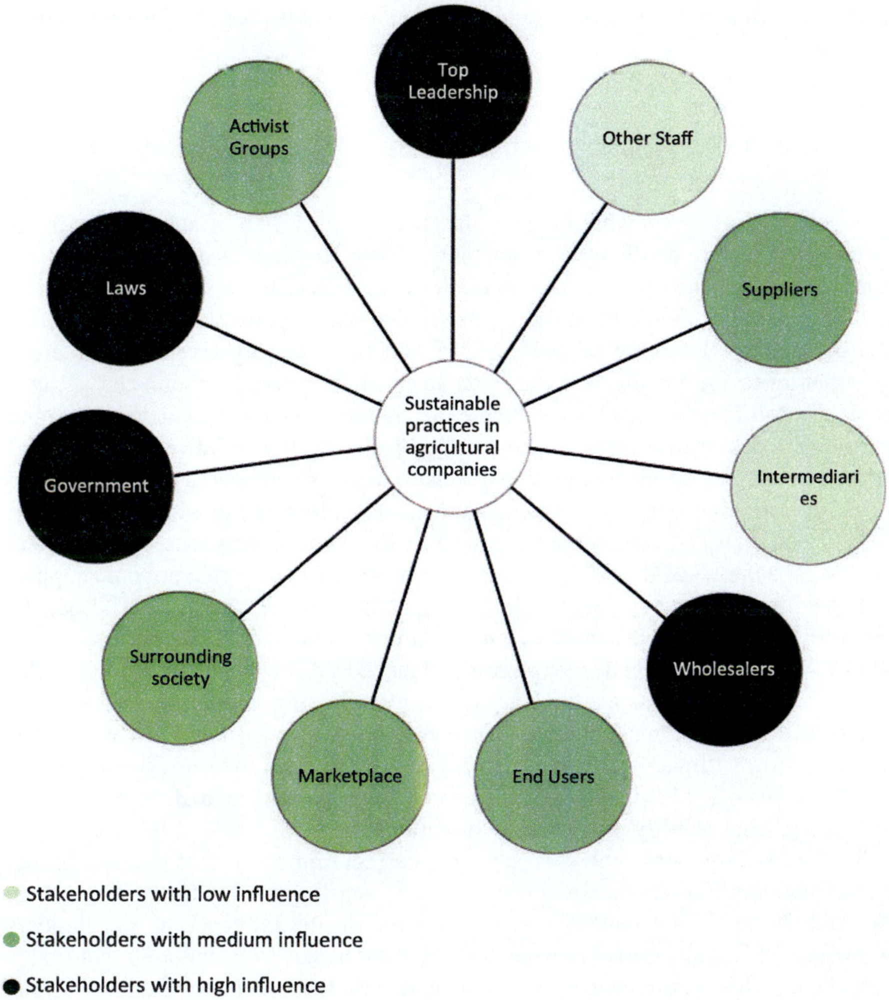

Fig. 1 Comparison between the levels of influence of stakeholders

The discussion and Fig. 1 described earlier have led the authors to submit the following propositions.

Proposition 1. Engaging stakeholders allows agricultural companies to implement sustainable practices beyond its own boundaries.
Proposition 2. Top leadership and management, wholesalers, governments and laws and regulations exert high influence in the intention of agricultural companies to implement sustainable practices beyond its own boundaries.
Proposition 3. Suppliers, end users, marketplace, the surrounding society, and activist groups exert medium influence in the intention of agricultural companies to implement sustainable practices beyond its own boundaries.
Proposition 4. Other staff and intermediaries exert low influence in the intention of agricultural companies to implement sustainable practices beyond its own boundaries.

8.2 Conclusion, Limitations, and Future Research Lines

This research offers valuable insights into the complex web of stakeholder influence on the intentions of agri-businesses in Spain to adopt sustainable practices. Particularly, this study provides a comprehensive understanding of the multifaceted dynamics at play in the Spanish agri-business sector concerning sustainability. It highlights the interplay of stakeholders and their varying levels of influence, contributing to the broader discourse on sustainable practices within the agricultural industry. The findings offer valuable insights for businesses, policymakers, and researchers seeking to promote and understand sustainability initiatives in this critical sector. For businesses, the recognition of stakeholders' multifaceted influence underscores the importance of proactive engagement and collaboration with diverse stakeholders to align sustainability goals with their values and expectations. Policymakers can leverage these insights to design regulations and incentives that not only support sustainability but also consider the unique challenges faced by agri-food companies, fostering an environment conducive to sustainable practices. Researchers, on the other hand, are encouraged to explore the nuanced power dynamics among stakeholders, conduct region-specific analyses, and employ diverse research methods to enrich our understanding of stakeholder roles in advancing sustainability within this critical sector. Ultimately, this research highlights the need for a holistic approach, where businesses, policymakers, and researchers work in tandem to drive meaningful and lasting sustainability initiatives in agriculture.

This study, like many others, is subject to certain limitations that warrant discussion. Three key limitations stand out, each offering insights for future research in the field. Firstly, the research primarily focuses on the influence of stakeholders within specific segments of companies' production chains. While this approach yields valuable insights, future investigations could benefit from a more granular analysis, honing in on specific areas of the production process where stakeholder influence is most pronounced. This level of detail could provide even more concrete and targeted results, shedding light on the precise mechanisms by which stakeholders impact sustainability practices. Secondly, despite our efforts to maintain objectivity during data collection, it's important to acknowledge that the data analysis is inevitably influenced by the perspectives and opinions of the interviewees. This inherent subjectivity underscores the need for future research to employ additional methods, such as surveys or external assessments, to triangulate findings and mitigate potential biases. Thirdly, the sample size of four companies used for analysis, while providing valuable insights, may limit the generalizability of the results. Expanding the scope of future research to include a larger and more diverse set of agri-food companies could yield a more comprehensive understanding of stakeholder influence within the sector. Additionally, exploring variations in stakeholder dynamics among different regions or agricultural subsectors could offer valuable comparative insights.

9 Collaboration

SAT 9989 Peregrín.
Agrícola Navarro de Haro S.L.
Murgiverde S. Coop.
Bio Campojoyma S.L.

References

Alonso-Martinez, D., De Marchi, V., & Di Maria, E. (2021). The sustainability performances of sustainable business models. *Journal of Cleaner Production, 323*, 129–145. https://doi.org/10.1016/j.jclepro.2021.129145

Andersson, S., Svensson, G., Molina-Castillo, F.-J., Otero-Neira, C., Lindgren, J., Karlsson, N. P. E., & Laurell, H. (2022). Sustainable development-Direct and indirect effects between economic, social, and environmental dimensions in business practices: A triangular comparison. *Corporate Social Responsibility and Environmental Management, 29*(5), 1158–1172. https://doi.org/10.1002/csr.2261

Avila, L. V., Germann Avila, A., Kieling, D. L., Cavalcante, D. M., & Lima, D. S. (2023). Sustainable recovery in small businesses: Analysis of sustainable practices and the goals for sustainable development. *Environmental Quality Management, 33*(1), 441–455. https://doi.org/10.1002/tqem.22070

Bowman, E. H., & Haire, M. (1975). A strategic posture toward corporate social responsibility. *California Management Review, 18*(2), 49–58. https://doi.org/10.2307/41164638/ASSET/41164638

Carroll, A. B. (1979). A three-dimensional conceptual model of corporate performance. *Source: The Academy of Management Review, 4*(4), 497–505.

Castillo-Díaz, F. J., Belmonte-Ureña, L. J., López-Serrano, M. J., & Camacho-Ferre, F. (2023). Assessment of the sustainability of the European agri-food sector in the context of the circular economy. *Sustainable Production and Consumption, 40*, 398–411. https://doi.org/10.1016/J.SPC.2023.07.010

Céspedes-Lorente, J., & Galdeano-Gómez, E. (2004). Environmental practices and the value added of horticultural firms. *Business Strategy and the Environment, 13*(6), 403–414. https://doi.org/10.1002/bse.420

Damke, L. I., Gomes, C. M., Kneipp, J. M., Godoy, T. P., & Motke, F. D. (2021). Sustainable management practices and innovation capacity in family agribusinesses. *Environmental Quality Management, 30*(4), 5–20. https://doi.org/10.1002/tqem.21724

Dansky, K. H., & Gamm, L. S. (2004). Accountability framework for managing stakeholders of health programs. *Journal of Health Organization and Management, 18*(4), 290–304. https://doi.org/10.1108/14777260410554296

De Falco, S. E., Scandurra, G., & Thomas, A. (2021). How stakeholders affect the pursuit of the environmental, social, and governance: Evidence from innovative small and medium enterprises. *Corporate Social Responsibility and Environmental Management, 28*(5), 1528–1539. https://doi.org/10.1002/csr.2183

De Massis, A., & Kotlar, J. (2014). The case study method in family business research: Guidelines for qualitative scholarship. *Journal of Family Business Strategy, 5*(1), 15–29. https://doi.org/10.1016/j.jfbs.2014.01.007

Donaldson, T., & Preston, L. E. (1995). The stakeholder theory of the corporations: Concepts, evidence, and implications. *The Academy of Management Review, 20*(1), 65–91.

Duque-Acevedo, M., Belmonte-Ureña, L. J., Terán-Yépez, E., & Camacho-Ferre, F. (2022). Sustainability and circularity in fruit and vegetable production. Perceptions and practices of reduction and valorization of agricultural waste biomass in south-eastern Spain. *Journal of Environmental Management, 316*, 115270. https://doi.org/10.1016/j.jenvman.2022.115270

Du Plessis, A. (2019). Current and future water scarcity and stress. In: Water as an inescapable risk (pp. 13–25). Springer. https://doi.org/10.1007/978-3-030-03186-2_2

Engert, S., & Baumgartner, R. J. (2016). Corporate sustainability strategy: Bridging the gap between formulation and implementation. *Journal of Cleaner Production, 113*, 822–834. https://doi.org/10.1016/J.JCLEPRO.2015.11.094

European Commission. (2023). Agriculture and rural development. Organic production and products. https://agriculture.ec.europa.eu/farming/organic-farming/organic-production-and-products_es

Evans, N., & Sawyer, J. (2010). CSR and stakeholders of small businesses in regional South Australia. *Social Responsibility Journal, 6*(3), 433–451. https://doi.org/10.1108/17471111011064799

Fereres, E., & Villalobos, F. J. (2016). Principles of agronomy for sustainable agriculture. In: Principles of agronomy for sustainable agriculture. Springer International Publishing. https://doi.org/10.1007/978-3-319-46116-8

Fernández, J. L., & Bajo, A. (2012). The stakeholder on interest groups theory, a key to CSR, business success and sustainability. *Adresearch ESIC International Journal of Communication Research, 6*(6), 130–143. https://doi.org/10.7263/adr.rsc.006.07

Freeman, R.E. (1984). Strategic management: A stakeholder approach. Pitman

Global Reporting Initiative. (2022). GRI 13: Agriculture, aquaculture and fishing sectors

Góes, H. A. A., Fatima, G., Santos Jhunior, R. O., & Boaventura, J. M. G. (2023). Managing for stakeholders towards corporate environmental sustainability. *Corporate Social Responsibility and Environmental Management, 30*(4), 1561–1572. https://doi.org/10.1002/csr.2448

Greenwood, M. (2007). Stakeholder engagement: Beyond the Myth of corporate responsibility. *Journal of Business Ethics, 74*, 315–327. https://doi.org/10.1007/s10551-007-9509-y

Gupta, A. K., & Gupta, N. (2020). Effect of corporate environmental sustainability on dimensions of firm performance: Towards sustainable development—evidence from India. *Journal of Cleaner Production, 253*, 119948. https://doi.org/10.1016/j.jclepro.2019.119948

Hillis, V., Lubell, M., & Hoffman, M. (2018). Sustainability partnerships and viticulture management in California. *Journal of Environmental Management, 217*, 214–225. https://doi.org/10.1016/j.jenvman.2018.03.033

Jaworski, C. C., Krzywoszynska, A., Leake, J. R., & Dicks, L. V. (2023). Sustainable soil management in the United Kingdom: A survey of current practices and how they relate to the principles of regenerative agriculture. *Soil Use and Management, 24*, 12908. https://doi.org/10.1111/SUM.12908

Knaggård, Å., Slunge, D., Ekbom, A., Göthberg, M., & Sahlin, U. (2019). Researchers' approaches to stakeholders: Interaction or transfer of knowledge? *Environmental Science and Policy, 97*, 25–35. https://doi.org/10.1016/j.envsci.2019.03.008

Kneipp, J. M., Gomes, C. M., Bichueti, R. S., Müller, L. O., & Motke, F. D. (2018). Strategic management of sustainable innovation: a case study in Brazilian industrial enterprises. *Organizations in Context Magazine, 14*(27), 131–185.

Liang, Y., Lee, M. J., & Jung, J. S. (2022). Dynamic capabilities and an ESG strategy for sustainable management performance. *Frontiers in Psychology, 13*, 7776. https://doi.org/10.3389/fpsyg.2022.887776

Loh, L., & Tan, S. (2020). Impact of sustainability reporting on brand value: An examination of 100 leading brands in Singapore. *Sustainability, 12*(18), 7392. https://doi.org/10.3390/SU12187392

Lozano, R., Nummert, B., & Ceulemans, K. (2016). Elucidating the relationship between sustainability reporting and organisational change management for sustainability. *Journal of Cleaner Production, 125*, 168–188. https://doi.org/10.1016/j.jclepro.2016.03.021

Lüdeke-Freund, F., Carroux, S., Joyce, A., Massa, L., & Breuer, H. (2018). The sustainable business model pattern taxonomy-45 patterns to support sustainability-oriented business model innovation. *Sustainable Production and Consumption, 15*, 145–162. https://doi.org/10.1016/j.spc.2018.06.004

MacLeod, C. J., Brandt, A. J., Collins, K., & Dicks, L. V. (2022). Giving stakeholders a voice in governance: Biodiversity priorities for New Zealand's agriculture. *People and Nature, 4*(2), 330–350. https://doi.org/10.1002/PAN3.10285

Massoud, M. A., Terkawi, M., & Nakkashz, R. (2019). Water reuse as an incentive to promote sustainable agriculture in Lebanon: Stakeholders' perspectives. *Integrated Environmental Assessment and Management, 15*(3), 412–421. https://doi.org/10.1002/ieam.4131

Mili, S., & Arovuori, K. (2023). The struggle to sustain agriculture in EU Southern neighbourhood partners: Improving data and policies. *EuroChoices, 22*(1), 50–57. https://doi.org/10.1111/1746-692X.12381

Mitchell, R. K., Agle, B. R., & Wood, D. J. (1997). Toward a theory of stakeholder identification and salience: Defining the principle of who and what really counts. *The Academy of Management Review, 22*(4), 853–886. https://doi.org/10.2307/259247

Moermond, C. T. A., & de Rooy, M. (2022). The Dutch chain approach on pharmaceuticals in water: Stakeholders acting together to reduce the environmental impact of pharmaceuticals. *British Journal of Clinical Pharmacology, 88*(12), 5074–5082. https://doi.org/10.1111/BCP.15509

Mohammadi-Nasrabadi, F., Salmani, Y., Banihashemi, S. M., Haghighian Roudsari, A., Zargaraan, A., & Esfarjani, F. (2020). Policy challenges of food advertisements from the viewpoints of Stakeholders: A qualitative study. *Food Science and Nutrition, 8*(4), 1949–1956. https://doi.org/10.1002/FSN3.1482

Mora-Córdova, D. E., Lituma-Loja, A. A., & González-Illescas, M. L. (2020). Certifications as a strategy for the competitiveness of exporting companies. *INNOVA Research Journal, 5*(2), 113–132.

Ordonez-Ponce, E., Clarke, A., & Macdonald, A. (2021). Business contributions to the sustainable development goals through community sustainability partnerships. *Management and Policy Journal, 12*(6), 2040–8021. https://doi.org/10.1108/SAMPJ-03-2020-0068

Punch K. (2005). Introduction to social research: Quantitative and qualitative approaches (2nd ed.). Sage

Rashid, Y., Rashid, A., Warraich, M. A., Sabir, S. S., & Waseem, A. (2019). Case study method: A step-by-step guide for business researchers. *International Journal of Qualitative Methods, 18*, 1609406919862424.

Rawlinson, C., Lesage, S., Gilles, I., & Peytremann-Bridevaux, I. (2022). Healthcare stakeholders' perspective on barriers to integrated care in Switzerland: Results from the open-ended question of a nationwide survey. *Journal of Evaluation in Clinical Practice, 28*(1), 129–134. https://doi.org/10.1111/JEP.13605

Reimer, A., Doll, J. E., Boring, T. J., & Zimnicki, T. (2023). Scaling up conservation agriculture: An exploration of challenges and opportunities through a stakeholder engagement process. *Journal of Environmental Quality, 52*(3), 465–475. https://doi.org/10.1002/JEQ2.20317

Riegler, M., Burton, A. M., Scholz, M., & de Melo, K. (2023). Why companies team up for sustainable development: Antecedents of company engagement in business partnerships for sustainability. *Business Strategy and the Environment*. https://doi.org/10.1002/BSE.3392

Seuring, S., & Mü Ller, M. (2008). From a literature review to a conceptual framework for sustainable supply chain management. *Journal of Cleaner Production, 16*(15), 1699–1710. https://doi.org/10.1016/j.jclepro.2008.04.020

Spanish Ministry of Agriculture, F. F. (2023). Environmental module for the agricultural sector

Svensson, G., Ferro, C., Hogevold, N., Padin, C., & Sosa Varela, J. C. (2018). Developing a theory of focal company business sustainability efforts in connection with supply chain stakeholders. *Supply Chain Management: An International Journal, 23*(1), 16–32. https://doi.org/10.1108/SCM-12-2015-0461

Svensson, G., Høgevold, N. M., Petzer, D., Padin, C., Ferro, C., Klopper, H. B., Carlos, J., Varela, S., Rico, P., & Wagner, B. (2016). Framing stakeholder considerations and business sustainability efforts: A construct, its dimensions and items. *Journal of Business and Industrial Marketing, 31*(2), 287–300. https://doi.org/10.1108/JBIM-05-2014-0094

Teece, D. J. (2007). Explicating dynamic capabilities: The nature and microfoundations of (sustainable) enterprise performance. *Strategic Management Journal, 28*(13), 1319–1350. https://doi.org/10.1002/SMJ.640

The Food and Agriculture Organization (FAO). (2023). Retrieved September 5, 2023, from https://www.fao.org/sustainability/es/

Valentinov, V., & Chia, R. (2022). Stakeholder theory: A process-ontological perspective. *Business Ethics, Environment and Responsibility, 31*(3), 762–776. https://doi.org/10.1111/BEER.12441

Van den Broek, K. L. (2019). Stakeholders' perceptions of the socio-economic and environmental challenges at Lake Victoria. *Lakes and Reservoirs: Research and Management, 24*(3), 239–245. https://doi.org/10.1111/LRE.12275

Van Gorp, B., & Van Der Goot, M. J. (2012). Communication, culture and critique sustainable food and agriculture: Stakeholder's frames. *Communication, Culture and Critique, 5*, 127–148. https://doi.org/10.1111/j.1753-9137.2012.01135.x

Vargas-Amelin, E., & Pindado, P. (2013). The challenge of climate change in Spain: Water resources, agriculture and land. *Journal of Hydrology, 518*(B), 243–249. https://doi.org/10.1016/j.jhydrol.2013.11.035

Walker, K., & Laplume, A. (2014). Sustainability fellowships: The potential for collective stakeholder influence. *European Business Review, 26*(2), 149–168. https://doi.org/10.1108/EBR-09-2013-0119

White, P. (2009). Building a sustainability strategy into the business. *Corporate Governance: The International Journal of Business in Society, 9*(4), 386–394. https://doi.org/10.1108/14720700910984936

Zander, K., Risius, A., Feucht, Y., Janssen, M., & Hamm, U. (2018). Sustainable aquaculture products: Implications of consumer awareness and of consumer preferences for promising market communication in Germany. *Journal of Aquatic Food Product Technology, 27*(1), 5–20. https://doi.org/10.1080/10498850.2017.1390028

Ana Isabel García-Agüero has a degree in Economics from the University of Almería, with 1 year of studies at the Jean Monnet University of Saint-Etienne (France), and has also completed three university master's degrees, one in SAP consulting, another in local development and co-development sustainable and the third is a master's degree qualifying for secondary education teaching, with a specialty in economics. She has worked as a university professor at the Changchun International University (China) and at the University of Almería (Spain). She is currently a doctoral student at the University of Almería. Her field of study focuses on agricultural competitiveness and the influence of stakeholders on sustainability and the agribusiness sector.

Eduardo Terán-Yépez Ph.D., is an Assistant Lecturer at the University of Almeria, Spain. His current research interests include international entrepreneurship, international marketing, tourist consumer behavior, sharing economy, bibliometric methodology, and sustainable management and education. His works have been published in journals such *International Journal of Hospitality Management, Current Issues in Tourism, Review of Managerial Science, Journal of International Entrepreneurship, Oeconomia Copernicana, Journal of Cleaner Production, Journal of Environmental Management, Journal of Family Business Management, Anais da Academia Brasileira de Ciências*, among others. He has also contributed with one book, five book chapters, and more than 30 papers at national and international conferences. In addition, he has participated in three competitive research projects funded by public entities and in four research contracts financed by public and private entities. He has been reviewer of more than 45 papers in journals such as *Journal of Business Venturing, Business Strategy and the Environment, Corporate Social Responsibility and Environmental Management, Small Business Economics*, and *Management Review Quarterly*.

Luis J. Belmonte-Ureña is Professor at the Economics and Business Department of the University of Almería, with more than 20 years of teaching experience. His main lines of research are related to sustainable development, social economy, circular economy, and agriculture. According to the Scopus database, he is the first most prolific author in the publication of articles related to the "circular economy" and "agriculture". Currently, Professor Belmonte-Ureña has 99 articles (69 indexed in Scopus and 51 indexed in the Web of Science), 22 books or book chapters and 26 national and international conferences. According to the classification of the journals by quartiles, Professor Belmonte-Ureña presents 23 Q1 articles, 19 Q2 articles (3 of them from Elsevier or Wiley publishers), 5 Q3 articles and 4 Q4 articles. He has also participated in 13 competitive research projects: European, national, and regional, acting as principal investigator in one of them. He is also a member of two international research networks: in France (ERASME-Centre d´excellence Jean Monnet sur le Dévéloppment Durable, coordinated by the University of Clermont-Ferrand) and in Colombia (CIRIEC-Colombia).

Francisco Camacho-Ferre is University Professor (Intensive Horticulture) at the School of Engineering of the University of Almeria. He has contributed as a speaker, poster presenter, or organizer in more than 200 courses, conferences, symposiums, etc. in Europe and America. He has directed or co-directed 174 Final Projects at the University of Almeria, as well as 20 Doctoral Theses. He was appointed expert in intensive horticultural crops by the Instituto de Cooperación Iberoamericana, working for four months in CORDECRUZ (Bolivia) in May 1983. Subsequently, he has carried out technical consultancies in Venezuela, Peru, and Argentina. From 2003 to 2015, development work in horticultural grafting with engineers and producers in Honduras, Guatemala, Dominican Republic, Cuba, Chile, Argentina, and Mexico through the United Nations Industrial Development Organization (UNIDO). He is the director of the University of Almeria´s own Master´s Degree in Horticultural Production, of which four editions have been offered for Latin America. He has participated and participates as a researcher or principal investigator in different projects sponsored by national and European organizations and private companies on different aspects of production and protection of horticultural crops (80 projects and research contracts).

He is author or co-author of 23 books, has written chapters in 38 other books and 130 articles in national and international journals. In May 2003, on the occasion of the feast of San Isidro, he was awarded the Encomienda de número de la Orden del Mérito Agrario Pesquero y Alimentario which was imposed on him by the Minister of Agriculture, Fisheries and Food. In November 2015, he was distinguished by the Government of Mexico and by the Deputy Director of the Montreal Protocol Division (UNIDO) for his participation in the National Plan for the Elimination of Methyl Bromide.

The Role of Family Farming in Socio-Economic Sustainability: An Exploratory Analysis of Rural Development in Southeast Spain

Emilio Galdeano-Gómez, Laura Piedra-Muñoz, María del Carmen García-Barranco, Yolanda Sorroche-del-Rey, Jesús Hernández-Rubio, and Javier Sánchez-García

1 Introduction

The significance of family farming in agricultural systems is becoming more apparent, serving as a vital link between economic, social, and environmental objectives (HLPE, 2013; Wuepper et al., 2020). Family farms stand at the heart of rural economies, rooted in their deep understanding of the local context and their ability to adapt. Their ambitions extend beyond profit, encompassing the well-being of the community and environmental preservation (Ikerd, 2013; Schwab do Nascimento et al., 2020; UPA, 2022).

The versatility of these farmers, evident in their provision of a range of products and services, is recognized in sustainable development strategies, including the rural

E. Galdeano-Gómez (✉) · L. Piedra-Muñoz · M. C. García-Barranco · Y. Sorroche-del-Rey · J. Hernández-Rubio · J. Sánchez-García
Mediterranean Research Center on Economics and Sustainable Development (CIMEDES), University of Almería, Almería, Spain
e-mail: galdeano@ual.es

L. Piedra-Muñoz
e-mail: lapiedra@ual.es

M. C. García-Barranco
e-mail: maricarmengarcia@ual.es

Y. Sorroche-del-Rey
e-mail: ysd813@ual.es

J. Hernández-Rubio
e-mail: jhernandez@ual.es

J. Sánchez-García
e-mail: jsg608@ual.es

M. d. C. Valls Martínez and J. M. Santos-Jaén (eds.), *Environmentally Sustainable Production*, https://doi.org/10.1007/978-3-031-52656-5_6

policies of the European Union, i.e., the concept of multifunctionality (Fuller et al., 2021; Mölders, 2013). While the environmental aspect often takes center stage, there is a pressing need for more research on their socio-economic achievements.

Sustainable development broadly aims for a long-term quality of life, balanced with cultural, social, and environmental harmony (Galdeano-Gómez et al., 2017; Suess-Reyes & Fuetsch, 2016). From a socio-economic perspective, family farms are indispensable in sustaining employment and economic resilience in rural regions, fostering entrepreneurial spirit and building intergenerational social capital (Piedra-Muñoz et al., 2016; Schwab do Nascimento et al., 2020). In other words, they play an important role in terms of business management and entrepreneurship in this economic context, for instance, as members of agricultural and livestock cooperatives or professional associations. In this way, these farmers and their networks can generate social capital and promote equity in social well-being, participation and cohesion across generations (Galdeano-Gómez et al., 2013; Holloway et al., 2021; Wu et al., 2023).

Traditional rural agriculture has typically been straightforward and low-tech, with structures that are not profitable and/or irrelevant on economic activity, but recent research underscores the significance of family farms, highlighting their economic influence and their role in preserving specific agricultural practices (see e.g., Suess-Reyes & Fuetsch, 2016; Fuller et al., 2021; Ge & Li, 2023). In diverse regions such as North America, Asia, and Europe, family-operated farms dominate the agricultural landscape, occupying extensive land areas and providing employment to millions, estimating over 80% of agricultural production and farmed land (UPA, 2022).

Despite their pivotal role, research on how family farming shapes regional socio-economic development remains scarce. Family-managed farms possess unique traits enabling them to achieve social and economic objectives (Ikerd, 2013; Piedra-Muñoz et al., 2016).

Our study is focused on the agri-food system of southeast Spain, a region steeped in a rich tradition of family farming. Production in this area is based on a small-scale family growing system that has evolved over six decades and is considerably endogenous, that is, there has been no external public planning nor political support (Galdeano-Gómez et al., 2017). We analyze how socio-cultural and economic factors related to family farms influence various indicators of socio-economic sustainability, such as demographic age distribution, income levels, employment rates, and cultural diversity.

This research work aims to illuminate the vital role of family farming in rural sustainability, contributing to the academic discourse on this subject by means of:

(i) Reviewing the involvement of the family component in rural sustainable development goals;
(ii) Exploring empirically the impact of factors related to the organization of the family farm and its multifunctional agrarian characteristics on reaching social and economic objectives.

2 Methodology

2.1 Description of Family Farming in Southeast Spain

In Spain, the structure of agricultural sector is woven largely by family-owned farms, representing a staggering 70% of the national farmland (UPA, 2022). A prime example of this agricultural model can be observed in the southeastern stretches, particularly along the coastal belts of Granada and Almeria provinces. Here, it can find over 15,000 small-scale family farms, each spanning an average of two and a half hectares. Predominantly horticultural, these farms have sowed the seeds for profound socioeconomic growth in the region over the past six decades. This progress was achieved with minimal to no governmental or Common Agricultural Policy (CAP) support in the more recent years (Galdeano-Gómez et al., 2017). The expansive family-oriented framework has anchored equitable local growth, marked by a uniform distribution of income and general well-being (Downward & Taylor, 2007; Piedra-Muñoz et al., 2016) (Fig. 1).

In this belt, the pulse of the regional economy beats with the rhythms of agriculture. Notably, the flow of provincial income and job prospects are directly tied to the agricultural seasons. This situation challenges the conventional economic view, which elevates industrialization as the beacon of development and relegates agriculture to an emblem of backwardness (Galdeano-Gómez et al., 2013). The family-farming ethos in this region is deeply rooted in its sociocultural basis. It nurtures generational

Fig. 1 Southeast Spain: location of family horticulture

Table 1 Production structure in southeast Spain and Spain (% of GDP)

Activities	1970		2020	
	Southeast Spain	Spain's National average	Southeast Spain	Spain's National average
Agriculture	29.1	12.8	24.2	4.1
Industry	13.8	30.9	14.1	19.4
Construction	9.2	7.8	9.3	9.8
Services	47.9	48.5	52.4	66.7

Source Galdeano-Gómez et al. (2017) and Cajamar (2022)

ties, ensuring that knowledge, traditions, and practices flow seamlessly across generations. The community spirit is vibrant, with many opting for collective platforms like cooperatives, emphasizing the close-knit social framework and networks underpinned by mutual trust and enduring commitments (Fuller et al., 2021; Wu et al., 2023; Wuepper et al., 2020).

Echoing characteristics identified in various global rural studies (Block & Spiegel, 2013; Schwab do Nascimento et al., 2020; Suess-Reyes & Fuetsch, 2016), these Spanish farm owners showcase a profound sense of belonging to their locale. Their roots run deep, enriched by longstanding personal bonds. Over the years, these connections have cemented both familial and regional ties, catalyzing the rural economic engine where they invest and engage (Wuepper et al., 2020). That is, these families of the farmers contribute by strengthening the local rural economy where they shop, spend money and participate in business activities (Galdeano-Gómez et al., 2017; Van der Ploeg, 2014).

The farming sector in this corner of Spain commands a robust 27% of employment and contributes 24% to the local GDP (Gross Domestic Product). An auxiliary boost comes from associated services and industries, accounting for 32% of the regional GDP (Cajamar, 2022) (Table 1).

The recent years have also witnessed a mosaic of cultures entering the farming tapestry, with a notable influx of foreign workers from countries like Morocco, Romania, and Ecuador, many of whom now helm their own agricultural ventures and are active cooperative members (García-Lorca et al., 2010).

2.2 *Specifications of Socio-Economic Sustainability*

To evaluate the influence of family farming on socioeconomic sustainability, we must navigate through a plethora of indicators, even when our lens is focused on a specific locale (Galdeano-Gómez et al., 2017; Gómez-Limón & Sánchez-Fernández, 2010). Building on the foundation laid by prior research, we have narrowed our focus to four pivotal indicators intertwined with family farming:

- Related to demographics of the farming community. This encompasses the **age structure of farmers**. It not only serves as a beacon for population stability and regional migratory patterns (Gómez-Limón & Sánchez-Fernández, 2010) but also speaks volumes about the legacy of agriculture passed down through generations (Tonts et al., 2012).
- Related to income metrics. The average **income** of a farm stand in relation to the GDP per capita, shedding light on financial equitability (Piedra-Muñoz et al., 2016). A comparative assessment with the GDP per capita of other Spanish regions provides deeper insights, especially concerning per capita social expenditure (Cajamar, 2022).
- Related to employment landscape. The **employment rate** holds paramount importance as it is a direct reflection of socioeconomic health in any rural vocation. Both the sheer number of jobs and the percentage of the population employed serve as crucial barometers (Galdeano-Gómez et al., 2013; Gómez-Limón & Sánchez-Fernández, 2010).
- Related to cultural heterogeneity of agriculture. This delves into the realm of **multiculturalism** within the farming sector, symbolizing sociocultural sustainability and cohesion (García-Lorca et al., 2010; Galdeano-Gómez et al., 2017). It is noteworthy to highlight the influx and integration of international workers and their families into the agrarian framework of southeastern Spain, adding diversity to its structures and organizations (Table 2).

Table 2 Socio-economic sustainability indicators and measurements

Indicators	Description of the measurement
Age structure	Average age of the farmer
Income	Worker income (either hired or a family member) over the interprofessional minimum salary of the country[a]
Employment rate	Average number of workers per basic crop unit (either hired or a family member) over the national average[b]
Multiculturalism	Number of nationalities per family farm (either hired or family members)

[a]This measurement is the difference between the national minimum salary (965.00 euros per month) and the average salary for family workers (net income of the farm) on one hand, and the average salary for hired workers in the sector (1096.35 euros per month) on the other

[b]The average employment rate in Spanish agriculture is 0.93 workers/year per farm (Agricultural technical unit—“Unidad técnica agrícola” in Spanish). Yet, the rate for the horticultural sector in southeast Spain is 2.5 per farm. In the present study, this variable is measured as job per hectare (1.82 workers)

Source Own elaboration

2.3 Sample

Chosen by cluster sampling, a total of 58 family farms were surveyed during the 2021–2022 fruit and vegetables growing season. Our survey delved into three principal areas:

- Social dimension within family farm management: education level, age, family relationships, family business transition, participants in decision making, number of workers (whether family members or hired), sex and nationality.
- Economic nuances shaping these farms: income, size, crop specialization, innovativeness and influence of other companies in the agricultural and livestock sector.
- Environmental stewardship and practices: agroecological practices, environmental innovation, agroecological practices and efficiency management of natural resources.

Grouped according to the survey design described above, the results obtained are the following:

a. Social dimension. Men dominate ownership, accounting for nearly 89.62%, with women representing a small fraction at 10.38%. Each owner is an autonomous, licensed entrepreneur. A 90.37% of farms are managed by heads of family, while the remaining 9.63% involve the younger generation co-steering the farm alongside their parents. The average age of decision-makers stands at 44. Women have carved out a respectable role in decision-making processes, contributing to 37.28% of such decisions. Education-wise, 42.14% boast secondary education with vocational training, while 13.22% have reached university or higher. A mere 3.08% have no formal education. As we gaze into the future, a heartening 88% of owners wish to bequeath their legacy to kin. Employment dynamics reveal an average of five steady workers, predominantly male (79%), with hired hands constituting 66.04%. The tapestry of nationalities within these farms is rich, averaging around five per farm.
b. Economic aspects. With an average of 3.6 ha under cultivation per farm, the 2021–2022 season saw a high yield of 8.1 tons per hectare, fetching an average of €44,256.17 per hectare. Using a 5-point Likert scale, farmers evaluated their industry relationships. Local businesses, marketing cooperatives, and auxiliary industries received a favorable 4.6 rating. Financial institutions earned a decent 3.7, while academia and research collaborations were valued at 3.6. However, government support trailed behind with a mere 1.4. Many farms showed a penchant for specializing in select crops and major part, 84%, were either already embracing or keen on adopting technological advancements, particularly environmentally-centric ones.
c. Environmental issues. An elevate percentage, 96%, of the surveyed farms in Almeria lean into environmental and quality management protocols like UNE 155400, UNE-EN-ISO 14001, Integrated Pest Management (IPM) and GLOBAL GAP. These are not just green labels, but also they enhance on-farm working

conditions. Furthermore, the major part, 73.09%, of these farms incorporates at least one environmental badge. Water conservation is paramount: 80.5% of interviewees have optimized their water usage via enhanced distribution and irrigation mechanisms. Other eco-initiatives orbit around land rejuvenation, waste management, and energy optimization, often in synergy with research institutions.

2.4 Description of the Variables

Drawing from our previously outlined sample, we derived measurements for the following socioeconomic performance indicators: age structure, income, employment, and multiculturalism. We have neatly compiled the descriptive statistics in Table 3 for your reference, categorizing them as dependent variables.

Likewise, to deepen the exploratory analysis and the multifunctional nature of family farming, based on our previous section, we have identified several explanatory variables, segmented into social, economic and environmental areas. Here is a breakdown:

- Decision makers. Number of people that make decisions on the family farm.

Table 3 Descriptive statistics of the variables

Variable	Mean	SD	Minimum	Maximum
Age	43.57	11.4806	21	67
Income	11,082.38	10,229.07	− 7166.4	42,812.16
Employment	1.8204	1.3612	0.5308	7.2
Multiculturalism	3.7216	2.0942	3	8
Decision makers	2.0819	1.1071	1	4
Women	0.8155	0.6830	0	2
Education	3.2180	1.2073	1	5
Generation	1.9508	0.8322	1	4
Business transition	0.9165	0.3267	0	1
Scale	3.7140	2.8519	0.65	19
Specialization	1.8516	1.0023	1	4
Sec_sector	3.7620	0.8129	2.3500	5
R + D proactivity	3.4099	0.9055	1	5
Eco-certification[a]	93.351	22.203	18.403	149.863
Eco-innovation	3.7058	1.2013	2	5

[a]Thousand kilograms
Source Own elaboration

- Women. The number of women contributing to major decisions. For example, Farmar-Bowers (2010) and also Piedra-Muñoz et al. (2016) have highlighted the role female farmers can play in steering sustainable development strategies.
- Education. This gauges the average educational background of our decision-makers. We have graded each individual on a scale: 1 (none), 2 (primary), 3 (secondary), 4 (high school or vocational training), and 5 (tertiary).
- Generation. Number of generations that have nurtured the family farm. This also shows accumulated expertise and tradition.
- Family business transition. Fictitious variable which scores 1 if the farmer envisions the next generation helming the farm, and 0 if not.
- Scale. Refers to the farmed expanse, measured in hectares, giving an idea of the farm size.
- Specialization. Assessed by the diversity of crops. A lower score here suggests a more specialized agricultural approach.
- Sec_sector. On a scale of 1–5, this is the farmer evaluation of how marketing cooperatives and other secondary services fare.
- R + D proactivity. Measures how actively a farmer collaborates with research institutions and universities to innovate and elevate their farm's competitive edge, rated from 1 to 5.
- Eco-certification. Reflects the farm's commitment to sustainable practices, like Integrated Pest Management or other environmental quality certification. Calculations were done considering the production weightage across various crops.
- Eco-innovation. This variable measures the family farm's eco-consciousness regarding the efficient use of natural resources and its openness to environmental innovation. A score of 0–5 evaluates if the family farm had implemented any innovative solutions or new technology to reduce environmental footprints.

2.5 *Specifications of the Empirical Model*

Crafting models that bridge a selection of dependent variables (socio-economic performance indicators) with a group of explanatory variables (traits of multifunctionality in family farms) brings its own challenges, notably issues with ambiguity and imprecise specification (Harrel, 2015). This research, though rooted in a theoretical framework that connects specific family farming attributes with socio-economic benchmarks, requires the crafting of customized models following thorough statistical-econometric evaluations (for instance, see Tonts et al., 2012).

In this way, we proceed to an initial regression analysis, exploring the four equations linked to the socio-economic markers, while encompassing all explanatory variables. Therefore, we begin with a general model as follows:

$$Y_i = \beta_0 + \beta_1 X_{1i} + \beta_2 X_{2i} + \beta_3 X_{3i} + \cdots + \varepsilon_i; \quad \varepsilon_i : N(0, \sigma^2) \tag{1}$$

where Y_i represents each one of the socioeconomic indicators i, X_{ji} is the value of each one of the explanatory variables and β_j is the parameter to be estimated based on the data obtained.

3 Results

Assuming that the residuals (ε_i) are normally distributed with consistent variance, Ordinary Least Squares (OLS) method stands out as the top choice for linearly estimating unknown parameters without any bias (Harrel, 2015). Initial checks, like the Breusch-Pagan test, confirmed no heteroskedasticity issues in Models 1 (related to age) and 2 (tied to income). However, we did find heteroskedasticity in Models 3 (linked to employment) and 4 (connected to multiculturalism). To tackle this, we used robust regression to find a reliable estimator that can handle the variance–covariance even when heteroskedasticity is present. It can see the results in Table 4.

In our analysis, Model 1 shows that as a farmer gets older, there's a positive relationship with several factors: their level of education, women in leadership roles, wanting to keep the farm in the family, the variety of crops (though more specialization tends to show a negative trend), a push towards innovation, efficiency in the secondary sector, the amount of certified produce, and a focus on using natural resources sustainably. In contrast, the variables corresponding to greater scale, more generations involved, and more people making decisions seem to have a negative impact. This model captures 34% of the variance, with major contributions from efficiency in the secondary sector, generational aspects, plans for keeping the business in the family, educational levels, and a push for competitive innovation. Women in leadership and a focus on environmental innovation also play crucial roles.

For Model 2, income seems to go up when women are in charge, the farmer is well-educated, the family is heavily involved in the farm, innovation is a priority, there is a strong link in the production processes, there are efforts towards environmental innovation, and the secondary sector is efficient. However, income drops when there is less crop specialization, more generations involved, more decision-makers, and somewhat surprisingly, when there's an intention to pass the farm to the next generation. This last point, though, does not have much of an impact. This model explains 45% of the variance, with significant factors being environmental certifications, farm size, the number of decision-makers, crop specialization, efficiency in the secondary sector, and education.

Model 3 points out that more workers per hectare on family farms are associated with fewer decision-makers, more quality certifications for crops, and a positive view of the secondary sector, even though it is only moderately significant. On the other hand, larger farms, more women making decisions, plans for business succession, and longer generational history of farm ownership all tend to decrease the number of workers per hectare. This model has a slightly lower fit, explaining 28% of the variance.

Table 4 Model estimations

Independent variables	Model 1. Dependent variable: age	Model 2. Dependent variable: income	Model 3. Dependent variable: employment	Model 4. Dependent variable: multiculturalism
Decision makers	− 0.08374 (2.5170) P > |t| 0.974	− 3.5849 ** (2.3306) P > |t| 0.013	− 0.2097* (0.1746) P > |t| 0.073	0.1327 (0.2643) P > |t| 0.618
Women	1.6311* (1.3936) P > |t| 0.061	0.8115 (3.6239) P > |t| 0.524	− 0.4285 (0.3526) P > |t| 0.231	0.1511 (0.3707) P > |t| 0.685
Education	1.4988** (1.2711) P > |t| 0.024	0.7568 * (1.3279) P > |t| 0.057	− 0.0224 (0.1564) P > |t| 0.887	0.3763* (0.1596) P > |t| 0.023
Generation	− 4.7597* (1.9699) P > |t| 0.020	− 0.3295 (1.8161) P > |t| 0.857	− 0.1596 (0.2110) P > |t| 0.453	0.0034 (0.1835) P > |t| 0.985
Business transition	7.0059** (5.9656) P > |t| 0.027	− 1.3148 (5.8023) P > |t| 0.822	− 1.2593 (1.4112) P > |t| 0.377	0.2288 (0.5051) P > |t| 0.653
Scale	− 0.0409 (0.4559) P > |t| 0.929	1.1281** (0.4203) P > |t| 0.010	− 0.1233** (0.0500) P > |t| 0.018	0.6098*** (0.0721) P > |t| 0.000
Specialization	2.1994 (1.8111) P > |t| 0.231	− 2.2852** (1.7282) P > |t| 0.019	− 0.0681 (0.2140) P > |t| 0.752	0.0921 (0.2568) P > |t| 0.722
Sec_sector	6.2589*** (1.5187) P > |t| 0.000	1.1939* (1.7415) P > |t| 0.049	0.0643 (0.2153) P > |t| 0.767	0.1021 (0.1648) P > |t| 0.539
R + D proactivity	0.4607** (1.6916) P > |t| 0.042	0.9666 (1.5693) P > |t| 0.054	− 0.0085 (0.1926) P > |t| 0.965	− 0.0492 (0.1668) P > |t|0.769
Eco-certification	0.1115* (0.6058) P > |t| 0.065	1.6544*** (1.052) P > |t| 0.003	1.5006** (1.2306) P > |t| 0.008	− 1.10e-06 (6.01e-06) P > |t| 0.855
Eco-innovation	0.06618* (0.1217) P > |t| 0.059	0.38402 (1.5693) P > |t| 0.141	0.1003 (0.1275) P > |t| 0.436	− 0.0313 (0.1272) P > |t| 0.807
Constant	28.6821 (11.3079) P > |t| 0.015	− 6.9317 (11.0316) P > |t| 0.533	3.5467 (1.5849) P > |t| 0.030	− 0.2355 (0.9543) P > |t| 0.806
R^2	0.3402	0.4515	0.2826	0.8216
F	0.0505	0.0029	0.1579	0.0000

Note Standard errors in parentheses. Level of significance: * $p < 0.1$; ** $p < 0.05$; *** $p < 0.01$
Source Own elaboration

Model 4 suggests that almost all factors positively influence multiculturalism, except for environmental certifications, innovations, and a push for R + D, though these have limited significance. So, larger farms with knowledgeable decision-makers tend to have a more diverse workforce. The fit of this model is high, explaining 82% of the variance.

Digging deeper with a secondary analysis focused on empirical models, our goal was to estimate parameters that align more closely with the unique characteristics of family farms in Southeast Spain. We looked at four models:

$$\text{Age} = f(\text{women, education, generation, business transition,} \\ \text{sec\_sector, } R + D \text{ proactivity, eco - certification, eco - innovation}) \quad (2)$$

$$\text{Income} = f(\text{dec\_makers, education, scale, specialization,} \\ \text{sec\_sector, } R + D \text{ proactivity, eco - certification}) \quad (3)$$

$$\text{Employment} = f(\text{decision makers, women, generation,} \\ \text{business transition, scale, specialization, eco - certification}) \quad (4)$$

$$\text{Multiculturalism} = f(\text{decision makers, education, scale, specialization,} \\ \text{sec\_sector, } R + D \text{ proactivity}) \quad (5)$$

The results are shown in Table 5.

From our findings, several factors influence the average age of farmers. A higher level of education, plans for the next generation to take over, and a strong inclination towards innovation stand out. Interestingly, efficiency in the secondary sector is also a big deal. The generational aspect is a bit tricky, possibly because many of the farmers we talked to still run the farms with their children. Women in decision-making roles and a commitment to the environment, including certifications and initiatives for sustainable resource use, also have a positive influence.

Looking at average income on family farms, the size of the farm is a major player, suggesting that economies of scale might be worth exploring. The certifications for their produce are also key, as well as how specialized their crops are and their connections with product marketing and secondary sector businesses. However, having more decision-makers seems to spread the earnings thin, especially as more family members get involved in running the farm. On the bright side, steps towards competitiveness and the farmer's education level boost income.

For employment, environmental certifications seem to bring in more specialized skills and uphold traditional practices, although this trend decreases as the farm gets bigger. Family dynamics, including participation from family members, women in leadership, and the number of decision-makers, as well as plans for succession, seem to lessen the need for additional employees. This might be because family members end up doing more of the work themselves.

Table 5 Estimations of the definitive models

Independent variables	Model 1. Dependent variable: age	Model 2. Dependent variable: income	Model 3. Dependent variable: employment	Model 4. Dependent variable: multiculturalism
Decision makers		− 3.1416** (1.2971) P > |t| 0.019	− 0.1749* (0.1640) P > |t| 0.092	0.2172* (0.1575) P > |t| 0.074
Women	1.1830* (1.2049) P > |t| 0.057		− 0.3762* (0.3340) P > |t| 0.066	
Education	1.5085** (1.2039) P > |t| 0.021	0.6819* (1.2518) P > |t| 0.068		0.3540** (0.1632) P > |t| 0.035
Generation	− 3.9785** (1.7896) P > |t| 0.031		− 0.1852 (0.2023) P > |t| 0.365	
Business transition	8.3671** (5.7774) P > |t| 0.017		− 1.2739 (1.3279) P > |t| 0.142	
Scale		1.1295*** (0.3982) P > |t| 0.007	− 0.1291** (0.0474) P > |t| 0.009	0.6135*** (0.0635) P > |t| 0.000
Specialization		− 2.5236** (1.5339) P > |t| 0.017	− 0.1071 (0.1206) P > |t| 0.379	0.1560* (0.1763) P > |t| 0.053
Sec_sector	6.7060*** (1.4100) P > |t| 0.000	1.3655** (1.6351) P > |t| 0.040		0.1091** (0.1585) P > |t| 0.041
I + D_proactivity	0.8374** (1.5978) P > |t| 0.036	0.8054* (1.4281) P > |t| 0.057		0.0534 (0.1646) P > |t| 0.747
Eco-certification	0.1112* (0.6013) P > |t| 0.067	0.1624*** (0.0480) P > |t| 0.001	2.7407*** (1.1706) P > |t| 0.002	
Eco-innovation	0.0863* (0.0116) P > |t| 0.074			
Constant	30.1513 (10.3409) P > |t| 0.005	− 7.6948 (9.9346) P > |t| 0.442	3.6202 (1.4121) P > |t| 0.014	0.0316 (0.9737) P > |t| 0.974
R^2	0.3260	0.4480	0.2790	0.8195
F	0.0068	0.0001	0.0505	0.0000

Note Standard errors in parentheses. Level of significance: *$p < 0.1$; **$p < 0.05$; ***$p < 0.01$
Source Own elaboration

On the multiculturalism front, bigger and less specialized farms attract a more diverse workforce. Farms run by well-educated people and those with more decision-makers also see more diversity. The secondary sector again plays a role, possibly offering jobs to people from different backgrounds, especially in secondary and marketing companies.

4 Discussion and Conclusions

Over recent years, the dialogue surrounding sustainable development has been notably intensified within rural policy planning. More pronouncedly, the agrarian sphere, especially the structure of family-owned farms, is being viewed as a potential pillar in forwarding sustainable agendas (Fuller et al., 2021; HLPE, 2013). Family farms, distinct from other institutional setups, uniquely champion multifunctionality. Their operations cater to market-driven products while concurrently fostering non-market outcomes, emphasizing ecological conservation. Yet, there is an underrepresented dimension to this conversation—the socioeconomic advantages of family farming. Their intrinsic understanding of local agriculture, adaptability, and generational knowledge transfer stand as pivotal assets to rural economies (Wuepper et al., 2020). The interplay between multifunctionality and sustainability within the family farm discourse suggests an intertwined relationship between their innate capabilities and the broader objectives of sustainable development. This connection, however, requires empirical validation, emphasizing region-specific nuances in rural sectors. Our study endeavors to bridge this gap, delving into these dynamics using southeast Spain's agricultural framework as a template, examining indicators such as farmer age, family and worker income, employment dynamics, and multiculturalism.

The study discerned several multifaceted determinants shaping family farms:

- Multicultural Elements: Noteworthy multicultural attributes are discerned (Cajamar, 2022), buoyed by educational prowess and decision-making breadth. Economic dynamics, farm scalability, and auxiliary industries introduce a rich tapestry of foreign workforce affiliations, fostering multiculturalism (Piedra-Muñoz et al., 2016).
- Farmer age. Predominantly shaped by the sector's vitality (Cajamar, 2022), factors like innovative inclination, regional agrifood influence, and educational reach of farmers emerge as decisive. Elements rooted in family lineage, particularly the prominence of women and generational farm inheritance, also weigh in (Holloway et al., 2021). Proactive environmental initiatives further catalyze a youthful farmer demographic (Gómez-Limón & Sánchez-Fernández, 2010; Suess-Reyes & Fuetsch, 2016).
- Income dynamics. Economic scales, specialized output, and ecologically certified products drive farm incomes, echoing past sectoral insights (Valera et al., 2014). Pivotal too are farmer education and an innovative drive aimed at bolstering

competitiveness, aligning with evolving agricultural practices (Galdeano-Gómez et al., 2017).

- Employment rate. Larger farms reflect reduced employment metrics, yet environmentally certified crops necessitate heightened manual labor (Cajamar, 2022). Heightened family participation, particularly female involvement, inversely affects external employment, with core farm operations shouldered predominantly by the familial nucleus.
- Multicultural elements. Noteworthy multicultural attributes are discerned (Cajamar, 2022), buoyed by educational prowess and decision-making breadth. Economic dynamics, farm scalability, and auxiliary industries introduce a rich tapestry of foreign workforce affiliations, fostering multiculturalism (Piedra-Muñoz et al., 2016).

Conclusively, our findings underline the juxtaposition of economic scalability with innovative vibrancy, ecological leanings, educational progression, and generational farm continuity in positively impacting the agrarian landscape's age profile, income spectrum, employment distribution, and multicultural essence in the agricultural area studied.

Notably, our study possesses certain constraints and overcoming them would imply carrying out future lines of research. Our lens predominantly focuses on southeast Spain's agricultural sector, characterized by petite family farms with minimal external European policy influences. Thus, extrapolating these findings across varied international terrains and farming categories warrants exploration. While our data orbits socioeconomic progression, a comprehensive sustainability perspective remains to be charted. The cross-sectional data offers a snapshot, urging a longitudinal perspective to gauge enduring relational dynamics.

Broadly, our research illuminates the socioeconomic sustainability ramifications of family farming traits, offering pivotal insights for rural agrarian frameworks underpinned by family-driven agriculture.

References

Block, J. H., & Spiegel, F. (2013). Family firm density and regional innovation output: An exploratory analysis. *Journal of Family Business Strategy, 4*, 270–280.

Cajamar. (2022). Análisis de la Campaña Hortofrutícola de Almería, Campaña 2021/2022. Cajamar Caja Rural, Almería, España

Downward, S. R., & Taylor, R. (2007). An assessment of Spain's program AGUA and its implications for sustainable water management in the province of Almería, southeast Spain. *Journal of Environmental Management, 82*, 277–289.

Farmar-Bowers, Q. (2010). Understanding the strategic decision women make in farming families. *Journal of Rural Studies, 26*, 141–151.

Fuller, A. M., Xu, S., Sutherland, L.-A., & Escher, F. (2021). Land to the tiller: The sustainability of family farms. *Sustainability, 13*, 11452.

Galdeano-Gómez, E., Aznar-Sánchez, J. A., & Pérez-Mesa, J. C. (2013). Sustainability dimensions related to agricultural-based development: The experience of 50 years of intensive farming in Almería (Spain). *International Journal of Agricultural Sustainability, 11*(2), 125–143.

Galdeano-Gómez, E., Aznar-Sánchez, J. A., Pérez-Mesa, J. C., & Piedra-Muñoz, L. (2017). Exploring synergies among agricultural sustainability dimensions: An empirical study on farming system in Almería (southeast Spain). *Ecological Economics, 140*, 99–109.

García Lorca, A. M., et al. (2010). Agriculture in drylands: Experience in Almería. In H. G. Brauch (Ed.), *Coping with global environmental change, disasters and security* (pp. 921–934). Springer.

Ge, J., & Li, L. (2023). Dual innovation of the business model: The regulatory role of entrepreneurial orientation in family firms. *Journal of Business and Industrial Marketing, 38*(7), 1587–1599.

Gómez-Limón, J. A., & Sánchez-Fernández, G. (2010). Empirical evaluation of agricultural sustainability using composite indicators. *Ecological Economics, 69*, 1062–1075.

Harrel, F. E. (2015). *Regression modeling strategies*. Vanderbilt University.

HLPE (2013). Investing in smallholder agriculture for food security. Report by the High Level Panel of Experts on Food Security and Nutrition, Rome

Holloway, L. A., Catney, G., Stockdale, A., & Nelson, R. (2021). Sustainable family farming futures: Exploring the challenges of family farm decision making through an emotional lens of 'belonging.' *Sustainability, 13*, 12271.

Ikerd, J. (2013). Family farms: Our promise for a sustainable future. Presentation at the Pennsylvania farmers Union annual convention, Dixon University, Harrisburg, PA, December 7, 2013. http://web.missouri.edu/ikerdj/papers/Pennsylvania%20Farmers%20Union%20Family%20Farms.htm

Mölders, T. (2013). Multifunctional agricultural policies: Pathways towards sustainable rural development? *International Journal of Sociology of Agriculture and Food, 21*(1), 97–114.

Piedra-Muñoz, L., Galdeano-Gómez, E., & Pérez-Mesa, J. C. (2016). Is sustainability compatible with profitability? An empirical analysis on family farming activity. *Sustainability, 8*, 893.

Schwab do Nascimento, F., Calle-Collado, A., Muñoz Benito, R. (2020). Economía social, solidaria y agroecología en cooperativas de agricultura familiar en Brasil como forma de Desarrollo de una agricultura sostenible. CIRIEC-España, Revista de Economía Pública, Social y Cooperativa, 98, 189–211

Suess-Reyes, J., & Fuetsch, E. (2016). The future of family farming: A literature review on innovative, sustainable and succession-oriented strategies. *Journal of Rural Studies, 47*(A), 117–140.

Tonts, M., Plummer, P., & Lawrie, M. (2012). Socio-economic wellbeing in Australian mining towns: A comparative analysis. *Journal of Rural Studies, 28*, 288–301.

UPA, Unión de Pequeños Agricultores y Ganaderos (2022). Anuario de la Agricultura y Ganadería Familiar en España 2022. UPA, Madrid

Valera, D. L., Belmonte, L. J., Molina, F. D., & López, A. (2014). Los Invernaderos de Almería. Análisis de su Tecnología y Rentabilidad. Cajamar Caja Rural, Almería, España

Van der Ploeg, J. D. (2014). Peasant-driven agricultural growth and food sovereignty. *Journal of Peasant Studies, 41*(6), 999–1030.

Wu, F., Guo, X., & Guo, X. (2023). The impact of cooperative membership on family farms' income: Evidence from China. *Applied Economics, 12*, 1–18.

Wuepper, D., Wimmer, S., & Sauer, J. (2020). Is small family farming more environmentally sustainable? Evidence from a spatial regression discontinuity design in Germany. *Land Use Policy, 90*, 104360. https://doi.org/10.1016/j.landusepol.2019.104360

Emilio Galdeano-Gómez (male) is Full Professor in Economics at Universidad de Almería. His research topics are innovation, environmental economics, supply chain networks, internationalization, and exports. He has participated in 36 R&D projects and contracts, with national and international companies and institutions. He has more than 50 papers published in SSCI journals (his work has 1742 cites and his H index is 26), such as Business Strategy and the Environment, Supply Chain Management: An International Journal, Journal of Business Economics and Management, World Economy, Environment and Resource Economics, Ecological Economics, and Research Policy.

Laura Piedra-Muñoz (female) works as a Full Professor in the Department of Economics and Business at the University of Almería in Spain. She has been an invited professor in Georgia State University (USA) in 2006 and the University of Manchester (United Kingdom) in 2022. She has 24 years of experience in studies on environmental economics, sustainable development and agricultural economics. She has participated in 28 R&D projects and contracts, with national and international companies and institutions. She has published 38 scientific articles in international peer-reviewed journals (h > 19; > 776 citations). She is the Editor-in-Chief of the International Journal of Migration Studies (RIEM, ISSN: 2173–1950) since 2013.

María del Carmen García-Barranco (female) holds a PhD in Economics, Business and Law (2020) and is an acting substitute lecturer in the Department of Economics and Business Studies at the University of Almeria (Business Organisation). Her area of expertise is supply chain applied to the agribusiness and tourism sectors. She is the author of several articles in high impact academic journals, books and book chapters related to this line of research.

Yolanda Sorroche-del-Rey (female) has undergraduate degree in Business Administration and Management and master's degree in Teaching Compulsory Secondary and Pre-university Education at the University of Almería (Spain). She is currently researching for her PhD in Economics. Research experience and interests are topics related to the interactions between environmental performance, international trade and business productivity, specifically in the agri-food sector. She has published one article in the journal Business Strategy and the Environment, which is indexed in the first decile (D1) of the Journal Citation Report.

Jesús Hernández-Rubio (male) holds a PhD degree in Economics by the University of Almeria, Spain, where he currently serves as an interim professor of Business Administration in the Department of Economics and Business. He has published in journals in Economics and Finance and participated in two R&D projects and contracts. His research interests include food safety, agribusiness, biotechnometrics and organizational behavior.

Javier Sánchez-García (male) holds a PhD degree in Economics from the University of Almeria, Spain, where he currently serves as an interim professor of Econometrics in the Department of Economics and Business. He has published in top SSCI journals in Economics and Finance, such as Finance Research Letters and the International Journal of Finance and Economics. He has participated in two R&D projects and contracts. His research interest includes time series econometrics, financial stability and macroeconomics.

Innovation Oriented Towards Sustainability in the Value Chain of Agri-Food Cooperatives

Sepide Mehrabi, Juan Carlos Pérez-Mesa, María del Carmen García-Barranco, and Cynthia Giagnocavo

1 Introduction

In the present day, sustainability is widely recognized as a catalyst for innovation. A substantial body of literature focuses on sustainability-driven innovation, sustainable development innovation, and sustainability-related innovation, particularly in the agri-food sector. However, a significant portion of this literature may not be directly relevant to this review for several reasons: (i) it predominantly pertains to specific case studies involving minority or niche products. (ii) it involves cases from less developed countries that are unlikely to be applicable to business models within the European Union (EU). (iii) it discusses "innovations" that are already considered standard practices or legal requirements in the European agricultural sector; (iv) it includes numerous technical papers, particularly in the fields of environment and agronomy, which do not contribute significantly to understanding the causal relationship between sustainability and innovation. It is essential to emphasize that the relationship between innovation and sustainability in agri-food systems is more intricate compared to other sectors. This complexity arises from differing and conflicting viewpoints, as noted by (El Bilali, 2018). On one hand, some perspectives advocate that future agricultural innovation must address not only technical matters but also

S. Mehrabi · J. C. Pérez-Mesa (✉) · M. C. García-Barranco · C. Giagnocavo
Department of Economics and Business, University of Almería (Agrifood Campus of International Excellence, ceiA3; Centre for Research on Intensive Mediterranean Agrosystems and Agri-Food Biotechnology, CIAMBITAL), Almería, Spain
e-mail: juancarl@ual.es

S. Mehrabi
e-mail: sepidemehrabi@ual.es

M. C. García-Barranco
e-mail: maricarmengarcia@ual.es

C. Giagnocavo
e-mail: cgiagnocavo@ual.es

M. d. C. Valls Martínez and J. M. Santos-Jaén (eds.), *Environmentally Sustainable Production*, https://doi.org/10.1007/978-3-031-52656-5_7

social issues to effectively confront challenges such as climate change, food security, and rural development. In this context, innovation in rural development, particularly in the European Union (EU), is viewed through the lens of social innovation, recognizing that innovation and technology in agriculture can have negative effects on the environment and rural livelihoods (EU SCAR, 2016). On the other hand, opposing viewpoints stress the significance of traditional technologies, such as fertilizers and crop protection products, in addressing the challenge of food security by increasing productivity. They contend that the stagnation or plateauing of crop yields poses a significant threat to global food security, necessitating breakthrough innovations for the "sustainable intensification" of production, as argued by Garnett et al. (2013).

In this context, this chapter aims to analyze the interplay between innovation and sustainability, focusing on the concept of SOI as applied to the fruit and vegetable sector, with a particular emphasis on the producer organization's pivotal role within the supply chain.

2 Definition of SOI in the Literature on Agri-Food Sector

The discourse surrounding companies steering their business activities towards sustainability through innovation initially centered on eco-innovations. Environmental concerns were identified as catalysts for strategic transformation. Eco-innovations encompass novel or improved processes, organizational structures, products, or technologies that benefit the environment by reducing or mitigating adverse environmental impacts. Over time, the notion of eco-innovations has transformed into SOI. Although several similar concepts already exist, such as sustainability-driven innovation, sustainable development innovation, CSR-driven innovation, and sustainability-related innovation (Klewitz & Hansen, 2014), SOI offers a more comprehensive perspective on sustainability, encompassing both environmental and social dimensions. It represents a deliberate, long-term approach towards sustainability. SOI diverges from conventional innovation in several significant ways (Cagliano et al., 2016):

- Strategic Objectives with Innovation (SOIs) typically encompass multidimensional goals. In certain cases, SOIs are framed within the context of radical innovations, as noted by Meynard et al. (2017).
- SOIs are inherently riskier endeavors that need adjustments to the supply chain's configuration, its member organizations, and the management of practices and activities.
- SOIs introduce a heightenned level of complexity and uncertainty due to the unknown outcomes resulting from the integration or trade-offs among social, economic, and environmental dimensions. For instance, shortening the supply chain may effectively increase social and environmental benefits, but demonstrating economic sustainability can be challenging, as highlighted by Chiffoleau and Dourian (2020).

- SOIs typically demand more collaborative and open-system approaches. Given their complexity, the diverse types of knowledge required, and the associated risks and uncertainties, particularly in terms of market acceptance, SOIs require the engagement of a variety of stakeholders, including non-conventional actors such as customers, NGOs, and other stakeholders like local communities.

Also, SOI implies modification (reorganization) of standard innovation classifications (technological, organizational, institutional, social, managerial, and marketing). In this regard, the concept of SOI requires a new terminology (Neutzling et al., 2018):

- Operational optimization: Companies have the objective of optimization resorting to incremental innovations based on "technical-fixes" (both to processes or products) in order to reduce impacts.
- Organizational transformation: sustainability becomes more deeply embedded in the company's strategy and culture.
- Systems building perspective: Engagement and collaboration with internal and external stakeholders.

Consequently, it is necessary to improve the structural characteristics and internal resources of the supply chain such as the degree of integration between the various stages of the supply chain and the cooperation between the subjects of the same supply chain in order to implement open innovation models (Meynard et al., 2017; Stanco et al., 2020). Open innovation has the potential to reduce innovation costs, expedite time-to-market, and share risk with other stakeholders (Arcese et al., 2015).

SOI implies that sustainability is deliberately used as a driver of innovation in ad hoc strategy from various points of view (Panigrahi et al., 2019):

- Environmental perspective (green packaging distribution, warehousing and transportation, implementing eco-friendly technologies, etc.).
- Social perspective (social perspectives code of conduct; employee rights, welfare and working conditions, etc.).
- Economic perspective (strategic collaboration and information sharing; business transparency, etc.).
- Governance perspective (governance structures, mechanisms and linkages; collaboration and formalisation, etc.).

In the same way, indicators of their status can be used: (e.g. eco-innovations, certifications, etc.). In this sense, there is a positive relationship between the SOI application and profitability (Piedra-Muñoz et al., 2016).

In all aspects of SOI strategy, technological support (such as digitalization) will have a relevant role (Krishnan et al., 2020; Renda, 2019). Also, the implementation of a higher degree of inter-organizational collaboration would be needed to include multiple variables for cooperating companies (Neutzling et al., 2018). For example, control and communication systems, inter-organizational learning capacities, resource sharing for specific objectives, standardized procedures of operation, etc. In addition, internal self-diagnosis would be necessary (Kumar et al., 2020):

(i) dynamic leadership, (ii) an effective communications structure, (iii) trust among employees; and (iv) performance evaluation system and reward facility, etc.

3 SOI in Literature of Value/Supply Chains

By taking the SOI perspective into account when examining supply chains, as suggested by (Geissdoerfer et al., 2018), a novel concept of Sustainability-Oriented Innovation Supply Chain (SOI-SC) can be formulated. In this definition, SOI-SC includes all the features of the green supply chain, circular supply chain, sustainable supply chain, and even Bioeconomy. As illustrated in Fig. 1, SOI-SC could create a core value proposition of the business as a competitive advantage, particularly in business models based on collaboration for SOI (see Fig. 1). For example, it is notable how the literature considers "short supply chains", especially in perishables, as a paradigmatic example to illustrate this new system (Volpentesta et al., 2013).

In the context of the SOI-SC concept, as outlined by (Gao et al., 2017), Sustainable Supply Chain Innovation (SSCI) emerges from Supply Chain Innovation(SCI). SSCI represents a comprehensive shift from incremental to radical changes, encompassing various forms of innovation, including product, process, organizational structure, and business models. It affects all functions within the supply chain and generates value for all stakeholders. Within the realm of SSCI, it is evident that innovation acts as the driving force for sustainability and not the result of it. In other words, sustainability is a consequence of innovation, not the other way around. SSCI operates within a business ecosystem framework, which is essentially an economic community

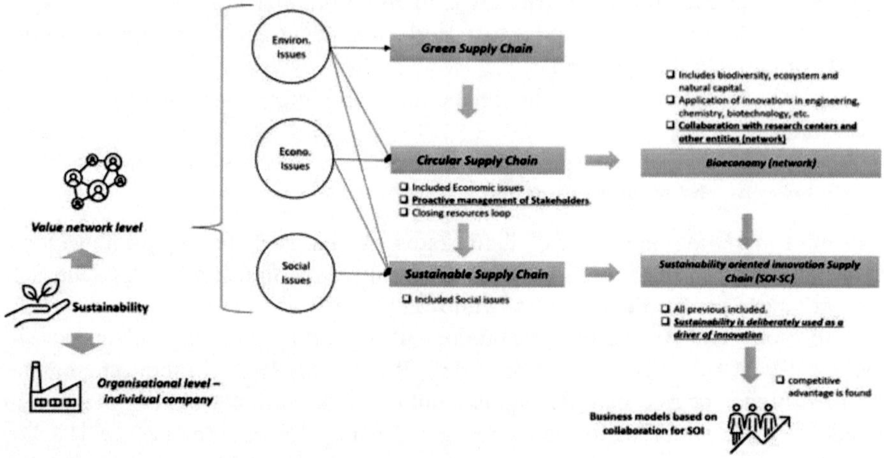

Fig. 1 New concept of Sustainability-Oriented innovation Supply Chain (SOI-SC). *Source* Own elaboration

composed of interconnected organizations and individuals. This ecosystem encompasses economic and non-economic agents, as well as relationships involving technology, institutions, sociological interactions, and culture. Over time, the participants within the business ecosystem develop their capabilities and adapt their roles, aligning themselves with the strategic objectives established by one or more central companies. Moreover, the position of these central companies as a leader is highly valued within the business ecosystem. They play a role in enabling other members towards shared visions, such as aligning investment strategies and mutually supportive roles etc. This perspective expands the conventional supply chain into a broader network of stakeholders and indirect business collaborators, fostering a dynamic and interdependent environment in which open innovation plays a central role (Liu & Stephens, 2019).

Business model innovation within the agri-food sector has been categorized by (Barth ct al., 2017) including three main innovation types: (1) Technological innovations that aim to maximize material and energy efficiency, create value from waste, and employ renewable substitutes and natural processes. (2) Social innovations that prioritize delivering functionality over ownership, adopting a stewardship role, and encouraging sufficiency. (3) Organizational innovations that involve repurposing society and the environment while developing scalable solutions. Moreover, sustainable transition studies emphasized the role of SBMIs in the agri-food sector. The successful transitions to sustainability in this sector rely on collaboration, a well-defined narrative and vision, continuous innovation, a strong sustainability foundation, profitability, and the potential influence of serendipitous external events. An entrepreneurial mindset is also crucial in this context (Ulvenblad et al., 2019).

4 Specificities of SOI in Fruit and Vegetables (F&V) Value Chains

4.1 Description of the F&V Chain and Its Relationships

The agri-food market is characterised by dynamic shifts in consumer demands, technological advancements, and socioeconomic influences. Within this complex landscape, a company's ability to develop capabilities and gain competitive advantages hinges on its adept management of the supply chain (SC). Supply chain management should be viewed as a sequence of operations designed to ensure the production and distribution of goods in the right quantities, delivered to chosen locations with high efficiency all with the primary aim of meeting consumer needs (Flynn et al., 2010). In order to ensure profitability in this sector which is characterized by volatility and ongoing transformations, efficiently coordinating and overseeing all the entities engaged in this process is crucial.

Growers occupy a pivotal position within a network responsible for delivering products to the end consumer. With the increasing trend toward consolidation in certain phases of the supply chain, such as retail distribution or the supply of seeds, phytosanitary products, and biotechnology, and the expanding international reach of these components, small-scale growers have been compelled to adopt organizational strategies. Typically, these growers have turned to a traditional approach for both horizontal expansion and market access, leading to the establishment of social economy firm associations such a cooperatives or producer organizations.

Over time, the organizational systems of growers have diversified in terms of roles, sizes, and modes of association, differing according to the country of origin and the product in question. Even within a single supply chainlarge enterprises with significant export capabilities can coexist with smaller, locally-focused ones. Furthermore, commercial chains can vary in length, with long chains involving intermediate entities before reaching the consumer, including agrifood industry members, wholesale firms, buying centers, and retailers (see Fig. 2). In contrast, shorter chains, while simpler, may still feature retailers as part of their strategy, as large distribution brands seek local suppliers for direct provision. This sector displays a complex supply network that encompasses both horizontal relationships, which are more common in the production phase and involve collaboration with other companies to enhance market access and foster coordinated innovation activities, as well as vertical relationships designed to optimize joint processes like transportation, provisioning, and marketing.

It's worth noting that small-scale agri-food product distribution is subject to continual restructuring (Fischer et al., 2010; Pérez-Mesa & Galdeano-Gómez, 2015). A significant change is the ascendancy of large-scale distribution as the dominant business model, significantly altering relationships within the supply chain (Arzu & Erman, 2002) and influencing the level of integration and collaboration among its members (Galdeano-Gómez et al., 2017). In this context, the agri-food supplier of a retailer, typically a producer organization, stands as a vulnerable link that must adapt to meet the demands of its customers (Dobson et al., 2013; Hingley, 2005). It's

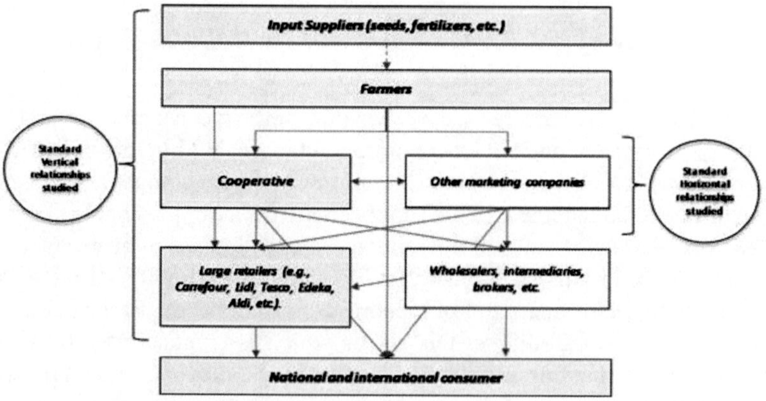

Fig. 2 Standard supply chain for F&V and classic relationships analysed. In grey, priority channel. *Source* Adaptation from Pérez-Mesa et al. (2021)

essential to also acknowledge the preceding link to the grower, consisting of firms responsible for developing, producing, and supplying growers with all necessary inputs, such as seeds, biotechnology, fertilizers, and phytosanitary products. These companies, largely multinational entities, are gaining significance due to their high concentration within the supply chain.

Agri-food companies, as integral components of the supply chain, must make critical decisions regarding the nature of the relationships they establish, whether vertical or horizontal in orientation. These decisions encompass the determination of necessary supplier-supplier collaborations and the extent of customer integration required. As a consequence, it is imperative for companies to comprehend the implications of these connections and structure their networks effectively to achieve mutually beneficial outcomes for both suppliers and customers (Brandenburger & Nalebuff, 1996). The supply chain can essentially be visualized as an inter-organizational network composed of multiple supplier-supplier-customer combinations. In this interconnected system, the advancement of individual companies can only be realized through the optimization of the entire network.

The analysis of collaboration within supply chains is pursued along two fundamental research avenues. The first concentrates solely on cooperation within specific segments and relationships in the supply chain (Dagnino & Mariani, 2010). The second approach takes into account the phenomenon in its entirety, considering all internal interactions within the operational network (Zerbini & Castaldo, 2007). Nevertheless, both these approaches are harmonious (Wilhelm, 2011), as vertical (supplier–buyer) and horizontal (supplier-supplier) relationships and strategies cannot be understood in isolation; they are intricately intertwined.

Producer organisations and cooperatives, in particular, face the challenge of managing relationships, owing to the presence of a multitude of members, each with distinct roles. This complexity must be factored into the design of a company's strategy within the supply chain (García Pérez et al., 2016). This approach aligns with the netchain concept (Lazzarini et al., 2001), which denotes the horizontal and vertical network established by cooperative partner members and their distributors. Within these networks, customer–supplier relationships, primarily from the buyers' perspective, have been extensively explored. Efficient collaboration with suppliers generally equips companies to adapt to unforeseen changes, implement well-conceived solutions to organizational issues, reduce logistics and inventory costs, and, ultimately, enhance their financial performance (Yu et al., 2007). Despite the potential for conflicts, these relationships tend to be characterized by complementarity. The literature underscores the advantages of dividing the supply chain so that suppliers can specialize in actions favoring quality while customers focus on promotion and sales (Gurnani et al., 2007).

In practice, this collaboration represents the initial step in an integration process, which involves closer connections and coordination among network members, culminating in increased commercial interactions (Droge et al., 2009). In the context of agricultural products, integration between suppliers and retailers can manifest in various forms, but fundamentally such integration aims to establish alliances that enhance scheduling and supply reliability through long-term relationships, with the ultimate goal of streamlining processes and ensuring quality, food safety, and sustainability (Rong et al., 2011).

One notable characteristic of supply chains is that one or several members, usually retailers, exert influence over the rest, and are referred to as the "hub company" (Wassermann & Faust, 1994) or "channel master" (Rice & Hope, 2011). To maintain their market share, suppliers often acquiesce to the demands of their customers, accepting this power imbalance (Peterson, 2002). Notably, in the agricultural supplies sector (including seeds, biotechnology, fertilizers, and phytosanitary products), the power and activity in vertical relationships has increased.

Interestingly, a supply chain dominated by one or several channel masters may prove inefficient, as it can demotivate other members in areas such aw innovation or marketing. Ideally, the relationship would evolve to create genuinely integrated chain entities (Fawcett et al., 2008), where all involved companies participate in decision-making and are intrinsically interdependent, sharing knowledge to resolve any issues that may arise.

In addition, incorporating the supplier-supplier relationship into this analytical framework is a crucial step in understanding how supply chains operate, particularly within cooperatives and producer organisations. This relationship is considered the most complex and dynamic (Hingley, 2005), involving the traditional challenges of structuring and coordinating processes among companies competing for the same customers (Choi & Hong, 2002). This complexity may explain the initial ambiguity surrounding the benefits of this collaboration.

A contentious issue revolves around the degree of integration between suppliers and their customers. Strategies in this context vary and are contingent on the pros and cons of having a single supplier-customer relationship versus multiple supplier-customer relationships. From the customer's perspective, the benefits of exclusive suppliers are readily apparent. Such benefits include enhanced negotiation capabilities, increased transparency, improved interaction, better investment planning, information exchange, and economies of scale. Nevertheless, multiple suppliers also bring advantages to the table, such as flexibility, price competition, and access to diverse information sources. Presently, many retailers are pursuing a hybrid approach that combines supplier integration for strategic items, often leaving smaller suppliers in a vulnerable position due to price pressures.

4.2 Sustainability-Innovation in F&V Chains

The current paradigm for binomial sustainability-innovation in F&V chains is between circular chains (Aznar-Sánchez et al., 2020; Augustin et al., 2020) and sustainable chains (Fruit-Logistica, 2020). The environmental aspect (eco-innovation) of sustainability is preferred in the literature over the social and economic one (García-Granero et al., 2020). The improvement of quality and health, although it is relates to standards, is also considered another relevant aspect (Chan et al., 2020; Peano et al., 2017).

In short, there is no clear guidance to apply SOI at the company level: a specific strategy in innovation based in sustainability is need (Trienekens et al., 2008). In this sense, the needs recognized by the professional sector are (Fruit-Logistica, 2020):

- The fresh produce sector is shifting towards circular business models that involve repurposing waste streams, maximizing the efficient use of renewable resources, and minimizing their environmental and social footprints (see Fig. 3).
- In the upcoming years, the industry must tackle concerns related to water, food waste, packaging, chemical utilization, and energy consumption (see Fig. 4).
- Prioritizing concerns related to labor, especially in light of the challenges faced by the horticulture sector regarding workforce issues, is of utmost importance.

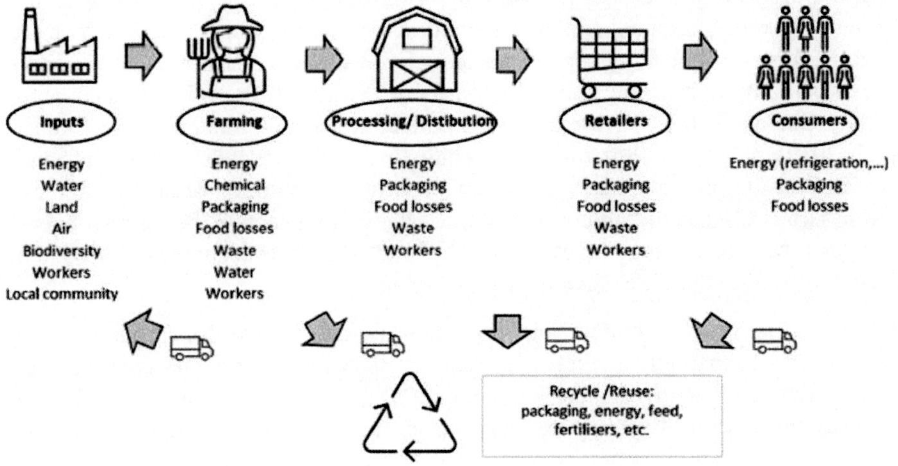

Fig. 3 Sustainability impacts in fresh F&V supply chain. 2020 *Source* Fruit Logistica ()

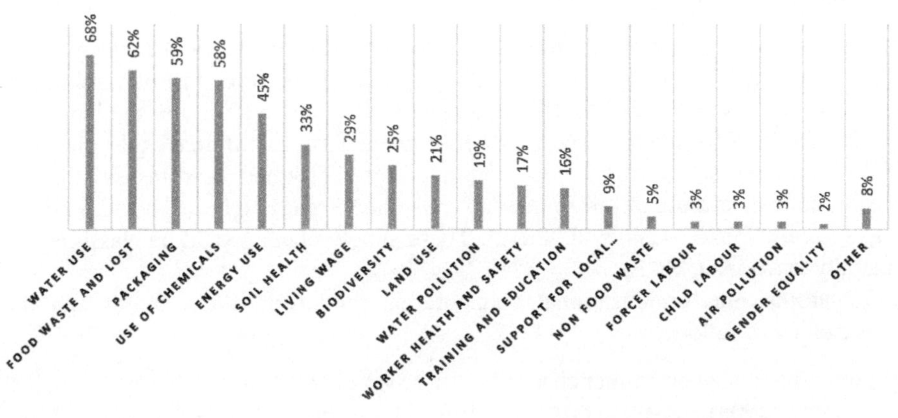

Fig. 4 Top sustainability issues for fresh F&V industry. 2020 *Source* Fruit Logistica ()

- The significance of technology and biology is evident. These two critical domains provide solutions to sustainability challenges. Technology encompasses closed greenhouse production systems, precision agriculture technologies like sensors, and robotic weeding solutions. Biology involves harnessing heat from geothermal sources and employing natural crop protection methods.
- Certification plays a pivotal role in addressing the gaps that governments may not fully cover. However, it should be complemented with metrics, incorporating relevant tools and footprinting. In the field of horticulture, the development and application of metrics is still in its early stage.

On the other hand, there are two somewhat contradictory tendencies between (i) long (or export-oriented) supply chains, more concerned with technological innovation focused on increasing productivity within a sustainable supply chain; and (ii) a more "retro" oriented tendency towards short supply chains (SSC), which are more focused on social and social-economic innovations. In this sense, it is important to ask: Can short supply chains be considered Sustainability-oriented innovation Supply Chains (SOI-SC)? Some authors defend the second tendency (Kuokkanen et al., 2019): "Sustainability transitions require innovations that disrupt both production and consumption practices. However, the original definition of disruptive innovation is too limited to capture a systemic change perspective, being also biased towards radical technology innovation at the production side". There is an argument (EIP-Agri, 2019) that SSC have important benefits:

- Pioneering novel products and processes (adding value for producers and processors, expanding market access, and diversifying consumer product choices).
- Investigating inventive business and marketing models (streamlining logistics and distribution, cost and resource sharing, and nurturing close relationships with consumers).
- Bridging the gap between urban and rural areas (ensuring fresh food access for city consumers, maintaining stable supply chains, and establishing procurement contracts for public institutions).

However, others authors express doubts about SSC economic and even environmental sustainability (Cagliano et al., 2016). In any case, SSC will be a minority option, whose model is difficult to implement for the vast majority of the industry, particulary those focused on exports. When examined from a different angle, the term "short" may not necessarily imply geographic proximity but could instead signify the thorough disclosure of information to consumers regarding location and production systems (Pérez-Mesa et al., 2019). These supply chains could be classified as spatially extended (SSCE).

Regarding inter-organizational collaboration models in F&V, trends can be summarized as follows:

- Adapting to final consumer changes is the driver of collaboration in supply chains (Verdow, 2008), mostly in order to improve the agility and timely operations (due to the perishability of product).

- The companies in the chain avoid the responsibility of applying SOI because there is a "channel master" or "hub company" (usually retailers, but currently also biotech companies) that dominates all relationships (Pérez-Mesa et al., 2019).
- The ascendance of large-scale distribution as the predominant business model, responsible for acquiring around 70% of fresh F&V production, triggers shifts in the dynamics of supply chain relationships. These alterations have notable implications for the extent of collaboration among its members, especially concerning innovation. Within this evolving landscape, the F&V supplier of a retailer, as mentioned above, finds itself in a vulnerable position, necessitating adaptation to meet the demands of its customer, who, in turn, channels the needs of the ultimate consumer. Furthermore, the link preceding the grower, comprising companies engaged in the development, production, and sale of vital inputs to growers, such as seeds, biotechnology, fertilizers, and phytosanitary products in many instances, wield influence, imposing their models of sustainable innovation (Pérez-Mesa et al., 2021).
- There has been a trend towards studying classic relationships (e.g. coopetition, Fig. 2) focused on economic objectives (Wiśniewska-Paluszak & Paluszak, 2019). The need to expand relationships is recognized (upstream and downstream, but also with research centers, consumers, NGOs, etc.). In general, there is a need to promote open innovation in a more participatory and proactive collaboration framework (Wiersinga et al., 2010). It is necessary to involve the heterogeneous actors that make up the entirety of the value chain in the development and creation of innovations.

Business models for sustainability require a transition from defensive to proactive and transformative models. The existence of an internal innovation strategy is key, but so is obtaining a minimum profitability. The current strong market competition makes the appearance of this type of model difficult (Long et al., 2018). This raises the question if the existence of a central company is completely necessary (obligatory), at least in the more classic supply chains.

5 Conclusion

The interplay between innovation and sustainability within agri-food systems presents a unique complexity compared to other sectors, as it grapples with diverse and conflicting perspectives. One perspective argues that upcoming agricultural innovations should not solely address technical challenges but also encompass social considerations. Conversely, opposing perspectives emphasize the continued importance of traditional technologies, such as fertilizers and crop protection products, in addressing food security concerns through increased productivity. The discourse on companies orienting their business activities toward sustainability via innovation initially centered on eco-innovations, wherein environmental concerns were identified as drivers of strategic change. Over time, the concept of eco-innovations has evolved into SOI. Various related terms like sustainability-driven innovation, sustainable development innovation, corporate social responsibility-driven innovation, and sustainability-related innovation have emerged. These approaches collectively offer a more holistic view of sustainability, encompassing both environmental and social dimensions. They represent a journey or orientation towards sustainability that demands a long-term vision in management.

The current paradigm for binomial sustainability-innovation in F&V chains is between circular chains and sustainable chains. The environmental aspects (eco-innovation) of sustainability is preferred in the literature over the social and economic one. The improvement of quality and health, although it is related to standards, is also considered another relevant aspect. In short, there is no clear guidance to apply SOI at the company level. In this sense, a specific strategy in innovation based on sustainability is needed. Regarding inter-organizational collaboration models, producer organizations in the middle of the chain avoid the responsibility of applying SOI because there is a "channel master" or "hub company" who dominates relationships. In addition, there are trend towards studying classic relationships focused on economic objectives. It is necessary to expand relationships upstream and downstream, but also with other actors (research centers, consumers, NGOs, etc.). In general, it is essential to encourage open innovation within a more engaging and forward-thinking collaborative framework.

Acknowledgements This study has been funded by the project Smart-Red UAL2020-SEJ-D2026 [UAL/CTEICU/ FEDER] "Adapting intermodal transport of perishables within an intelligent supply chain".

References

Arcese, G., Flammini, S., Lucchetti, M. C., & Martucci, O. (2015). Evidence and experience of open sustainability innovation practices in the food sector. *Sustainability (Switzerland), 7*(7), 8067–8090. https://doi.org/10.3390/su7078067

Arzu, G., & Erman, T. (2002). Supply chain performance measurement: A literature review. *International Journal of Production Research, 48*(17), 5137–5155.

Augustin, M. A., Sanguansri, L., Fox, E. M., Cobiac, L., & Cole, M. B. (2020, September). Recovery of wasted fruit and vegetables for improving sustainable diets. *Trends in Food Science and Technology, 95*, 75–85. https://doi.org/10.1016/j.tifs.2019.11.010

Aznar-Sánchez, J. A., Velasco-Muñoz, J. F., García-Arca, D., & López-Felices, B. (2020). Identification of opportunities for applying the circular economy to intensive agriculture in Almería (South-East Spain). *Agronomy, 10*(10). https://doi.org/10.3390/agronomy10101499

Barth, H., Ulvenblad, P. O., & Ulvenblad, P. (2017). Towards a conceptual framework of sustainable business model innovation in the agri-food sector: A systematic literature review. *Sustainability, 9*(9), 1620. https://doi.org/10.3390/su9091620

El Bilali, H. (2018). Relation between innovation and sustainability in the agro-food system. *Italian Journal of Food Science, 30*(2).

Brandenburger, A. M., & Nalebuff, B. J. (1996). *Co-opetition.* Doubleday.

Cagliano, R., Worley, C. G., & Caniato, F. F. A. (2016). The challenge of sustainable innovation in agri-food supply chains. In *Organizing for sustainable effectiveness* (Vol. 5, pp. 1–30). Emerald Group Publishing Ltd. https://doi.org/10.1108/S2045-060520160000005009

Chan, H. Y., Halim-Lim, S. A., Tan, T. B., Kamarulzaman, N. H., Jamaludin, A. A., & Wan-Mohtar, W. A. A. Q. I. (2020). Exploring the drivers and the interventions towards sustainable food security in the food supply chain. *Sustainability (Switzerland), 12*(19). https://doi.org/10.3390/SU12197890

Chiffoleau, Y., & Dourian, T. (2020). Sustainable food supply chains: Is shortening the answer? A literature review for a research and innovation agenda. *Sustainability (Switzerland), 12*(23), 1–21. https://doi.org/10.3390/su12239831

Choi, T. Y., & Hong, Y. (2002). Unveiling the structure of supply networks: Case studies in Honda, Acura, Daimler Chrysler. *Journal of Operations Management, 20*(5), 469–493.

Dagnino, G. B., & Mariani, M. (2010). Co-opetitive value creation in entrepreneurial contexts: An applied interpretative framework. In S. Yami, S. Castaldo, G. B. Dagnino, F. Le Roy, & W. Czakon (Eds.), *Co-opetition: Winning Strategies for the 21st Century* (pp. 101–123). Edward Elgar.

Dobson, P., Waterson, M. Y., & Davies, S. (2013). The patterns and implications of increasing concentration in European food retailing. *Journal of Agricultural Economics, 54*(1), 111–125.

Droge, C., Vickery, S. K., & Jacobs, M. A. (2009). Does supply chain integration mediate the relationships between product/process strategy and service performance? An empirical study. *International Journal of Production Economics, 137*(2), 250–262.

EIP-agri. (2019). *Innovation in short food supply chains.*

EU SCAR. (2016). Agricultural knowledge and innovation systems towards the future. In *European commission.*

Fawcett, S. E., Magnan, G. M., & Mccarter, M. W. (2008). Benefits, barriers, and bridges to effective supply chain management. *Supply Chain Management: An International Journal, 13*(1), 35–48.

Fischer, C., Hartmann, M., Reynolds, N., Leat, P., Revoredo-Giha, C., Henchion, M., Albisu, L. M. Y., & Gracia, A. (2010). Factors influencing contractual choice and sustainable relationships in European agrifood supply chains. *European Review of Agricultural Economics, 36*(4), 541–569.

Flynn, B., Huo, B., & Zhao, X. (2010). The impact of supply chain integration on performance: A contingency and configuration approach. *Journal of Operations Management, 28*(1), 58–71.

Fruit-Logistica. (2020). Do the right thing. *FruitLogistica.* https://doi.org/5

Galdeano-Gómez, E., Pérez-Mesa, J. C., & Aznar-Sánchez, J. (2017). Internationalisation of SMEs and simultaneous strategies of cooperation and competition: An exploratory analysis. *Journal of Business Economics and Management, 17*(6), 1114–1132.

Gao, D., Xu, Z., Ruan, Y. Z., & Lu, H. (2017). From a systematic literature review to integrated definition for sustainable supply chain innovation (SSCI). *Journal of Cleaner Production, 142*, 1518–1538. https://doi.org/10.1016/j.jclepro.2016.11.153

García Pérez, A. M., Román Cervantes, C., & García Martínez, M. (2016). De la cooperativa agroalimentaria a la learning netchain. Hacia un planteamiento teórico interorganizativo e interpersonal. *Revesco, 121*, 114–144.

García-Granero, E. M., Piedra-Muñoz, L., & Galdeano-Gómez, E. (2020). Multidimensional assessment of eco-innovation implementation: Evidence from spanish agri-food sector. *International Journal of Environmental Research and Public Health, 17*(4). https://doi.org/10.3390/ijerph17041432

Garnett, T., Appleby, M. C., Balmford, A., Bateman, I. J., Benton, T. G., Bloomer, P., Burlingame, B., Dawkins, M., Dolan, L., Fraser, D., Herrero, M., Hoffmann, I., Smith, P., Thornton, P. K., Toulmin, C., Vermeulen, S. J., & Godfray, H. C. J. (2013). Sustainable intensification in agriculture: Premises and policies. *Science, 341*(6141), 33–34. https://doi.org/10.1126/science.1234485

Geissdoerfer, M., Morioka, S. N., de Carvalho, M. M., & Evans, S. (2018). Business models and supply chains for the circular economy. *Journal of Cleaner Production, 190*, 712–721. https://doi.org/10.1016/j.jclepro.2018.04.159

Gurnani, H., Erkoc, M., & Luo, Y. (2007). Impact of product pricing and timing of investment decisions on supply chain co-opetition. *European Journal of Operational Research, 180*(1), 228–248.

Hingley, M. (2005). Power imbalance in UK agrifood supply channels: Learning to live with the supermarkets? *Journal of Marketing Management, 21*(1/2), 63–88.

Klewitz, J., & Hansen, E. G. (2014). Sustainability-oriented innovation of SMEs: A systematic review. *Journal of Cleaner Production, 65*, 57–75. https://doi.org/10.1016/j.jclepro.2013.07.017

Krishnan, R., Yen, P., Agarwal, R., Arshinder, K., & Bajada, C. (2020, June). Collaborative innovation and sustainability in the food supply chain- evidence from farmer producer organisations. *Resources, Conservation and Recycling*, 105253. https://doi.org/10.1016/j.resconrec.2020.105253

Kumar, A., Moktadir, M. A., Khan, S. A. R., Garza-Reyes, J. A., Tyagi, M., & Kazançoğlu, Y. (2020). Behavioral factors on the adoption of sustainable supply chain practices. *Resources, Conservation and Recycling, 158*(March), 104818. https://doi.org/10.1016/j.resconrec.2020.104818

Kuokkanen, A., Uusitalo, V., & Koistinen, K. (2019). A framework of disruptive sustainable innovation: An example of the Finnish food system. *Technology Analysis and Strategic Management, 31*(7), 749–764. https://doi.org/10.1080/09537325.2018.1550254

Lazzarini, S., Chaddad, F., & Cook, M. (2001). Integrating supply chain and network analyses: The study of netchains. *Journal on Chain and Network Science, 1*(1), 7–22.

Liu, Z., & Stephens, V. (2019). Exploring Innovation Ecosystem from the perspective of sustainability: Towards a conceptual framework. *Journal of Open Innovation.*

Long, T. B., Looijen, A., & Blok, V. (2018). Critical success factors for the transition to business models for sustainability in the food and beverage industry in the Netherlands. *Journal of Cleaner Production, 175*, 82–95. https://doi.org/10.1016/j.jclepro.2017.11.067

Meynard, J. M., Jeuffroy, M. H., Le Bail, M., Lefèvre, A., Magrini, M. B., & Michon, C. (2017). Designing coupled innovations for the sustainability transition of agrifood systems. *Agricultural Systems, 157*, 330–339. https://doi.org/10.1016/j.agsy.2016.08.002

Neutzling, D. M., Land, A., Seuring, S., & Nascimento, L. F. M. (2018). Linking sustainability-oriented innovation to supply chain relationship integration. *Journal of Cleaner Production, 172*, 3448–3458. https://doi.org/10.1016/j.jclepro.2017.11.091

Panigrahi, S. S., Bahinipati, B., & Jain, V. (2019). Sustainable supply chain management: A review of literature and implications for future research. *Management of Environmental Quality: An International Journal, 30*(5), 1001–1049. https://doi.org/10.1108/MEQ-01-2018-0003

Peano, C., Girgenti, V., Baudino, C., & Giuggioli, N. R. (2017). Blueberry supply chain in Italy: Management, innovation and sustainability. *Sustainability (Switzerland), 9*(2). https://doi.org/10.3390/su9020261

Pérez-Mesa, J. C., & Galdeano-Gómez, E. (2015). Collaborative firms managing perishable products in a complex supply network: An empirical analysis of performance. *Supply Chain Management: An International Journal, 20*(2), 128–138.

Pérez-Mesa, J. C., Piedra-Muñoz, L., García-Barranco, M. C., & Giagnocavo, C. (2019). Response of fresh food suppliers to sustainable supply chain management of large European retailers. *Sustainability (Switzerland), 11*(14). https://doi.org/10.3390/su11143885

Pérez-Mesa, J. C., Piedra-Muñoz, L., Galdeano-Gómez, E., & Giagnocavo, C. (2021). Management strategies and collaborative relationships for sustainability in the agrifood supply chain. *Sustainability (Switzerland), 13*(2), 1–16. https://doi.org/10.3390/su13020749

Peterson, C. (2002). The Learning Supply Chain: Pipeline or Pipedream? *American Journal of Agricultural Economics, 84*(5), 1329–1336.

Piedra-Muñoz, L., Galdeano-Gómez, E., & Pérez-Mesa, J. C. (2016). Is sustainability compatible with profitability? An empirical analysis on family farming activity. *Sustainability (Switzerland), 8*(9). https://doi.org/10.3390/su8090893

Renda, A. (2019). The age of Foodtech: Optimizing the agri-food chain with digital technologies. In R. Valentini, J. L. Sievenpiper, M. Antonelli, & K. Dembska (Eds.), *Achieving the sustainable development goals through sustainable food systems.*

Rice, J. B., & Hope, R.M. (2011). Supply chain vs. Supply chain: The hype and the reality. *Supply Chain Management Review, 5*(5), 47–54.

Rong, A., Akkerman, R., & Grunow, M. (2011). An optimization approach for managing fresh food quality throughout the supply chain. *International Journal of Production Economics, 131*(1), 421–429.

Stanco, M., Nazzaro, C., Lerro, M., & Marotta, G. (2020). Sustainable collective innovation in the agri-food value chain: The case of the "Aureo" wheat supply chain. *Sustainability*. https://doi.org/10.3390/su12145642

Trienekens, J., Van Uffelen, R., Debaire, J., & Omta, O. (2008). Assessment of innovation and performance in the fruit chain: The innovation-performance matrix. *British Food Journal, 110*(1), 98–127. https://doi.org/10.1108/00070700810844812

Ulvenblad, P., Ulvenblad, P., & Tell, J. (2019). An overview of sustainable business models for innovation in Swedish agri-food production. *Journal of Integrative Environmental Sciences, 16*(1), 1–22. https://doi.org/10.1080/1943815X.2018.1554590

Verdow, C. (2008, January). Description of the theoretical framework for a "Consumer driven and responsive supply chain", 1–56.

Volpentesta, A. P., Ammirato, S., & Della Gala, M. (2013). Classifying short agrifood supply chains under a knowledge and social learning perspective. *Rural Society, 22*(3), 217–229. https://doi.org/10.5172/rsj.2013.22.3.217

Wassermann, S., & Faust, K. (1994). *Social network analysis: Methods and applications.* Cambridge University Press.

Wiersinga, R. C., Hiller, S. R. C. H., Zimmermann, K. L., Verdouw, C., & Van Der Lans, I. A. (2010). Consumer driven and innovative fruit supply chains. *Towards Effective Food Chains: Models and Applications*, 113–136. https://doi.org/10.3921/978-90-8686-705-9

Wilhelm, M. M. (2011). Managing co-opetition through horizontal supply chain relations: Linking dyadic and network levels of analysis. *Journal of Operations Management, 29*(7–8), 663–676.

Wiśniewska-Paluszak, J. A., & Paluszak, G. T. (2019). The role of inter-organisational relations and networks in agribusiness: The case for the Polish fruit and vegetable industry. *International Journal on Food System Dynamics, 10*(2), 176–194. https://doi.org/10.18461/ijfsd.v10i2.11

Yu, S., Park, J. E., & Min, S. (2007). Factors of determining long-term orientation in interfirm relationships. *Journal of Business Research, 60*(12), 1225–1233.

Zerbini, F., & Castaldo, S. (2007). Stay in or get out the Janus? The maintenance of multiplex relationship between buyers and sellers. *Industrial Marketing Management, 36*(7), 941–954.

Sepide Mehrabi has diverse professional and academic experience in food industry engineering, business and management studies. With an engineering degree in food industry, MBA and international business studies, Sepide has an extensive understanding of the industry from both technical and managerial perspectives. Currently, Sepide is a Ph.D. student researching sustainable business model innovations in food value chains, also has over three years of experience working as a researcher on European projects related to sustainable-oriented innovations in agri-food sector. Their work focuses on developing new and innovative business models to make the food industry more sustainable.

Juan Carlos Pérez-Mesa is Full Professor in the Economics and Business Studies Department at the University of Almeria. He specializes in the agri-business field and supply chain management. He is the author of several studies published in high impact journals and has participated in several international research projects related to agri-food He currently is the Secretary of the Mediterranean Research Center of Economics and Sustainable Development (CIMEDES). Also, he is the coordinator of the Market Intelligence Department for the Andalusian Association of Organizations of Fruit and Vegetable Producers (APROA) and collaborator with the Spanish Interprofessional Organization of Greenhouse Crops (HortiEspaña). He is also a member of the Directing Board fo the Spanish Association of Agricultural Economics (AEEA).

María del Carmen García Barranco holds a Ph.D. in Economics and Business Management and is an acting substitute lecturer in the Department of Economics and Business Studies at the University of Almeria (Business Organisation). Her area of expertise is supply chain applied to the agribusiness and tourism sectors. She is the author of several articles in high impact academic journals, books and book chapters related to this line of research.

Cynthia Giagnocavo holds a Ph.D. in Economics and Business Management, a Masters of Environmental Studies, degrees in law, literature and philosophy. She is a professor in the department of Business Management in the University of Almería and is the director of the Cátedra Coexphal-UAL in agricultural, cooperative studies and sustainable development. Research areas focus on the intersection of organisational, technical, social, economic and institutional innovation and impact, with a particular interest in agri-food, social enterprise and cooperative and collaborative business models. She is also the Global Chair of the International Cooperative Alliance Committee on Cooperative Research and sits on the social research committee of CIRIEC International. She has worked in business and finance in London, Madrid and Toronto and is qualified as a Barrister, solicitor and attorney in three jurisdictions.

Public Sector and Sustainable Development

Budget Policies Versus Sustainable Development Goal in the Main Spanish Municipalities

Check for updates

Pedro Gil-García, Natalia Alonso-Morales, Arturo Haro-de-Rosario, and Alejandro Sáez-Martín

1 Introduction

The Sustainable Development Goals (SDGs) address all countries and aim at reconciling economic and social objectives with ecological goals. Now the SDGs face the challenge of achieving sustainable development (Brundtland et al., 1987), that is, continuing the improvement of living conditions for those in need and, at the same time, preserve the ecological integrity of the planet for future generations (Eisenmenger et al., 2020).

In this context, Bebbington and Unerman (2018) argue that the SDGs open new avenues for sustainability research. Particularly and in line with the aim of study of this work, Bose & Khan (2022) state that the role of accountants and financial management tasks continues to be a neglected area in research on sustainable development.

The SDGs cover all aspects of human life, addressing one of the most important issues the world faces today, the reduction of inequalities. This objective, under the name of SDG 10, has been the objective of analysis by numerous works (Cojocaru et al., 2022; Sánchez & Valera, 2021; Úbeda et al., 2022) and it has been observed that it is an extremely difficult and complex objective (Cojocaru et al., 2022). In addition, Úbeda et al. (2022) consider SDG 10 one of the most relevant, due to the fact that inequality is not only based on income inequality, but also on inequality of opportunities, whether due to different access to education, different health coverage or access to other services, that is, what is known as the Welfare State.

The main objective of the SDGs is to improve the coherence of the different government plans and policies, and for this, budget management is key (Hege &

P. Gil-García · N. Alonso-Morales · A. Haro-de-Rosario (✉) · A. Sáez-Martín
University of Almeria, Almeria, Spain
e-mail: arturo.haro@ual.es

A. Sáez-Martín
e-mail: alejandro.saez@ual.es

M. d. C. Valls Martínez and J. M. Santos-Jaén (eds.), *Environmentally Sustainable Production*, https://doi.org/10.1007/978-3-031-52656-5_8

Brimont, 2018). Furthermore, this consistency between the SDGs and budgets can increase accountability, as it can reveal country's progress towards the SDGs and help assess government performance. However, most countries make their budgets based on their performance, the SDGs could add an additional factor to assess the sustainability on the budget. In fact, the meetings of the United Nations Development Program agreed to incorporate the SDGs into budgets as a push to achieve the SDGs by 2030 (Sisto et al., 2020).

In this context, greater budget management is summed up in a higher quality of the social security system, where introducing and expanding social security plans can eliminate key problems of inequality (Cojocaru et al., 2022). Where SDG 10 is especially significant, which promotes the idea of a society in which all people enjoy the benefits of economic growth, although economic growth can sometimes contribute to growing income inequality, and therefore opportunities. Due to this, the concepts of poverty, inequality and greater social exclusion need to be analyzed and justified from various points of view. This is because there are a set of factors, especially linked to the economic dimension, which are often problematic and influence the way in which social problems are addressed (Madanipour et al., 2015).

Among the different member states, the government of Spain has been one of the pioneers in presenting its budgets aligned with the spending that applies to each SDG, where they recognize that aligning the objective of the SDGs with budget items and policies makes their synergy difficult. Despite this, more countries are considering integrating the SDGs into their budget programs, so that the evaluation of the impact of budget allocation also becomes increasingly influential (Sisto et al., 2020).

In addition, budgetary policies depend to a large extent on the political structure of the different levels of government, and particularly of local governments, affecting both decision-making and its implementation (Bisogno et al., 2023). In this sense, Bisogno et al. (2023) exposes how political culture has shifted from a concern with materialistic values (e.g. economic well-being, etc.) to post-materialistic values (e.g. environmental protection, quality of life, etc.), the latter of which are reflected in the SDGs. Considering political ideology of our politicians as fundamental in sustainable consumption through beliefs and concerns (Dehghanpour & Rezvani, 2021).

Having exposed the above, the objective of this work is to analyze how the budgetary policies of the municipalities influence in the degree of compliance with SDG 10. For this, we will analyze the budgetary policies associated with specific activities, a greater current expenditure associated with the social spending and fiscal pressure, and therefore, the budget's ability to collect. In addition, we can not forget the political color in budget management that is included in the study, as well as the social context, such as migration associated with inequality of resources.

The aim of this work is to contribute to improve and analyze which budgetary policies are related to the SDGs and which of these budgetary measures are the most optimal. In this way, public managers can be helped to elaborate and improve their budgetary policies that contribute to the sustainable development of cities. In addition, this work can improve the literature to date and our empirical understanding of how budgets, political ideology, and migration policies affect the achievement of the SDGs, along with opening the way for future lines of research.

In order to achieve the stated objective, we have structured this work into six sections. After this introduction, the state of the question is deepened and a review of the literature is carried out, followed by the methodology applied to carry out this work. In the fourth and fifth sections, the results obtained are analyzed and discussed, and finally, the most relevant conclusions are presented.

2 State of the Question and Review of the Literature

In recent years, citizens have demanded from governments greater efficiency in resources, besides that they provide a better understanding of the actions they carry out. These ideas are supported by international organizations, such as the G-20 (2013) and the International Federation of Auditors (IFAC, 2013), highlighting the importance of sustainable development (Ortiz-Rodríguez et al., 2018).

Among the different levels of government, the literature has focused on local governments to measure the sustainability and development of the SDGs (Benito et al., 2021; Hege & Brimont, 2018; Navarro-Galera et al., 2019; Ortiz-Rodríguez et al., 2018; Saha, 2009). These authors consider how local governments are essential to achieve the SDGs, since their budgetary actions are closely related to cover basic services for citizens and being evaluated them efficiently (Benito et al., 2021).

Cities are the human ecosystems where the best options and opportunities for social and economic development are offered and, in turn, are the places where significant inequalities and some of the greatest sustainability challenges are detected (Sánchez de Madariaga et al., 2018, 2020). For this reason, cities are the space where solutions must be put into practice, involving all comers. The Spanish Network for Sustainable Development (Sánchez de Madariaga et al., 2018, 2020) emphasizes how SDG 10 is directly related to municipal powers, where local action plays a fundamental role (Sánchez de Madariaga et al., 2018, 2020).

The main objective of the SDGs is to improve the coherence of the different government plans and policies, and for this, budget management is key (Hege & Brimont, 2018). In this regard, Hege and Brimont (2018) emphasize how coherence between the SDGs and budgets can increase accountability, since it can reveal a country's progress towards the SDGs and evaluate government actions, in short, in what is public money spent?

For this reason, in the meetings of the United Nations Program they agreed to incorporate the SDGs into budgets as a boost to achieve the completion of the SDGs by 2030. Despite this, most countries do not make their budget based on in the SDGs (Sisto et al., 2020). However, Spain has been one of the pioneering countries implementing the SDGs in the 2021 budgets, where they specify that it has been designed with a double-sided quantitative and qualitative approach to measure the budgetary effort that each spending policy made to achieve the goal of the SDGs (Spanish Ministry of Finance).

In this context, better budget management is summed up in a higher quality of the social security system, where introducing and expanding social security plans can eliminate key problems of inequality (Cojocaru et al., 2022). Where Cojocaru et al. (2022) emphasizes the importance of SDG 10, which promotes the idea of a society in which all people enjoy the benefits of economic growth, although economic growth can sometimes contribute to grow income inequality, and therefore of opportunities. Indeed, the concepts of poverty, inequality and greater social exclusion need to be analyzed and justified from various points of view.

The implementation of the SDGs not only requires a will or policies to implement plans, but it is also necessary to allocate resources to be able to implement them successfully (Hendriks, 2018), which is summarized in that sustainable development must be reflected in budgets to have a significant effect. In this line, Sisto et al. (2020) identify that there are links between the assignation of a budget program and the achievement of the SDGs in Spanish local governments. In this regard, it should be noted that the structure of the budgets of Spanish local governments is the same for all and they have the same programs to classify spending and establish the amount thereof (Orden EHA/3565/2008).

In the study by Sisto et al. (2020), addressing SDG 10 of our objective, only 28.4% of the budget programs are decisive for the achievement of the SDG. It should be noted that the significant budgetary programs are 211 and 231, where the first of these items refers to civil servants' pensions, and the second to a set of actions to guarantee primary care for basic needs.

Another aspect to take into account for the achievement of the SDGs is the spending in general and particular terms, as studied by Bisogno et al. (2023), how a higher or lower current expenditure per capita influences the achieving of SDGs. To this end, López (2011) highlights the relevance of knowing the level of efficiency of the public sector, since the more efficient they are, the more resources will be obtained and, therefore, less spending will be necessary to guarantee a certain amount of public services to the public society.

However, attempts to reduce public spending face up the setback of the growing demands for public spending by citizens. Consequently, due to these needs, improving the efficiency of public spending appears as a more appropriate and essential alternative to give continuity to the new public spending programs required by the new economic scenarios (López, 2011).

However, one must not forget the fact that all spending is financed with the total resources obtained by the public sector, which are called public incomes, and which come mainly from the payment of taxes of citizens (López, 2011). In addition, local authorities face considerable budgetary and institutional limitations, and tend to rely heavily on support from other levels of government to carry out effective sustainability actions in urban areas (Hickmann, 2021). However, a higher tax pressure can negatively affect citizens or even worsen the situation, increasing inequalities. In this context, Seelkopf and Bastiaens (2020) analyzed how tax revenues supported compliance with the SDGs.

In view of the previous studies and the aim of this work, which is to determine if the budgetary policies for the allocation of resources, current spending and fiscal pressure influence the degree of compliance with SDG 10. Our hypothesis is divided into three:

H1.a: The allocation of budgetary resources in spending programs influences the fulfillment of SDG 10.

H1.b: Budgeted current spending influences compliance with SDG 10.

H1.c: Fiscal pressure influences the fulfillment of SDG 10.

Another important issue for sustainable government development is political will, and with this, it is highlighted the importance of political ideology in the actions and achievement of the SDGs (Arora-Jonsson, 2023; Weber, 2017). In this sense, DeGroff and Cargo (2009) and Bardal et al. (2021) explain how the implementation of certain policies reflects a complex change process where government decisions are transformed into programs or practices with several objectives, including the SDGs. Along these lines, authors such as Bisogno et al. (2023) or Martínez-Córdoba et al. (2020), emphasize the need to analyze political ideology for the analysis of the SDGs and public management, however, their results are not conclusive. Therefore, taking these arguments into account, we propose the following hypothesis:

H2: Political ideology significantly affects compliance with SDG 10 in Spanish municipalities.

Along with political ideology, we cannot forget the importance of migration and its social inclusion, especially in a deeply globalized world like the one we live in. In this regard, Szymańska (2021) highlights how conflicts in unstable countries such as Syria, Afghanistan or Venezuela have caused an unprecedented migration in EU countries in recent years. Therefore, the integration of immigrants at the level of the EU Member States is a challenge that we cannot ignore, even decisive in terms of well-being and prosperity (Szymańska, 2021).

In this way, Guerrero-Gómez et al. (2021) analyze how the proportion of immigrants among the total population affects compliance with the SDGs, in the context in which the existence of a significant immigrant population could reflect a greater need for municipal services, both in quantity and quality. Therefore, we propose the following hypothesis:

H3: A greater presence of immigrants significantly affects compliance with SDG 10 in Spanish municipalities.

3 Metodology

In accordance with the proposed objective, the research is planned in two phases. In a first phase, a descriptive analysis of the dependent variable will be carried out—REDS Report compliance with SDG 10—and the independent ones -budgetary, political and migration—presented by local governments. And secondly, explanatory analysis will be carried out to determine how the independent variables influence the degree of compliance with the SDG.

The dependent variable considered has been Sustainable Development Goal 10 (Reduce inequalities). This variable has been obtained from the New REDS report:

The SDGs in 100 Spanish cities (Sánchez de Madariaga et al., 2018, 2020). This report is based on individual indicators to represent each SDG, where for their assessment, each one of them have been rated between 0 and 100, where 100 is the best situation and 0 the worst. Therefore, our sample will consist of the 103 Spanish municipalities with more than 80,000 inhabitants analyzed in 2020.

Our independent variables, which are explained in detail in Table 1, are budget items 211 and 231 proportional to the budget, current spending per capita, tax pressure, political ideology and immigration rate.

Table 1 Variables explicativas, método de medición e hipótesis prevista

Classification	Variables	Abbreviation	Unit of measurement	Source	Hypothesis
Budget policies	211 proportional	211 pro	Proportion of the budget amount assigned to item 211: pensions of civil servants among the total budget	Spanish Ministry of Finance and Civil Service	H1.a: The allocation of budgetary resources in spending programs influences the fulfillment of SDG 10
	231 proportional	231 pro	Proportion of the budget amount assigned to item 231: actions to guarantee primary care for basic needs between the total budget proportional to the total budget		
	Current expenditure per capita	CEPC	Current expenditure (expenditure Chaps. 1–5 of the budget)/Total population	Spanish Ministry of Finance and Civil Service	H1.b: Budgeted current spending influences compliance with SDG 10
	Tax pressure	TP	(Amount of income from Chaps. 1 to 2 of the budget)/Total population	Spanish Ministry of Finance and Civil Service	H1.c: The budgeted fiscal pressure influences the fulfillment of SDG 10
Political	Politic Ideology	PI	Political ideology of the parties: 1 = progressive; 0 = conservative		H2: Progressive ideologies have a positive influence
Social	Immigration rate	IR	(Number of immigrants/total population) * 100	Statistics National Institute	H3: A greater number of immigrants influences positively

Source Self made

For the explanatory analysis, a Pearson correlation matrix will be made first. Subsequently, a multiple linear regression will be carried out, using the STATA 15.0 program. The regression will follow the following model:

$$\begin{aligned} SDG_{ki} = \beta_0 + \beta_1 211\ pro_i + \beta_2 231\ pro_i + \beta_3 \text{Current spending per capita} \\ + \beta_4 \text{Tax pressure} + \beta_5 \text{Political Ideology} \\ + \beta_6 \text{Immigration rate} + \varepsilon_i \end{aligned} \quad (1)$$

where β is the variable to be estimated, ε is the probability of error, and the subscript i refers to each municipality in the sample.

The dependent variable (SDG_k)) refers to the degree of compliance with k (SDG 10).

4 Analysis of Results

4.1 Descriptive Analysis

In Table 2, we can observe how the average compliance with this SDG is 49,58, that is, in Spain more than half of the total score for this objective is not achieved. However, we find exceptions, such as Fuenlabrada (Madrid), with the highest score of 72,81, although we find local governments such as Pamplona (Navarra) with 19,76, the lowest score.

In relation to the budgetary variables −211 and 231 proportional-, we can observe that neither together reach an average of 10% of the total budget of the municipalities. Highlighting how a maximum of 8,48% of civil servants' pensions in Elche (Alicante) and a maximum of 22,96% are allocated to actions to guarantee primary care for basic needs in Pamplona (Navarra).

Regarding the current expenditure per capita-CEPC—we appreciate how, on average, 901,32€ is spent per inhabitant, despite this, the average does not even

Table 2 Descriptive analysis of the study variable

Variable	Average	Deviation	Máx	Min
ODS 10	49,58	11,03	72,81	19,76
211 pro	0,27%	0,01	8,48%	0,01%
231 pro	8,69%	0,04	22,96%	3,67%
CEPC	901,32	306,49	2310,62	635,84
TP	812,48	901,97	8867,25	131,29
PI	1	–	1,00	0,00
IR	11,71%	0,08	42,33%	1,41%

Source Self made

reach 40% of the maximum of some local governments. Having its maximum in Santiago de Compostela (A. Coruña), and the minimum in Parla (Madrid).

Following this line, analyzing the fiscal pressure-PI-, we see how each citizen contributes an average of 812,48 € to public coffers. We also find governments with a maximum fiscal pressure, as is the case of El Puerto de Santa María (Cádiz), and a minimum tax burden in Lorca (Murcia). It is interesting how, on average, citizens receive more than what they contribute, specifically of the 812,48 € that citizens contribute, they receive 901,32€ from the State.

Regarding the variable political ideology-PI-, on average we found values close to 1, indicating that more than half of the municipalities are governed by progressive governments, that is, 64 of the 103 municipalities are progressive. As for the immigration rate-IR—we appreciate that on average 11,71%, that is, 11,71% of the population is immigrant, although we can appreciate that there are some municipalities such as Torrevieja (Alicante) where 42, 33% of the population is foreign.

4.2 Explanatory Analysis

Regarding the correlations, in Table 3, few significant correlations are observed between the independent variables, although these correlations are of low degree, practically not exceeding 20% correlation. This fact is positive for building the model, since there is no multicollinearity problem between the regressors.

In Table 4 we see the results of the explanatory analysis, where all the independent variables are significant with the dependent variable, giving as a result that our regression model is conclusive. And therefore, all the variables affect compliance with the SDG with a probability between 95 and 99% probability. In addition, of all the observations, the variables explain 31,98% of compliance with SDG 10.

In relation to the budget variables −211 and 231 pro-, the results show that both variables are negatively significant by 99% and 95%, respectively, in compliance

Table 3 Correlations of the independent variables

	ods10	211 pro	231 pro	CEPC	TP	PI	IR
ods10	1						
pro 211	−0,2193**	1					
pro 231	−0,2304**	0,1096	1				
CEPC	−0,2327**	−0,0773	0,0766	1			
TP	−0,1702*	−0,0550	−0,208**	−0,221**	1		
PI	0,3628***	−0,0062	−0,0447	−0,0116	−0,1811*	1	
IR	−0,2704***	−0,051	0,0548	0,1412	0,0569	−0,159	1

Significant correlations at the 0,10 level*. Significant correlations at the 0,05 level**, Significant correlations at the 0,01 level***
Source Self made

Table 4 Results of the Regressive Analysis

Results of the regressive analysis		
F(6,92)	8,68**	
Adj R-squared	0,3198	
E	Coef	t
211 pro	−207	−2,91***
231 pro	−62	−2,33**
CEPC	−0,0,106,648	−3,26***
TP	−0,0,028,406	−2,58**
PI	6,06,862	3,04***
IR	−26,03,802	−2,09**

Significant correlations at the 0,10 level*. Significant correlations at the 0,05 level**. Significant correlations at the 0,01 level***
Source Self made

with SDG 10. In short, the greater amounts allocated to these items, worse is the result in compliance with the SDGs. Results that coincide with the study by Sisto et al. (2020), where in his study of compliance with the different SDGs, he shows that these budget items have a negative relationship with compliance with this SDG.

On the other hand, another budgetary variable such as current expenditure per capita -CEPC—is 99% significant with negative relationship. These results show that higher the current public spending are, less the fulfillment of the SDG is. These results are in line with Bisogno et al. (2023), who find similarities in Spanish and Italian local governments.

Regarding the tax pressure -TP-, the study shows that it is 95% significant, confirming that higher the tax pressure is, that is, higher taxes, lower is the compliance of the SDG. These results disagree with the study of Seelkopf and Bastiaens (2020), where they analyze how an increase in tax pressure is beneficial for the progress of the SDGs.

In relation to the political ideology variable -PI-, we observe that the variable is 99% significant with a positive relationship, that is, in progressive parties compliance with the SDG is greater. Results contrary to Bisogno et al. (2023) or Martínez-Córdoba et al. (2020), since this variable is not significant in their study.

Regarding the immigration rate-IR-, we appreciate that this variable is significant in the reduction of inequalities, although in a negative way. What is summarized in that the greater the number of immigrants compared to the residents, greater the inequalities are, results that agree with Guerrero-Gómez et al. (2021).

5 Conclusion

The fulfillment and success of the 2030 Agenda is based on the policies, plans and programs developed by all countries. However, these strategies require financing and resource mobilization plans (United Nations, 2020). The study, supported by the United Nations Office for Project Services (UNOPS) and published by The Economist Intelligence Unit in 2020, argues that effective public strategies allow governments to make significant progress towards achieving the SDGs. Due to this importance and relevance, the objective of our study is to demonstrate whether budgetary policies affect, to a greater or lesser extent, compliance with the SDGs. And, ultimately, our empirical results show that it is.

So measuring public budgets and how countries mobilize public resources to achieve the SDGs plays a fundamental role. However, the alignment of a budget policy to a particular SDG does not necessarily mean the first is having a direct or indirect impact on the second one (Sisto et al., 2020). Furthermore, in our study, the budget items aimed at supporting this objective are not positive, so we should focus our efforts on making better use of public funds allocated to these items, instead of seeking to increase their amount, that is, make them more efficient. From another point of view, these items are already high enough and it would be advisable to reduce them, and allocate these funds to other items that are of interest to achieve the SDGs. In short, an adequate budget allocation is essential to move towards sustainable development.

Along these lines, it is striking how greater the public spending is, worse the social inequalities are. On the one hand, an increase in public spending can be beneficial if it is directed at specific policies and programs that align with the SDGs. On the other hand, a reduction in public spending could lead to a reduction in the main social services, worsening social well-being. Therefore, the key to moving towards sustainable development is to ensure that public resources are efficiently used, in addition to focusing on addressing the key challenges that public institutions face on their path to achieving the SDGs. However, there are serious difficulties in valuing public production, since unlike what happens in the private sector, there is no definition of what a unit of public product is and there is no sales price for it. In the public sector, part of the production is intangible, which makes its quantification difficult (López., 2011).

Related to the previous variables, a greater fiscal pressure implies that the government has a greater spending capacity. Although following the theory of Alfred Laffer, who proposed a theoretical curve that relates tax collection to tax rates, in the shape of a bell (inverted U) and maintained that collection increases as tax rates increase, up to a certain point where any marginal increase would lead to a reduction in total collection (Cacay et al., 2021). In summary, a significant increase in taxes can negatively affect citizens' purchasing power and reduce consumption, negatively impacting citizens' quality of life and hindering access to essential services. The key is to find an appropriate balance that ensures that fiscal resources are used effectively and efficiently to move towards sustainable development. Although it is important to highlight how, despite the high Spanish fiscal pressure, public spending continues to be greater than what citizens contribute, once again highlighting the importance of greater public management to cover the main public services in the most efficient way.

On the one hand, this research completes the present literature on political ideology, demonstrating its influence and relevance in the fulfillment of SDG 10, since in previous works this variable is considered non-significant (Bisogno et al., 2023 or Martínez-Córdoba et al., 2020). However, this research shows how municipalities governed by progressive parties comply to a greater extent with the objective of reducing inequalities. Progressive leaders tend to pursue policies with the aim of redistributing wealth more equitably, seeking to address inequalities in terms of access to opportunities, services, and rights, that is, reducing disparities both economically and socially. Despite the increasing global pressure on politicians, who are encouraged to implement policies consistent with sustainable goals (Bisogno et al., 2023), progressive parties have shown a more positive influence on sustainability. This is because their aims and electoral strategies are based on policies which advocate for sustainable growth, resulting in greater equality of opportunities.

Regarding the immigration rate, it has been found that greater the number of immigrants are, lower the compliance of SDG is. Migration is a highly visible reflection of inequalities, since many decide to migrate with the aim of achieving a better life elsewhere. However, inequalities can also arise from migration itself. The existence of a significant immigrant population could reflect a greater need for municipal services, both in quantity and quality (Guerrero-Gómez et al., 2021). So if this is not well managed, it could negatively affect local citizens who already depend on these services. Furthermore, the type of immigration that a country receives directly affects its integration; the majority of immigrants from Spain come from underdeveloped countries and different cultures, making their integration into Western society difficult, increasing economic and social differences. Indeed, despite the open border policies of progressive governments, it does not seem that the effect of all of them is positive for the SDGs.

According to the results obtained, this work presents several contributions at an academic and practical level. At an academic level, it contributes to the lack of works that relate and examine how budgetary policies influence compliance with the SDGs in governments and in particular with local governments and SDG10. On the other hand, this work has a great contribution for public managers, and in particular for those responsible for municipal budgets. Specifically, our results show that budgetary policy is decisively for the development of the SDGs, in the same way as the control of current spending and the fiscal pressure of citizens, since the efficiency of both factors is key to sustainability from a country. Regarding political ideology and immigration rate, we can show how progressive policies are on the same line with SDG, although their open border policies don´t seem to affect positively on SDG, despite of their intentions to correct inequalities.

This work is not without limitations, because this study has selected municipalities with more than 80,000 inhabitants, so compliance with this SDG has not been able to be analyzed in small municipalities. In addition, this study only analyzes one SDG, leaving aside the other 16.

Furthermore, this work opens future lines of research, such as seeing if budget variables influence the rest of the SDGs and checking if our result coincides with the others. Another alternative would be to increase our sample, selecting a more heterogeneous sample of local governments. Another alternative would be to carry out a study of the local governments of the different EU countries, for a more global vision and under a supranational regulatory framework.

References

Arora-Jonsson, S. (2023). The sustainable development goals: A universalist promise for the future. *Futures, 146*, 103087.

Bardal, K. G., Reinar, M. B., Lundberg, A. K., & Bjørkan, M. (2021). Factors facilitating the implementation of the sustainable development goals in regional and local planning—experiences from Norway. *Sustainability, 13*(8), 4282.

Bebbington, J., & Unerman, J. (2018). Achieving the United Nations Sustainable Development Goals: an enabling role for accounting research. *Accounting, Auditing & Accountability Journal, 31*(1), 2–24.

Benito, B., Guillamón, M. D., Martínez-Córdoba, P. J., & Ríos, A. M. (2021). Influence of selected aspects of local governance on the efficiency of waste collection and street cleaning services. *Waste Management, 126*, 800–809.

Bisogno, M., Cuadrado-Ballesteros, B., Rossi, F. M., & Peña-Miguel, N. (2023). Sustainable development goals in public administrations: Enabling conditions in local governments. *International Review of Administrative Sciences*, 00208523221146458.

Bose, S., & Khan, H. Z. (2022). Sustainable development goals (SDGs) reporting and the role of country-level institutional factors: An international evidence. *Journal of Cleaner Production, 335*, 130290.

Brundtland, G., Khalid, M., Agnelli, S., Al-Athel, S., Chidzero, B., Fadika, L, & Okita, S. (1987). Our Common Future ('Brundtland report'). USA: Oxford University Press.

Cacay Cacay, J. C., Ramírez Chávez, G. D. R., & Campuzano Vásquez, J. A. (2021). Efecto del crecimiento económico y la presión fiscal sobre el impuesto al valor agregado. *Revista San Gregorio, 1*(47), 111–125.

Cojocaru, T. M., Ionescu, G. H., Firoiu, D., Cismaş, L. M., Oţil, M. D., & Toma, O. (2022). Reducing inequalities within and among EU Countries—Assessing the achievement of the 2030 agenda for sustainable development targets (SDG 10). *Sustainability, 14*(13), 7706.

Dehghanpour Farashah, A., & Rezvani, Z. (2021). *The role of political ideology (left vs right) on sustainable consumption.* ISDRS2021-the 27th Annual Conference, International Sustainable Development Research Society.

DeGroff, A., & Cargo, M. (2009). Implementación de políticas: Implicaciones para la evaluación. *nuevo directorio Eval*, 47–60.

Eisenmenger, N., Pichler, M., Krenmayr, N., Noll, D., Plank, B., Schalmann, E., Wandl, M.-T., & Gingrich, S. (2020). The Sustainable Development Goals prioritize economic growth over sustainable resource use: A critical reflection on the SDGs from a socio-ecological perspective. *Sustainability Science, 15*, 1101–1110.

English, L. M., & Mayo, P. (2019). Lifelong learning challenges: Responding to migration and the sustainable development goals. *International Review of Education, 65*, 213–231.

European Commission. (2014). Directive 2014/95/EU of the European Parliament and the Council of 22 October 2014 amending Directive 2013/34/EU as regards disclosure of non-financial and diversity information by certain large undertakings and groups. *ABlEU L, 330*, 1–9.

Galera, A. N., de los Ríos Berjillos, A., Lozano, M. R., & Valencia, P. T. (2014). Transparency of sustainability information in local governments: English-speaking and Nordic cross-country analysis. *Journal of Cleaner Production, 64*, 495–504.

Guerrero-Gómez, T., Navarro-Galera, A., & Ortiz-Rodríguez, D. (2021). Promoting online transparency to help achieve the Sustainable Development Goals: an empirical study of local governments in Latin America. *Sustainability, 13*(4), 1837.

Hege, E., & Brimont, L. (2018). Integrating SDGs into national budgetary processes. *Studies, 5*(18).

Hendriks, C. J. (2018). Municipal financing for sustainable development: A case of South Africa. *Local Economy, 33*(7), 757–774.

Hickmann, T. (2021). Locating cities and their governments in multi-level sustainability governance. *Politics and Governance, 9*(1), 211–220.

Kuhn, H. (2020). Reducing inequality within and among countries: Realizing SDG 10—A developmental perspective. *Sustainable Development Goals and Human Rights, 5*, 137–153.

López, N. R. (2011). La eficiencia y su importancia en el sector público. *Extoikos, 1*, 38–47.

Madanipour, A., Shucksmith, M., & Talbot, H. (2015). Concepts of poverty and social exclusion in Europe. *Local Economy, 30*(7), 721–741.

Martínez-Córdoba, P. J., Raimo, N., Vitolla, F., & Benito, B. (2020). Achieving sustainable development goals. Efficiency in the Spanish clean water and sanitation sector. *Sustainability, 12*(7), 3015.

Ministerio de Hacienda España. Informe de Alineamiento de los Presupuestos Generales del Estado con los Objetivos de Desarrollo Sostenible de la Agenda 2030; Gobierno de España: Madrid, España, 2020.

Navarro-Galera, A., Ortiz-Rodríguez, D., & Alcaraz-Quiles, F. J. (2019). Un impulso a la transparencia sobre sostenibilidad en gobiernos locales europeos mediante factores poblacionales, socioeconómicos, financieros y legales. *Spanish Journal of Finance and Accounting/revista Espanola De Financiacion y Contabilidad, 48*(4), 525–554.

Ortiz-Rodríguez, D., Navarro-Galera, A., & Alcaraz-Quiles, F. J. (2018). The influence of administrative culture on sustainability transparency in European local governments. *Administration & Society, 50*(4), 555–594.

Rassanjani, S. (2018). *Ending poverty: Factors that might influence the achievement of sustainable development goals (SDGs) in Indonesia.*

Saha, D. (2009). Factors influencing local government sustainability efforts. *State and Local Government Review, 41*(1), 39–48.

Sánchez de Madariaga, I., García López, J., Sisto, R. (2018). Los Objetivos de Desarrollo Sostenible en 100 ciudades españolas. Madrid: Red Española para el Desarrollo Sostenible (REDS).

Sánchez De Madariaga, I., Benayas Del Álamo, J., García López, J., Sisto, R., Urquijo Reguera, J., & García Haro, M. (2020). Los Objetivos de Desarrollo Sostenible en 100 ciudades españolas ¿Cómo está avanzando la Agenda 2030 a nivel local? Una mirada práctica. Red Española para el Desarrollo Sostenible.

Sánchez, L. E. H., & Valera, A. (2021). Desafíos actuales de la extensión en relación al ODS 10. *Masquedós-Revista De Extensión Universitaria, 6*(6), 15–15.

Schmidt-Traub, G., Kroll, C., Teksoz, K., Durand-Delacre, D., & Sachs, J. D. (2017). National baselines for the Sustainable Development Goals assessed in the SDG Index and Dashboards. *Nature Geoscience, 10*(8), 547–555.

Schramade, W. (2017). Investing in the UN sustainable development goals: Opportunities for companies and investors. *Journal of Applied Corporate Finance, 29*(2), 87–99.

Seelkopf, L., & Bastiaens, I. (2020). Achieving sustainable development goal 17? An empirical investigation of the effectiveness of aid given to boost developing countries' tax revenue and capacity. *International Studies Quarterly, 64*(4), 991–1004.

Sisto, R., García López, J., Quintanilla, A., de Juanes, Á., Mendoza, D., Lumbreras, J., & Mataix, C. (2020). Quantitative analysis of the impact of public policies on the sustainable development goals through budget allocation and indicators. *Sustainability, 12*(24), 10583.
Szymańska, A. (2021). Reducing socioeconomic inequalities in the European Union in the context of the 2030 agenda for sustainable development. *Sustainability, 13*(13), 7409.
Úbeda, F., Forcadell, F. J., Aracil, E., & Mendez, A. (2022). How sustainable banking fosters the SDG 10 in weak institutional environments. *Journal of Business Research, 146*, 277–287.
United Nations (2020). SDG Indicators [online] https://unstats.un.org/sdgs/indicators/database/.
Weber, H. (2017). Politics of 'leaving no one behind': Contesting the 2030 sustainable development goals agenda. *Globalizations, 14*(3), 399–414.

Pedro Gil-García is a master's student in Accounting Auditing, graduated in Finance and Accounting, both from the University of Almería (Spain). He is currently pursuing a doctorate in the Department of Economics and Business. His research interests are focused on public accounting, management system and financial information in the federal and local governments. In addition, he has participated in some international conferences, such as the 42th and 45th EBES Conference (Lisbon/Budapest) in 2023.

Natalia Alonso-Morales has a degree in Finance and Accounting from the University of Almería (Spain), where she is currently studying to acquire a Master's Degree in Auditing. She is also a Ph.D. student in the Department of Economics and Business. Her research interests are focused on the management system and financial sustainability in the local governments and the sustainable development. She has participated in the international 45th EBES Conference (Budapest) in 2023.

Arturo Haro-de-Rosario is a Senior lecturer in finance, in the Department of Economics and Business at the University of Almería (Spain). His research interests are focused on the online financial information (web and social media), the management system and financial information in the federal and local governments, online citizen engagement, corporate social responsibility and business ethics.

He is author of 31 articles in national and international journals, including Online Information Review, New Media and Society, Information, Communication & Society, Small Business Economics, Social Science Computer Review, Transforming Government: People, Process and Policy, Local Government Studies, etc. Also, he has written 10 book chapters in international publishing companies (IGI Global, Springer and Taylor & Francis, among others). In addition, he has participated in numerous national and international congress/conferences. Finally, note that he participates as a reviewer in relevant journals, such as International Review of Administrative Sciences and Romanian Journal of Political Science.

Alejandro Sáez-Martín is a Senior lecturer in accounting, in the Department of Economics and Business at the University of Almería (Spain). His research interests are focused on the financial information disclosures on the Web and social media (e-government), on the management system and financial information in the federal and local governments. He is author of numerous articles in national and international journals, including Public Management Review, Local Government Studies, New Media & Society, Social Science Computer Review, etc. Also, he has written some book chapters (IGI Global and Springer).

Analysis of the Influence of Financial Sustainability on the Achievement of SDG 2 in Public Sector

Natalia Alonso-Morales, Pedro Gil-García, Alejandro Sáez-Martín, and Arturo Haro-de-Rosario

1 Introduction

The European Union (EU) represents the starting point for the budgetary stability and financial sustainability that we know today and it is also a requirement and fundamental pillar for countries that want to be part of the EU (Martínez, 2018). However, the 2008 crisis revealed the ineffectiveness of the measures that had been working thanks to the economic boom, which is why, following various EU regulations and directives, the European constitutions were modified (López-Montoto, 2012).

The aforementioned crisis particularly affected Spain, with an exceptional increase in unemployment and a collapse in GDP (Meléndez, 2012). Therefore, following the Council Directive /85/EU, Article 135 of the Spanish Constitution was reformed to limit deficit and indebtedness (López-Montoto, 2012), in order to ensure the sustainability of public finances (Council Directive, 2011/85/EU). This constitutional reform required the development of an Organic Law, and therefore the Organic Law 2/2012, of 27 April, on Budgetary Stability and Financial Sustainability (OLBSFS) was finally approved in 2012 (OL, 2/2012).

The OLBSFS introduces the principle of financial sustainability, which is implemented by limiting public debt. The so-called expenditure rule establishes that the current credit operations of Local Entities may not exceed 110% of current revenues settled or accrued (RDL, 2/2004). Furthermore, the average payment period, which cannot exceed 30 days, is fundamental for the economic-financial analysis of local governments (Rebollo, 2017).

The financial sustainability of municipalities conditions the capacity of local governments to provide public services to their citizens (Lopez-Hernandez et al.,

N. Alonso-Morales · P. Gil-García · A. Sáez-Martín · A. Haro-de-Rosario (✉)
University of Almeria, Almeria, Spain
e-mail: arturo.haro@ual.es

A. Sáez-Martín
e-mail: alejandro.saez@ual.es

M. d. C. Valls Martínez and J. M. Santos-Jaén (eds.), *Environmentally Sustainable Production*, https://doi.org/10.1007/978-3-031-52656-5_9

2012), and, therefore, their capacity to act in favour of Sustainable Development (Bisogno et al., 2023). In this sense, after various conferences held by the United Nations to solve the sustainability problems that were occurring around the world, the Sustainable Development Goals (SDGs) were finally approved in 2015 (Benito et al., 2023). Among the various SDGs is Goal 2, Zero Hunger, which aims to ensure food security and nutrition while implementing sustainable agriculture (United Nations, 2015). In this regard, there has been a worrying increase in the number of people suffering from hunger since 2014 (United Nations, 2020). In particular, in Spain, the problems of poverty, inequality, precariousness and helplessness continue to be very real, despite the progress made in achieving the SDGs (Government of Spain, 2020).

Several authors have tried to analyse the relationship between financial sustainability and sustainable development, including Ziolo et al. (2018) who analyse how GDP affects the financing of sustainable development and Benito et al. (2023), who study how the implementation of the SDGs affects financial sustainability. Ríos et al. (2022) focus on other aspects of the public sector and analyse the relationship between efficiency in public service delivery and the achievement of the SDGs. The study by Bisogno et al. (2023) aims to examine how balanced budgets, among other economic and political variables, affect SDG implementation.

However, there is little empirical evidence to help explain how the fulfilment of financial sustainability influences sustainable development and in particular the fulfilment of the SDGs at the local level. Therefore, the objective of this research is to examine how municipalities' compliance with financial sustainability influences levels of sustainable development through the achievement of SDG 2. In addition, this research aims to determine whether other variables such as population density, state fiscal support, tax burden, financial autonomy and investment are behind local sustainable development, which help to understand the extent to which local governments comply with Goal 2 of Zero Hunger.

Given the above, we believe that this research can be of great relevance due to the lack of studies on this subject and the current problematic situation, facilitating knowledge of the importance of sustainable development and, particularly, of SDG 2. In addition, it is expected to allow Local Entities to know the relationship between their financial situation and the fulfilment of Goal 2, encouraging them to improve in both aspects. On the other hand, it can be useful not only for governments, but also for different stakeholders, such as citizens and businesses, motivating them to take action in favour of the goals of the 2030 Agenda and to demand that public entities implement measures in this regard. From an academic perspective, it can facilitate further research on the relationship between financial sustainability and the other SDGs.

The paper is structured in 6 sections. Following the introductory section, Sects. 2 and 3 present a review of both the OLBSFS and the 2030 Agenda, respectively. Section 4 explains the methodology used. In Sect. 5 we explain the results obtained from the descriptive and explanatory analysis. Finally, the last section summarises the main conclusions and possible future lines of research.

2 Financial Sustainability in European Public Entities

The introduction of European financial sustainability was one of the main axes for the current EU members to join, as was the case of Spain (Martínez, 2018). The Treaty establishing the European Economic Community, drawn up in 1957, simply established that the economic policies of the member states should respond to a matter of common interest; however, it did not require genuine budgetary discipline. Such budgetary discipline came about with the adoption of the Treaty on European Union (TEU) in 1992. Subsequently, the Stability and Growth Pact (SGP) was drawn up to ensure control of public deficits (López-Montoto, 2012). In accordance with the SGP, Spain passed Law 18/2001 of 12 December 2001 on General Budgetary Stability (LGBS) and Organic Law 5/2001 of 13 December 2001 to contain public deficits (Paredes & Jiménez, 2015).

The evolution of public finances in European countries as a result of the SGP was favourable, however, it was favoured by the economic growth experienced until 2001. By 2002 and 2003 the situation worsened, questioning whether the SGP works in times of economic recession (Serrador, 2004). This forced a vital reform of the SGP to introduce objectives adapted to the economic cycle. This reform led to the elaboration in Spain of Law 15/2006 and Organic Law 3/2006, both of 26 May, to amend the LGBS and its complementary law (López-Montoto, 2012).

Spain's entry into the European Union, with its respective reduction of public debt, allowed for a long period of economic growth until 2008, when a global crisis began that particularly affected Spain, manifesting the incapacity of the previous Budgetary Stability Law. In this context, a reform of Article 135 of the Spanish Constitution was carried out in 2011 to limit structural deficits and public debt. This article established the mandate to develop an Organic Law, with the subsequent approval of the Organic Law on Budgetary Stability and Financial Sustainability in 2012. Moreover, the reform was also motivated by Spain's commitment to the European Union (OL, 2/2012).

Financial sustainability, one of the principles incorporated thanks to the recent law, is defined in the OLBSFS (Art. 4.2) as "the capacity to finance present and future spending commitments within the limits of deficit, public debt and commercial debt arrears". It is instrumented through the restriction of public debt to 3% of nominal GDP for Local Entities (Art 13, OL 2/2012). However, due to the difficulty of determining GDP for these entities, the debt limit established in the recast text of the Law Regulating Local Treasuries is applied (Chaparro & Herrera, 2023). Therefore, it is established that current credit operations, both short and long term, cannot exceed 110% of current revenues settled or accrued (RDL, 2/2004). On the other hand, there is the Average Payment Period, which is an irrefutable indicator of the financial situation of public administrations and fundamental in budgetary discipline (Olmo Vera et al., 2018). The limit established for this is 30 days, to prevent the debtor from enjoying additional liquidity at the expense of the creditor and to reduce late payment (L, 3/2004). Both restrictions are established as one of the fundamental magnitudes in the economic-financial analysis of Local Entities, due to the fact that, if they are not complied with, they will not be able to access various exceptional measures (Rebollo, 2017).

In the wake of the financial crisis in Spain, several authors have tried to assess the financial situation of local authorities, but in each case, they have opted for a different financial indicator (Lopez-Hernandez et al., 2012). This is due to the interdependence and complex relationships between the different elements of the financial system (Gadanecz & Jayaram, 2008). Gadanecz and Jayaram (2008) after a review of the existing literature summarise the most commonly used measures, which, in the public sector, are GDP growth, the fiscal position (the ability to find financing for its expenditures in excess of its revenues) and inflation. Although there is no consensus, all definitions of financial stability include debt, revenue and expenditure (Bisogno et al., 2017).

In terms of existing empirical studies related to the OLBSFS, we find works such as Olmo Vera et al. (2021) that analyse whether various political and economic factors influence the PMP of Local Entities in Spain. Benito et al. (2023) study how the implementation of the 2030 Agenda affects the financial sustainability of municipalities, considering the principles of the OLBSFS.

Despite the existence of previous studies that analyse the OLBSFS, given its relevance, we consider it necessary to continue advancing on this issue, how financial sustainability influences sustainable development. Likewise, there is a lack of previous work on the possible applications of the Law and, in particular, on its usefulness for sustainable development. For this reason, we will now set out the scope of such sustainable development and its relationship with the OLBSFS.

3 The Pursuit of Sustainable Development in the Public Sector: The 2030 Agenda at the Local Level

The term Sustainable Development was first included in the Brundtland Report in 1987, defined as the ability to meet the needs of the present generation without compromising those of future generations (García-Berlanga et al., 2020). Subsequently, in the 1990s, various United Nations Conferences were held with the aim of achieving this sustainable development, highlighting the approval of the International Development Goals, which were initially exclusive to donor countries, until they were renamed the Millennium Development Goals (MDGs) in 2000 (Sanahuja & Vázquez, 2017). The MDGs were to be achieved by 2015, so the 2030 Agenda was adopted that year in order to make further progress towards sustainability (Bisogno et al., 2023).

The 2030 Agenda aims to protect the environment, ensuring peace and prosperity for citizens and their cities. It is composed of the 17 Sustainable Development Goals (SDGs), which are integrated, so that actions to improve one aspect will condition the others (Ríos et al., 2022).

A consensus can be seen in the literature about the connection between sustainability and the public sector (Benito et al., 2023; Bisogno et al., 2023; Guarini et al., 2022). Therefore, there is a need for the public sector to develop long-term action plans and measure the sustainability achieved, but they have limited resources to do so (Benito et al., 2023). Bisogno et al. (2023) argue that fiscal deficits can prevent resources from being directed towards reducing inequalities, while stable financial conditions allow for the implementation of measures to achieve the SDGs.

In particular, local governments have an important responsibility to implement measures to achieve sustainability due to their proximity to citizens (Ríos et al., 2022). To this end, they have a number of tools at their disposal, such as urban development plans, traffic and urban transport regulations, waste collection, procurement, etc. (Guarini et al., 2022).

Among the various SDGs, SDG 2 on Zero Hunger is currently a very relevant issue because the historically consistent trend of decreasing numbers of people suffering from hunger peaked in 2014, when it started to increase. It is necessary to act quickly by providing humanitarian aid and making a profound change in the agri-food system (FAO, 2020).

SDG 2 aims to achieve the end of hunger by ensuring food security, improving nutrition and promoting sustainable agriculture. Specifically, it has 5 targets and 3 implementation mechanisms to achieve this goal (United Nations, 2015).

We found several works that empirically analyse the factors that influence sustainable development, Ziolo et al. (2018) conducted a research on how financial stability, measured by GDP, affects the financing of sustainable development. Bardal et al. (2021) carried out a survey among sustainability officers in Norwegian municipalities who ranked lack of resources as one of the main barriers to SDG implementation. Both Puertas and Marti (2022) and Bisogno et al. (2023) analysed the factors affecting SDG compliance by Spanish municipalities, but did not take into account the principles of the OLBSFS.

Given this background, we consider that, despite the importance that financial sustainability may have in sustainable development, there are no studies that analyse the impact of compliance with the OLBSFS principles on municipalities' implementation of the SDGs. For this reason, we will set out below the research methodology that aims to achieve this goal.

4 Research Methodology

4.1 Sample

To carry out this analysis, we used a sample of the 103 Spanish municipalities that appear in the second edition of the report "The SDGs in 100 Spanish cities" (municipalities with more than 80,000 inhabitants and provincial capitals), produced in 2020 by the Sustainable Development Solutions Network (SDSN). This report provides an overview of the status of the different SDGs in Spanish municipalities, using indicators adapted to that territory (Sánchez et al., 2020).

Spain and SDG2 have been used as a sample because the problems of poverty, inequality, precariousness and helplessness continue to be highly visible in Spain today, despite the progress made in achieving the SDGs (Government of Spain, 2020). With regard to the situation of SDG 2 in Spain, children living in poverty have higher rates of malnutrition and obesity (15.9%) than those from higher income households (4.6%), which has important consequences throughout their adult lives (Government of Spain, 2020). In addition, we are faced with a food production and distribution system concentrated in large corporations with great power to influence the prices of the food they speculate on (Isakson, 2014).

Furthermore, we focus our research on local governments because their proximity to citizens makes them ideally placed to promote sustainability (Benito et al., 2023). In this sense, several authors have argued that local governments should adopt a sustainable approach to achieve better results (Ji & Tate, 2021; Ríos et al., 2022). In particular, we analyse municipalities with larger populations, in line with numerous previous empirical studies on sustainable development (Benito et al., 2023; Ríos et al., 2022), because large municipalities have greater resources than small ones, so the sustainability analysis has greater scope and impact (Navarro-Galera et al., 2016).

4.2 *Methodology of the Analysis*

Prior to achieving the proposed objective of determining the influence of financial sustainability on SDG 2, we analysed the data on the variables set out in Table 1. Afterwards, we carried out an explanatory analysis of the factors that condition the fulfilment of this SDG.

4.2.1 Dependent Variable

As mentioned above, the SDSN published in 2020 the second edition of the Report "The SDGs in 100 Spanish cities" with a total of 106 indicators grouped according to the 17 SDGs of the 2030 Agenda. Each indicator has been normalised using a scale from 0 to 100, with a higher numerical value representing a better result. To obtain the overall result for each SDG, an arithmetic average of the respective indicators is used (Sánchez et al., 2020).

The dependent variable, SDG 2 of Zero Hunger, is constructed from a series of indicators that will also be used in the explanatory analysis, assuming a total of 5 targets. Target 2.1 is measured by the percentage of the area under agriculture that is organic by province. Target 2.2 reflects the Consumer Price Index Base 2018 to May 2020 per province. Target 2.3 shows employees in the primary sector over total employees. Target 2.4 uses descriptive data from the Spanish Urban Agenda prepared

Table 1 Variables

Abbreviation	Variable	Definition
Dependent		
SDG 2	Sustainable development goal 2	The value assigned to zero hunger compliance (arithmetic mean of the scores assigned to the targets)
Target 2.1	Target 2.1	% of the area under agriculture that is organic
Target 2.2	Target 2.2	Consumer Price Index Base 2018 to May 2020
Target 2.3	Target 2.3	Employees in the primary sector over total employees
Target 2.4	Target 2.4	Descriptive data from the Spanish urban agenda
Target 2.5	Target 2.5	% of the municipality's surface area devoted to woody crops, arable crops, meadows and olive groves and vineyards
Independent		
APP	Average payment period	No. of days of delay in the payment of commercial debt in economic terms
Ind	Indebtness	The % of indebtedness of each municipality (consolidated outstanding debt/consolidated current revenues)
Financial autonomy	Financial autonomy	Capacity of municipalities to finance themselves through taxes (Chaps. 1–3/total revenues)
State fiscal support	State fiscal support	Current transfers received as a percentage of total income (Chap. 4/total revenues)
Investment	Investment	The relevance of long-term projects (Capital expenditure / current expenditure)
Tax burden	Tax burden	Taxes received by the municipality per capita (Chaps. 1 and 2/ Population)
Population density	Population density	The number of inhabitants per km^2(Population/Km^2)

Source Own elaboration

by the Ministry of Transport, Mobility and Urban Agenda. Target 2.5 measures the percentage of the municipality's surface area devoted to woody crops, arable crops, meadows and olive groves and vineyards. Finally, depending on the fulfilment of these targets, a score is assigned to the SDG 2 of each municipality (REDS, 2020).

4.2.2 Independent Variables

Financial sustainability

Several authors have emphasised the need for good financial conditions in order to be able to carry out sustainability measures (Bisogno et al., 2023). Ponce et al. (2018) showed that municipalities' budget stability and social welfare sustainability were closely correlated.

We measure Financial Sustainability based on Spanish laws, which state that local governments are financially sustainable if they have an Average Payment Period (APP) of less than 30 days (L, 3/2004) and an outstanding capital of less than 110% of current revenues (RDL, 2/2004). Therefore, we use the independent variables of APP and Debt Level, in line with authors such as Benito et al. (2023), who considered these variables as dependent variables. However, Benito et al. (2023) concluded that only APP had a significant and negative relationship with SDG compliance. Ajili and Ayoub (2020) also considered the level of indebtedness as an independent variable, determining that the higher the indebtedness, the lower the quality of life of the population.

Based on this, we establish the following hypothesis:

H1: Financial sustainability negatively influences the fulfilment of SDG 2.

Financial autonomy

Financial autonomy is closely interlinked with the sustainability of social welfare (Ponce et al., 2018). Farmer (2022) argues that when local governments control their own finances, they have a greater capacity to respond to demands with public spending and thus to provide a high level of welfare through the provision of public services.

We find several empirical papers that relate financial autonomy to sustainability. First, Farmer (2022) concluded that it was not a significant variable, as did Bisogno et al. (2023) for 2020 data, although with 2018 data he finds a significant and negative relationship. In contrast, Mutiarani and Siswantoro (2020) and Ponce et al. (2018) claim that tax-financed municipalities are more socially responsible, so we hypothesise the following:

H2: Financial autonomy negatively influences the fulfilment of SDG 2.

State fiscal support

Municipalities can be financed not only through taxes, but also through transfers received from higher governments, with the source of funding (via taxes or state support) influencing the type of spending policies pursued and social accountability (Ponce et al., 2018). Saha (2009) links state support to an increase in the adoption of sustainable policies by local governments. This is because it reduces uncertainty about future resource flows, incentivising commitment to sustainability policies (Farmer, 2022).

The empirical study by Ponce et al. (2018) concluded that transfers were not significantly related to the sustainability of social welfare. On the other hand, Farmer (2022) showed that state support enables local authorities to make sustainability efforts. This leads to the following hypothesis:

H3: State fiscal support positively influences the fulfilment of SDG 2.

Public investment

A large part of sustainability measures requires long-term investments (Ji & Tate, 2021). Mutiarani and Siswantoro (2020) argue that the higher the capital expenditure, the higher the level of SDG compliance in local governments. Therefore, we establish investment as an independent variable. Authors such as Bisogno et al. (2023) concluded that it was not a significant variable in relation to the people-centred SDGs. However, Ajili and Ayoub (2020) did find a significant and positive relationship between investment and people's quality of life, so we establish a new hypothesis:

H4: Public investment positively influences the achievement of SDG 2.

Tax burden

A higher tax burden could increase the population's demand for quality services, although it could also lead to local governments using these resources without control (Martínez-Córdoba et al., 2020). It also implies greater citizen participation, so local authorities will be motivated to provide more welfare in return (Mutiarani & Siswantoro, 2020).

The analysis by Ponce et al. (2018) shows that the higher the tax burden, the greater the social promotion. On the other hand, Farmer (2022) found no relationship.

Based on the above, we propose the following hypothesis:

H5: The tax burden positively influences the fulfilment of SDG 2.

Population

Portney (2003) argues that although large cities have a greater need to pursue sustainability, the scale of that need can also be an obstacle. Other authors such as Puertas and Marti (2022) also mention the possibility of such density acting as an obstacle.

In terms of empirical work, both Portney (2003) and Puertas and Marti (2022) found no significant relationship between density and sustainability. However, Saha (2009) concluded that the higher the population density, the higher the sustainability measures.

This leads us to formulate the last hypothesis:

H6: Population density positively influences the fulfilment of SDG 2.

Table 1 shows a summary of the variables used in the analysis and their units of measurement. To conclude, it is worth noting that the explanatory analysis (Pearson correlation matrix and multiple linear regression) was carried out using STATA 15.0. In addition, 6 models are carried out with SDG 2 and each of its targets according to the following equation:

$$\begin{aligned}\text{SDG 2/Targets} = {} & + \text{APP}_i + \text{Indebtness}_i + \text{Financial autonomy}_i \\ & + \text{State fiscal support}_i + \text{Investment}_i + \text{Tax burden}_i \\ & + \text{Population density}_i\end{aligned}$$

5 Analysis of the Results

5.1 Results of the Descriptive Analysis

The results in Table 2 show a high variability of both the APP of the municipalities and their indebtedness, even finding one municipality with an indebtedness of 0 while others are highly indebted. The average APP is 47.89 days, which is in breach of the law as it is over 30 days. As for the average indebtedness, it is less than 110%, thus complying with the maximum limit. On the other hand, the municipalities have an average financial autonomy of 62.49% and a state fiscal support of 29.60%, which indicates that they are mainly financed by taxes and fees, hence their average tax burden of 827.67. The results also indicate that the municipalities do not tend to make large investments, since the average is 10.54%, with some municipalities even having 0% investment. Finally, population density has a high standard deviation, so that there will be highly populated municipalities compared to others with fewer inhabitants per km^2.

With regard to the dependent variable of SDG 2, we observe a relatively low average compliance of 27.50, the maximum value being 77.63. It can also be seen that the minimum value for most of the targets is around 0. Therefore, the results indicate that sustainable development has not yet been achieved in the area of Zero Hunger.

Table 2 Results obtained from the descriptive analysis of the variables

Variable	Average	Standard deviation	Max	Min
SDG 2	27.50	14.69	77.63	8.46
Target 2.1	5.14	6.42	26.59	0
Target 2.2	105.03	1.01	107.22	102.89
Target 2.3	1.39	2.31	13.92	0.03
Target 2.4	1.93	3.69	22.77	0
Target 2.5	27.41	21.46	78.23	0
APP	47.89	83.91	624.34	6.22
Indebtness	46.99%	68.17%	438.00%	0.00%
Financial autonomy	62.49%	10.20%	88.00%	32.00%
State fiscal support	29.60%	9.27%	61.00%	8.00%
Investment	10.54%	7.84%	49.00%	0.00%
Tax burden	827.67	256.78	2156.53	499.76
Population density	2855.81	3585.05	19,517.94	42.54

Source Own elaboration

5.2 Results of the Explanatory Analysis

Prior to the explanatory analysis, we analysed the Pearson correlation matrix to check the lack of significance between independent variables. Table 3 shows that there are significant correlations between the independent variables, both at 0.05 and 0.01, however, they are all lower than 0.8, so there is no multicollinearity problem that could affect the proposed model (Neter et al., 1996).

Table 4 shows the results of the multiple linear regression analysis, showing the factors that influence municipalities to meet the Sustainable Development Goal of Zero Hunger. All the proposed models are significant, i.e. they have the capacity to explain how certain factors affect the SDG and its targets, which is why they constitute the object of study, in addition to confirming their linearity thanks to Fisher's F test shown in the first row of the table.

According to the results, the average payment period is not significant, in contradiction with Benito et al. (2023) who state that there is a relationship between compliance with the SDGs and the APP, so that the greater the compliance with the SDGs, the lower the APP.

This result is contradictory to the one obtained by Ajili and Ayoub (2020), who claim that higher indebtedness is detrimental to the quality of life of the population, and to the one obtained by Benito et al. (2023), who did not observe a significant relationship between SDG compliance and indebtedness. However, it is only significant for SDG 2, but not for its targets.

Financial autonomy is significant in relation to SDG 2, affirming that the greater the financial autonomy of municipalities, the less the SDG will be met. In addition, it would also be significant in the same direction with targets 2.3 and 2.4. It disagrees

Table 3 Correlation between independent variables

	APP	Indebtness	Financial autonomy	State support	Investment	Tax burden	Population density
APP	1						
Indebtness	0.6934***	1					
Financial autonomy	−0.2109**	−0.1093	1				
State support	0.0383	−0.0076	−0.7566***	1			
Investment	−0.2218**	−0.0886	−0.3211***	−0.0905	1		
Tax burden	0.0069	−0.1156	−0.2035**	0.0922	0.1073	1	
Population Density	−0.0624	−0.0552	−0.3023***	0.4185***	−0.0321	0.0794	1

* Significant correlation at the 0.10 level. ** Significant correlation at the 0.05 level. *** Significant correlation at the 0.01 level

Source Own elaboration

Table 4 Regression analysis results (targets)

	SDG 2		Target 2.1		Target 2.2		Target 2.3		Target 2.4		Target 2.5	
	F(7,94)	10,50***	F(7,94)	3,25***	F (7,94)	2,37**	F(7,85)	8,27***	F(7,94)	4,52***	F(7,94)	17,34***
	Adj R-squared	0,4394	Adj R-squared	0,1981	Adj R-squared	0,1219	Adj R-squared	0,1851	Adj R-squared	0,1825	Adj R-squared	0,3851
E	**Coef**	**t**	**Coef**	**t**	**Coef**	**t**	**Coef**	**t**	**Coef**	**t**	**Coef**	**t**
APP	−0,0,094,343	−0,43	−0,0,003,981	−0,03	−0,0,010,459	−0,47	−0,0,020,839	−0,49	−0,0,031,769	−0,70	−0,0,067,213	0,19
Ind.	5,941,061	2,22**	2,772,545	1,60	−0,2,752,906	−1,02	0,4,336,699	0,87	0,3,983,403	0,68	4,291,591	0,99
Fin. Aut.	−41,72,069	−1,87*	−12,50,493	−0,73	−2,751,167	−1,45	−9,444,951	−2,63***	−11,2289	−2,35**	−54,4551	−1,65
St. support	−17,68,526	−0,70	−8,172,494	−0,53	−4,392,936	−2,17**	−7,23,943	−2,33**	−5,963,645	−1,08	−48,93,898	−1,46
Inv.	−55,53,299	−3,09***	−23,52,606	−2,11**	−1,329,736	−0,95	−7,038,139	−1,92*	−11,60,483	−2,50**	−63,36,842	−2,36**
Tax burd.	−0,0,126,074	−2,72***	−0,0,034,978	−1,91*	−0,0,005,778	−1,33	−0,001,343	−1,89*	−0,0,032,551	−2,28**	−0,0,231,003	−3,82***
Pop. Den.	−0,0,020,637	−5,19***	−0,000,027	−0,18	0,0,000,885	3,52***	−0,0,001,989	−3,21***	−0,0,003,299	−3,15***	−0,0,026,591	−5,47***

* Significant correlation at the 0.10 level. ** Significant correlation at the 0.05 level. *** Significant correlation at the 0.01 level

APP = Average Payment Period; Ind. = Indebtness; Fin. Aut. = Financial autonomy; St. support = State fiscal support; Inv. = Investment; Tax burd. = Tax burden; Pop. Den. = Population density

Source Own elaboration

with the study conducted by Farmer (2022), whose results indicated that fiscal decentralisation was not significant. It also disagrees with the results obtained for 2020 by Bisogno et al. (2023), but agrees with the results obtained for 2018 which state that the greater the financial autonomy, the lower the achievement of the people-related SDGs. Finally, it disagrees with Mutiarani and Siswantoro (2020) and Ponce et al. (2018) as they found that revenues from local sources have a positive effect on SDG compliance levels when using general indicators, but not significant when using local indicators, and on social accountability, respectively.

Regarding the variable State fiscal support, we did not find a significant relationship with the fulfilment of SDG 2. We are in line with Ponce et al. (2018), who concluded that transfers received by municipalities was not a significant variable. In contrast, the results of the study carried out by Farmer (2022) indicate that state fiscal commitments promote municipal participation in social sustainability policies. On the other hand, in Table 4 we observe that the variable State fiscal support is significant with goals 2.2 and 2.3, so that the higher the state support, the lower the fulfilment of these goals.

Investment, according to the results, is significant in the fulfilment of SDG 2, so as the investment increases, the fulfilment of SDG 2 decreases. This is in contradiction to the results obtained by Bisogno et al. (2023) who state that investment is not significant for the people-related SDGs. We also disagree with Ajili and Ayoub (2020), who assert that there is a significant and positive relationship between investment and quality of life. It is also significant in the achievement of targets 2.1, 2.3, 2.4 and 2.5.

As stated in the previous hypotheses, tax burden is significant for the fulfilment of SDG 2. According to the results, the higher the tax burden in the municipality, the lower the fulfilment of SDG 2. The same is true for targets 2.1, 2.3, 2.4 and 2.5. The results obtained by Ponce et al. (2018) affirm the opposite, i.e. the higher the tax burden, the greater the social promotion. On the other hand, we disagree with Farmer (2022), who concluded that per capita taxes were not significant.

Finally, according to the results obtained, population density is significant in relation to the fulfilment of SDG 2, so that the higher the density in the municipality, the lower the fulfilment of the SDG. These results differ from those obtained by Portney (2003) and Puertas and Marti (2022), who find no relationship between population density and sustainability. They also differ from those obtained by Saha (2009), who states that there is a significant and positive relationship between density and sustainability initiatives. In conclusion, we also find that the higher the population density, the lower the fulfilment of targets 2.3, 2.4 and 2.5, but the higher the fulfilment of target 2.2.

6 Conclusion

The implementation of the SDGs is an issue that concerns all levels of society and in particular local governments. However, the resources available to municipalities are limited and subject to financial sustainability regulations. This is why, in this study on the application of the Spanish law on financial sustainability, the aim is to find out whether the application of these regulations affects the achievement of SDG 2 in Spanish municipalities through the APP and the Debt.

The results of this study conclude that financial sustainability is partially determinant in the achievement of SDG 2. In particular, the APP has not been a factor to be considered in the implementation of SDG 2, but the level of indebtedness has had a positive effect.

An analysis of these results shows that SDG 2 is far from the ideal level. This may have been due to the fact that it is one of the SDGs with the highest number of factors involved, since the poverty of the population, the increase in inflation and food prices, the structure of food production systems, among others, make it difficult for local governments to take action. However, the existence of municipalities with a high level of compliance shows that there is always room for improvement.

The results of our study show that there are a number of factors to consider beyond the level of indebtedness. These factors are population density, tax burden, financial autonomy and investment. However, most of them have not turned out to affect in the same way as we suggested. This may have been because the dependent variables in the studies used to establish the hypotheses consisted of groupings of several SDGs (Bisogno et al., 2023; Mutiarani & Siswantoro, 2020; Puertas & Marti, 2022) and even sustainability measures other than the SDGs (Ajili & Ayoub, 2020; Farmer, 2022; Ponce et al., 2018; Saha, 2009).

As a general rule, municipalities do not make large investments, do not receive large fiscal support and have a high level of financial autonomy and tax burden. Therefore, we can conclude that Local Entities have a great decision-making capacity, which could represent a problem due to the lack of control by the central government.

With the elaboration of this study we contribute to the development of knowledge on the currently latent issue of sustainable development. With this, Local Entities can know what factors they must take into account when implementing the SDGs. It can even help central and regional governments to learn about possible ways to influence local governments to act in a sustainable way. Given that greater indebtedness of municipalities generates greater compliance with SDG 2, it could mean that indebtedness is useful when it is directed towards social spending. Regarding financial autonomy, as it has a negative significance, higher governments should consider influencing local authorities to comply with certain measures. On the other hand, municipalities that invest less show higher levels of sustainability, which could indicate that municipalities that invest are not targeting their resources correctly in order to be able to make a difference. Regarding the negative relationship of tax burden with the fulfilment of SDG 2, it should be taken into account that municipalities with a higher proportion of high-income earners have a higher tax burden, which means that being richer leads to greater inequality. The last variable to be considered by governments is population density, with the most populated municipalities being the least likely to meet the Goal. This may be due to the greater need for public services concentrated in one area, complicating their management, and thus reducing their efficiency and effectiveness in providing them.

On the other hand, the achievement of the sustainability goal is not only the responsibility of the public sector, so we can raise social awareness and motivate the private sector to act in favour of sustainability. It can also serve as a reference for future research focusing on other SDGs or using other countries or other levels of government as a sample.

Despite the important contributions of this work, it is not without limitations. Firstly, we do not take into account the entire Spanish population, only municipalities with more than 80,000 inhabitants and provincial capitals, leaving the situation of SDG 2 in several localities unknown. In addition, some of the targets have been measured at the provincial rather than municipal level, which may hide information due to the diversity of characteristics within a single province. Finally, we only analysed the year 2020, not taking into account the progress made by the municipalities in implementing the SDG, which is a very relevant issue given that its achievement requires a long period of time.

References

Ajili, W., & Ayoub, H. (2020). The social sustainability of public debt in the framework of Middle East and North African countries: Egypt, Lebanon, Morocco, Tunisia, and Turkey. *Journal of Sustainable Development, 13*(4), 251–267.

Benito, B., Guillamón, M. D., & Ríos, A. M. (2023). The sustainable development goals: How does their implementation affect the financial sustainability of the largest Spanish municipalities. *Sustainable Development.*

Bisogno, M., Cuadrado-Ballesteros, B., & García-Sánchez, I. M. (2017). *Financial sustainability in local governments: Definition, measurement and determinants* (pp. 57–83). Financial Sustainability in Public Administration.

Bisogno, M., Cuadrado-Ballesteros, B., Rossi, F. M., & Peña-Miguel, N. (2023). Sustainable development goals in public administrations: Enabling conditions in local governments. *International Review of Administrative Sciences*, 00208523221146458.

Cereceda, R. C., Hernández, S. R., & Rivera, E. D. O. (2016). Los Objetivos de Desarrollo Sostenible y los retos para su implementación. *Pluralidad y Consenso, 5*(26).

Chaparro, F. P., & Herrera, C. L. (2023). El cumplimiento de la LOEPSF en el sector local: ¿disciplina fiscal o elementos estructurales e institucionales diferenciadores? *Papeles De Economía Española, 175*, 97–187.

Directiva 2011/85/UE DEL CONSEJO, *sobre los requisitos aplicables a los marcos presupuestarios de los Estados miembros.* 8 de noviembre 2011.

FAO, FIDA, OMS, PMA y UNICEF. (2020). *Versión resumida de El estado de la seguridad alimentaria y la nutrición en el mundo 2020. Transformación de los sistemas alimentarios para que promuevan dietas asequibles y saludables.* Roma, FAO. https://doi.org/10.4060/ca9699es

Farmer, J. L. (2022). State interventions and interlocal collaborations across the three pillars of sustainability. *State and Local Government Review, 54*(2), 120–145.

Gadanecz, B., & Jayaram, K. (2008). Measures of financial stability-a review. *Irving Fisher Committee Bulletin, 31*(1), 365–383.

García-Berlanga, O. M., Desfilis, T. P., Martínez, A. R. E., & Peña, A. V. (2020). Objetivos de Desarrollo Sostenible (ODS). Escenario actual. *Objectius de desenvolupament sostenible en el territori valencià*, 25–40.

Gobierno de España. (2020). Reconstruir lo común. En *La implementación de la agenda 2030 en España. Informe de progreso 2020.*

Guarini, E., Mori, E., & Zuffada, E. (2022). Localizing the sustainable development goals: A managerial perspective. *Journal of Public Budgeting, Accounting & Financial Management, 34*(5), 583–601.

Gutiérrez Ponce, H., Chamizo González, J., & Cano Montero, E. I. (2018). Budget stability, financing and social responsibility in Spanish municipalities. *Contaduría y administración, 63*(3).

Isakson, S. R. (2014). Food and finance: The financial transformation of agro-food supply chains. *The Journal of Peasant Studies, 41*(5), 749–775.

Ji, H., & Tate, M. P. (2021). Spillover effects of central cities on sustainability efforts in a metropolitan area. *Policy Sciences, 54*, 95–121.

Ley 3/2004, *por la que se establecen medidas de lucha contra la morosidad en las operaciones comerciales.* 29 de diciembre. BOE núm. 314.

Ley Orgánica 2/2012. *Estabilidad Presupuestaria y Sostenibilidad Financiera.* 27 de abril 2012. BOE núm. 103.

Lopez-Hernandez, A. M., Zafra-Gomez, J. L., & Ortiz-Rodriguez, D. (2012). Effects of the crisis in Spanish municipalities' financial condition: An empirical evidence (2005–2008). *International Journal of Critical Accounting, 4*(5–6), 631–645.

López-Montoto, L. (2012). Estabilidad presupuestaria y sostenibilidad financiera: Nuevas reglas de consolidación fiscal de las haciendas locales. *Anuario Aragonés Del Gobierno Local, 4*, 323–360.

Martínez, E. G. (2018). *La estabilidad presupuestaria y el principio de especialidad presupuestaria en el ámbito de las entidades locales.*

Martínez-Córdoba, P. J., Raimo, N., Vitolla, F., & Benito, B. (2020). Achieving sustainable development goals. Efficiency in the Spanish clean water and sanitation sector. *Sustainability, 12*(7), 3015.

Meléndez, E. S. (2012). El mercado de trabajo español en la crisis económica (2008–2012): desempleo y reforma laboral. *Revista De Estudios Empresariales. Segunda Época, 2*, 29–57.

Mutiarani, N. D., & Siswantoro, D. (2020). The impact of local government characteristics on the accomplishment of Sustainable Development Goals (SDGs). *Cogent Business & Management, 7*(1), 1847751.

Naciones Unidas. (1993). *Informe de la Conferencia de las Naciones Unidas sobre el Medio Ambiente y el Desarrollo 1992.* Naciones Unidas, Nueva York. https://undocs.org/es/A/CONF.151/26/Rev.1(vol.I)

Naciones Unidas. (2002). *Informe de la Cumbre Mundial sobre el Desarrollo Sostenible 2002.* Naciones Unidas, Nueva York. https://undocs.org/es/A/CONF.199/20

Navarro-Galera, A., Rodríguez-Bolívar, M. P., Alcaide-Muñoz, L., & López-Subires, M. D. (2016). Measuring the financial sustainability and its influential factors in local governments. *Applied Economics, 48*(41), 3961–3975.

Neter, J., Kutner, M., Nachtsheim, C., & Wasserman, W. (1996). Applied linear statistical models 4th edition. New York: McGraw hill

Olmo Vera, J., Brusca Alijarde, M. I., & Labrador Barrafón, M. (2018). *La morosidad en el sector local español: factores de riesgo y prevención* (No. ART-2018–115155).

Olmo Vera, J., & Brusca, I. (2021). Determinantes del periodo medio de pago municipal y eficacia del principio de sostenibilidad de la deuda comercial: Determinants of the municipal average payment period and effectiveness of the commercial debt sustainability principle. *Revista De Contabilidad-Spanish Accounting Review, 24*(1), 1–18.

Paredes, Á. A., & Jiménez, Ó. R. (2015). Una reflexión sobre los resultados obtenidos por la Administración Local en el marco de la reforma del artículo 135 de la Constitución española. *Revista De Estudios De La Administración Local y Autonómica, 3*, 95–113.

Portney, K. E. (2003). *Taking sustainable cities seriously*. Massachusetts Institute of Technology.

Puertas, R., & Marti, L. (2022). Regional analysis of the sustainable development of two Mediterranean countries: Spain and Italy. *Sustainable Development*.

Real Decreto Legislativo 2/2004, *por el que se aprueba el texto refundido de la Ley Reguladora de las Haciendas Locales*. 5 de marzo. BOE núm. 59.

Rebollo, M. P. (2017). Estudio descriptivo del Periodo Medio de Pago (PMP) en las Entidades Locales para el período septiembre 2014 a septiembre 2016. *Enrique Benítez Palma, 69*, 115–128.

Ríos, A. M., Guillamón, M. D., Cifuentes-Faura, J., & Benito, B. (2022). Efficiency and sustainability in municipal social policies. *Social Policy & Administration, 56*(7), 1103–1118.

Saha, D. (2009). Factors influencing local government sustainability efforts. *State and Local Government Review, 41*(1), 39–48.

Sánchez de Madariaga, I., del Álamo, J.B., García López, J., Sisto, R., Urquijo, J. (2020). Los Objetivos de Desarrollo Sostenible en 100 Ciudades Españolas, 2nd ed.; REDS: Madrid, Spain; ISBN 978-84-09-25763-8.

Sanahuja, J. A., & Vázquez, S. T. (2017). *Del milenio a la sostenibilidad: retos y perspectivas de la Agenda 2030 para el desarrollo sostenible*.

Serrador, A. G. (2004). *Teoría General sobre reglas fiscales*. Revista electrònica. 2ª época, 8.

United Nations. (2015). *Transforming our world: the 2030 Agenda for Sustainable Development*. Naciones Unidas, Nueva York. https://documents-dds-ny.un.org/doc/UNDOC/GEN/N15/291/89/PDF/N1529189.pdf?OpenElement

Ziolo, M., Ghoul, M. B. G. B., & Aydın, H. İ. (2018). Financial stability vs. sustainable development and its financing. In *Regaining Global Stability After the Financial Crisis*, (p. 88–107). IGI Global.

Natalia Alonso Morales has a degree in Finance and Accounting from the University of Almería (Spain), where she is currently studying to acquire a Master's Degree in Auditing. She is also a PhD student in the Department of Economics and Business. Her research interests are focused on the management system and financial sustainability in the local governments and the sustainable development. She has participated in the international 45th EBES Conference (Budapest) in 2023.

Pedro Gil García is a master's student in Accounting Auditing, graduated in Finance and Accounting, both from the University of Almería (Spain). He is currently pursuing a doctorate in the Department of Economics and Business. His research interests are focused on public accounting, management system and financial information in the federal and local governments. In addition, he has participated in some international conferences, such as the 42th and 45th EBES Conference (Lisbon/Budapest) in 2023.

Alejandro Sáez Martín is a Senior lecturer in accounting, in the Department of Economics and Business at the University of Almería (Spain). His research interests are focused on the financial information disclosures on the Web and social media (e-government), on the management system and financial information in the federal and local governments. He is author of numerous articles in national and international journals, including Public Management Review, Local Government Studies, New Media & Society, Social Science Computer Review, etc. Also, he has written some book chapters (IGI Global and Springer).

Arturo Haro de Rosario is a Senior lecturer in finance, in the Department of Economics and Business at the University of Almería (Spain). His research interests are focused on the online financial information (web and social media), the management system and financial information in the federal and local governments, online citizen engagement, corporate social responsibility and business ethics.

He is author of 31 articles in national and international journals, including Online Information Review, New Media and Society, Information, Communication & Society, Small Business Economics, Social Science Computer Review, Transforming Government: People, Process and Policy, Local Government Studies, etc. Also, he has written 10 book chapters in international publishing companies (IGI Global, Springer and Taylor & Francis, among others). In addition, he has participated in numerous national and international congress/conferences. Finally, note that he participates as a reviewer in relevant journals, such as International Review of Administrative Sciences and Romanian Journal of Political Science.

How Did the Covid-19 Pandemic Affect the Structure of Energy Final Consumption of the Households Across EU Countries? Findings from Eurostat Data

Aleksandra Matuszewska-Janica

1 Introduction

Energy is one of the main drivers of economic growth and plays an important role in the (material) development of society (Farooqi et al., 2021; González-Eguino, 2015; Odum, 2007; Tutak & Brodny, 2022). Eurostat estimates (Eurostat, 2023a) that, as of the end of 2022, within the European Union (27 countries, EU) final energy consumption (FEC) was 939.89 Mtoe (megatonnes of oil equivalent) with households' final energy consumption at 261.77 Mtoe (representing 28% of total FEC). Eurostat data indicate that households are the second-largest sector in terms of electricity consumption (after the transport sector and before the industrial sector).

As part of its energy policy, the EU has set itself 5 main EU objectives (European Commission, 2015): (1) Security, solidarity and confidence, which includes ensuring energy security; (2) Climate action, decarbonizing the economy, leading EU countries towards a low-carbon economy; (3) Energy efficiency, where (increased energy efficiency) is understood as reducing dependence on imported energy and energy carriers, lower emissions and stimulating employment and economic growth; (4) A fully integrated internal energy market that enables the free movement of energy within the EU (infrastructure and appropriate regulation); (5) Research, innovation and competitiveness that promotes breakthroughs in low-carbon and clean energy technologies.

The targets listed are closely linked to actions to achieve the Sustainable Development Goals, in particular SDG7 affordable and clean Energy and SDG12 responsible consumption and production. These two goals are qualified by the European Green Deal (European Commission, 2019; European Union, 2023). With these targets in

A. Matuszewska-Janica (✉)
Institute of Economics and Finance, Warsaw University of Life Sciences, Warsaw, Poland
e-mail: aleksandra_matuszewska@sggw.edu.pl

M. d. C. Valls Martínez and J. M. Santos-Jaén (eds.), *Environmentally Sustainable Production*, https://doi.org/10.1007/978-3-031-52656-5_10

mind, the package Fit for 55 was prepared in 2021. This package is a set of proposals to revise and update EU legislation to put in place new initiatives, so that EU policies are in line with the climate targets set. The guidelines are to reduce net greenhouse gas emissions by at least 55% by 2030 and, in the longer term, to transform economies to zero-carbon.

Households, which consume almost a third of the EU's energy, are an important part of this transformation. Households in this transformation can be considered in two dimensions. The first dimension concerns the consumption of energy from renewable and clean sources. The second dimension is about access to energy and meeting basic energy needs (where they live and at a reasonable cost). Despite the high economic development of the EU area, which usually translates into quality of life for citizens, not all households have access to energy to the same extent. The phenomenon of no access or limited access to energy is called energy poverty. Energy poverty is usually defined as the inability of a household to provide the required (socially and materially) level of energy services (Biernat-Jarka et al., 2021; Bouzarovski, 2014). By treating energy poverty as one dimension of poverty, we can link the reduction of energy poverty to the next goal of SDG1—no poverty.

EU countries are taking a wide variety of actions in energy and climate policy. The European Commission and the European Parliament allow them considerable flexibility in creating these actions. They only oblige them to achieve defined goals. Therefore, this differentiated approach and different economic and social conditions result in different energy consumption patterns among countries. This has raised the question of the extent to which EU countries differ in their household energy final consumption structures.

In the Covid-19 period, when a social distance mechanism was introduced in the initial phase to limit the transmission of the SarsCov2 virus, the result was a significant reduction in the activity of certain economic areas. In 2020, when the greatest restrictions were in place, Energy final consumption fell by 5.6% across the EU economies. The most significant decreases were observed in the transport sector by 12.9% and in the service sector by 5.7% (see Eurostat data (Eurostat, 2023a)).

Although no change in the amount of energy consumption was recorded in households during this period, if we look through the prism of the energy products category (taken from Standard International Energy Product Classification—SIEC), in the electricity category households were the only sector to record an increase in consumption of 1.2% (the other sectors, apart from not elsewhere specified, recorded a decrease in electricity consumption).

Hence, the question arose whether, in the case of households, preventive measures during the Covid-19 pandemic may also have changed the structure of energy consumption and whether this affected differences between EU countries.

The aim of the analysis presented in this paper was to assess the similarities and differences in the consumption of different energy products in households across the EU, and to see whether the shutdown of economies during the Covid-19 pandemic had an impact on changing this structure.

The analysis used descriptive statistics (included in the exploratory data analysis), the Gini concentration coefficient and the k-means algorithm. The data used in the analysis came from publicly available Eurostat databases (Eurostat, 2023a).

2 EU Household Energy Consumption in Relation to Sustainable Development

The wealth and level of development of a region's population is closely linked to the type and degree of access to energy. The more available usable energy and more efficient energy conversion technologies, the better the conditions for the development of individuals, households, communities, society and its economy (Energypedia, 2023). Therefore, access to energy is one of the fundamentals of household functioning. It is important not only (energy) for household activities (such as heating, lighting and use of basic appliances) or transport, but also for the delivery of social services such as education and health (Energypedia, 2023). Therefore, access to energy is considered not only from the viewpoint of economic development but also in the context of poverty, specifically energy poverty.

As mentioned earlier, energy poverty is linked to a household's inability to provide an adequate level of energy services (Biernat-Jarka et al., 2021; Bouzarovski, 2014). In European Union countries, reliable and affordable access to clean energy is also included in this definition (Hasheminasab et al., 2023). 'Between 50 and 125 million people in the EU countries are confronting poor energy access', as reported (Hasheminasab et al., 2023) based on 2009 statistics. In turn (Pereira & Marques, 2023) quoting data from the European Energy Poverty Advisory Hub (EPAH) indicates that around 34 million European citizens can suffer from energy poverty. It should be noted here that estimating the number of people who fall into the category of energy poverty can be difficult, as it is difficult to define the term precisely (Pereira & Marques, 2023). Nevertheless, the numbers quoted indicate the substantial scale of the phenomenon. This requires appropriate action to be taken at various levels of decision-making.

The reduction of energy poverty is strictly in line with the implementation of the Sustainable Development Goals SDGs (United Nations, 2015), more precisely SDG7 - ensure universal access to affordable, reliable, sustainable and modern energy for all and SDG1 - end poverty in all its forms everywhere. It is worth noting that the SDGs at EU level have been formulated somewhat differently. In this case, SDG7 refers to affordable and clean energy. One of the tasks being undertaken within SDG7 is to ensure universal access to affordable, reliable and modern energy services. This has a direct relevance to energy poverty, as according to the definition quoted, those at risk do not adequately have such access.

According to (United Nations, 2018), current indicators for monitoring progress towards SDG7 include (1) Proportion of Population with Primary Reliance on Clean Cooking Fuels and Technology and (2) Proportion of Population with Access to Electricity. In this regard, the UN also recommends Recommended considering complementary indicators such as: (a) affordability, expressed as share of household income spent on energy and (b) energy poverty, encompassing access to and quality of heating and cooling. At the EU level, the indicator for population unable to keep home adequately warm is considered in this respect. According to this indicator, the proportion of such people increased during the Covid-19 pandemic from 6.9% in 2019 to 7.5% in 2020 in the EU (Eurostat, 2023b). It is likely that this indicator could be higher in 2020 were it not for significant government support to offset the negative effects of lockdown. The level of the indicator decreased in the next year (2021) to 6.9%. In turn, it should be added that in 2022 the indicator increased significantly to 9.3%. This was affected by increases in the price of energy carriers and restrictions in their supply, mainly as a result of the Russian-Ukrainian war.

Energy poverty is closely related to the general concept of poverty (Wang & Lin, 2022). This is mainly because households with insufficient income largely have unmet needs related to energy use. Therefore, improving the elimination of energy poverty is in line with the implementation of SDG1.

Household energy use is also analysed from the point of view of renewable and low-carbon sources and sustainable energy use (clean energy). The share of renewable energy sources in the energy mix consumed by households in the EU is systematically increasing, as shown by Eurostat statistics. This includes both direct consumption of such sources (e.g. through the use of photovoltaic panels) and indirect consumption (e.g. the use of electricity produced from renewable sources and then supplied to households). The paper (Pais-Magalhães et al., 2022) indicates that, in the EU, groups with the highest energy demand for heating and cooling showed better renewable outcomes and a higher propensity to invest in energy-saving measures.

Some studies (Simionescu et al., 2023) regarding to energy poverty (EP), have shown that, in the short term, the exploitation of renewable energy sources (RES) promotes an increase in energy prices at an early transitional stage, inducing higher energy poverty. However, when RES penetrates the system and becomes the dominant source of energy supply, it ultimately helps reduce EP. Thus, the use of RES as a primary source of energy requires long-term and well-considered actions, minimizing the negative effects on the final consumer, in particular households.

In 2021, households' energy final consumption from RES was 55.6 Mtoe, with a share of 21.2% in the energy mix of this sector. RES consumption was 10.2% higher than in 2020, with a 1.7% decrease in 2020 compared to 2019. (Eurostat, 2023a, see also data in Tab@@le 2) This slight decrease may have been due primarily to the milder winter season of 2019/2020, which was much warmer than the preceding and following seasons (ECMWF, 2022), and in households RES is used mainly for space heating. It is also worth mentioning that electricity consumed by households is largely produced from RES. It is also worth mentioning that electricity consumed by households is largely produced from RES. In 2021, production from this type of source reached 94.7 Mtoe (37.9% share of total electricity production), which was 1.3% higher than production in 2020 and 9.8% higher than in 2019. The progressive increase in the share of RES in the energy mix is mainly driven by actions taken as a result of the EU's energy policy. The current EU energy policy in a broader context is based on the provisions of the Energy Union (European Commission, 2015), which are listed in the Introduction section.

A more precise mandatory energy policy target for the use of renewable sources was formulated in a Regulation PE/55/2018/REV/1 (European Parliament, 2018d) and in Directive EU/2018/2001 (European Parliament, 2018b) where a renewable energy share of at least 32% was set. In subsequent years, this target was modified. In 2021, it was set at 40% and in March 2023 (under a provisional agreement (European Commission, 2023a) at 42.5%. Russia's invasion of Ukraine is not insignificant in this regard. Due to these developments, the European Commission has indicated the need to accelerate the EU's decoupling from fossil fuels (European Commission, 2023b). The findings at the global level are also worth mentioning here. According to the International Energy Agency (IEA) report (IEA, 2021b), electricity will account for almost 50% of total energy consumption in 2050. The report also assumes that by 2050, almost 90% of electricity generation will come from renewable sources, with wind and photovoltaics together accounting for almost 70%. Most of the remaining energy will come from nuclear power plants.

A further target is to reduce energy consumption by 11.7% EU-wide by 2030, compared to the 2020 reference scenario projections (European Commission, 2021; European Parliament, 2023b). As a large amount of energy consumption relates to space heating, considerable focus is placed on, among other things, the modernization of buildings.

According to (European Parliament, 2023b) member states should also ensure that 'at least 3% of public buildings are renovated each year into nearly zero energy buildings or zero-emission buildings'. The certification and modernization issue also applies to households. As indicated in (European Commission, 2021), detailed public sector regulation has a multiplier effect as the public sector acts as a role model. As a consequence, private consumers mimic the actions of the public sector in improving energy efficiency. In these measures, households are supported by national and municipal authorities. The basis for this is set out in the Energy Efficiency Directive (EED), Directive 2012/27/EU as amended by Directive (EU) 2018/2002 (European Parliament, 2018a) and Directive (EU) 2018/2002 (European Parliament, 2018c) and Directive (EU) 2018/844 (European Parliament, 2018a) and Directive (EU) 2023/1791 (European Parliament, 2023a).

The level of energy efficiency under EU legislation is subject to certification in accordance with Directive 2010/31/EU (European Parliament, 2010) updated by Directive (EU) 2018/844 (European Parliament, 2018a). The legislation is also binding for the residential sector (households). The European Commission indicates that modernization strategies to improve energy efficiency should also include measures that contribute to the alleviation of energy poverty. However, Member States are allowed to set such actions in this regard as they consider appropriate (European Commission, 2021). It is worth pointing out here that investments in renewable energy sources promote energy efficiency (Simionescu et al., 2023), particularly in the residential sector (Dato, 2018).

These actions related to changes in EU legislation are closely linked to the provisions of the 'Fit For 55' framework (European Commission, 2021), which sets a climate target of at least a 55% reduction in EU emissions by 2030 and a total reduction by 2050. As a result, energy policy actions are deeply and multifaceted linked to the achievement of the SDGs targets.

As previously mentioned, the first phase of the Covid-19 pandemic (2020) recorded a decrease in energy consumption in the main sectors (industry, transport and services) and the household sector showed a small increase, which is directly linked to the shift to remote working and learning (Rokicki et al., 2022), which according to the data resulted in an increase in the share of consumption by households in the final total consumption from 26.5% in 2019 to 28% (Eurostat, 2023a). In the second phase of the pandemic (2021), we observed an increase in energy consumption in the sectors mentioned. This is a result of the lifting of pandemic restrictions and the economic recovery. According to the EEA (2023), the largest increases in energy consumption were in the transport sector (by 9% in 2021 compared to 2020) and in buildings (by 6%, the household is the basic unit of the residential sector, see Chen et al., 2023), the latter of which was heavily influenced by cold weather. Here it is important to note that the reduction in mobility and the decrease in energy consumption during the lockdown period resulted in a reduction in gas emissions, which had a positive impact on the environment (Rita et al., 2021).

A number of factors influence households' energy consumption patterns, and thus also the structure of this consumption. (Chen et al., 2023) list the following groups of factors: demographic factors (related to households: household size and its structure, household income, household members' habits, age and education level. etc.); environmental factors (relating to broad economic and social issues and natural environment issues such as geographical location or climate, etc.).

As mentioned earlier, EU countries possess a level of autonomy in developing their energy policies at the domestic level. Consequently, there are varying approaches to handling energy matters across these countries. As highlighted in the study (Jedlińska & Olkuski, 2019) 'Member states, based on their individual requirements and objectives, interpret or implement the EU regulations in diverse ways.' This custom-tailored approach can yield dual outcomes.

Firstly, this leads individual countries to accelerate the achievement of their targets by using methods and strategies tailored to their specific conditions. Secondly, it can lead to significant divergence between countries, especially in crisis. Actions taken during such periods are usually aimed at mitigating the negative effects of the crisis, and the energy market is particularly vulnerable to these changes. This vulnerability was highlighted, among others, in a study presented in (Matuszewska-Janica et al., 2023), where the structure of electricity prices for households across the EU was analysed. Between 2019 and 2021, there was an increase in differentiation among countries based on this price structure.

Hence, the question arose as to how this variation manifests in other facets of the energy market. It's also a question whether we can anticipate an amplified divergence among EU coutries due to the impact of the Covid-19 pandemic in other energy market segments. This article delves into examining the pattern of energy usage, which is profoundly critical in accomplishing the sustainable development objectives. This objective aligns with the EU's energy and climate policy, aiming to transition

EU economies into zero-emission models by 2050. Given that households represent the second-largest consumer, alterations in the energy sector significantly impact their functionality, particularly concerning issues like energy poverty. This analysis focused on household energy consumption.

Empirical studies comparing the structure of final household energy consumption recently in all EU countries have not yet been published. Therefore, this article fills a research gap in this area.

3 Data and Methods

Eurostat presents the structure of electricity consumption using the Standard International Energy Product Classification SIEC (DESA UN, 2017). At level 1 (section), 10 variants are distinguished, with only 7 of them related to households (refer to Table 1). The subsequent analysis utilized the following variables pertaining to the types of energy carriers used in households across the EU.

- gas_i^t—share of natural gas in the total households' FEC of ith country households in period t (w %);
- res_i^t—share of renewables and biofuels in the total households' FEC of ith country households in period t (w %);
- $electr_i^t$—share of electricity in the total households' FEC of ith country households in period t (w %)

Table 1 EU27 household consumption of energy products, 1990 and 2019–2021

SIEC section	Final consumption (Mtoe)				Change (%)		
	1990	2019	2020	2021	2020/2019	2021/2020	2021/1990
Total	239.84	248.18	248.12	261.77	0.0	5.5	9.1
Solid fossil fuels	27.21	6.57	6.72	6.44	2.3	−4.3	−76.4
Peat and peat products	0.83	0.20	0.20	0.20	0.7	−3.8	−76.5
Natural gas	56.43	79.62	78.60	87.71	−1.3	11.6	55.4
Oil and petroleum products	58.30	29.07	30.46	24.95	4.8	−18.1	−57.2
Renewables and biofuels	23.08	51.34	50.46	55.61	−1.7	10.2	140.9
Electricity	44.27	60.53	61.28	64.27	1.2	4.9	45.2
Heat	27.48	20.84	20.40	22.60	−2.1	10.8	−17.8

Source Own elaboration based on Eurostat data (Eurostat, 2023a)

- $heat_i^t$—share of heat in the total households' FEC of ith country households in period t (w %)
- fpp_i^t—total share of emission-intensive sources in the total households' FECn of ith country households in period t (w %)

Three sections were classified as emission-intensive sources: (a) solid fossil fuels, (b) peat and peat products and (c) oil and petroleum products (excluding biofuel portion). The treatment of combining these three components was driven by several considerations. Firstly, it allowed the number of variables to be reduced. In the case of the k-means method, too many variables are not recommended, as this can make it difficult to achieve a breakdown of satisfactory quality. Secondly, some sources are not used in all EU countries. For example, peat and peat products are used in only four countries, and solid fossil fuels are marginal in the vast majority of countries, with a share of less than 2%. Thirdly, EU energy policy aims at reducing carbon-intensive sources, so a combination of these three fuel types seems natural.

The examination relied on data accessible to the public sourced from the Eurostat energy balances database (Eurostat, 2023a). This dataset comprises annual information, encompassing 27 EU countries (as per the 2020 configuration) and spans from 2019 to 2021.

The analysis is divided into two parts. Initially, we illustrate the fluctuations in average households' FEC across European Union countries from 2019 to 2021. Subsequently, the countries are categorized based on their households' FEC structure. We utilized the k-means technique as a research tool to group the EU countries. McQueen (1967) this algorithm. Its detailed explanation can be found in various publications, including (Han et al., 2012) and (Yadav & Sharma, 2013), among others. The process of implementing the k-means algorithm encompasses multiple steps, as specified (Bieszk-Stolorz & Dmytrów, 2019) they are: (1) the selection of variables and objects, (2) variable normalization, (3) the selection of a clustering method and distance measure, and (4) the selection of the number of clusters.

We incorporated three normalization methods into the analysis: standardization, positional standardization, and unitization with a zero minimum (Walesiak, 2018). When computing the distance between objects, we also explored various approaches, such as the commonly used Manhattan distance and Euclidean distance, along with the generalized distance measure (GDM, see: Jajuga et al., 2003; Walesiak, 2016). The final phase involved determining the number of classes (k).

We assessed different divisions with varying cluster numbers (k); however, due to the limited number of objects (27 countries), our considerations were restricted to k values ranging from 2 to 12. To select the best division, we employed the silhouette index (SI, (Kaufman & Rousseeuw, 1990) pp. 83–88), the details of which can also be found in (Dudek, 2020) and (Roszko-Wójtowicz & Grzelak, 2021). The highest SI value indicates the optimal number of clusters. According to literature, acceptable divisions typically exhibit an SI of at least 0.5 (suggesting a reasonable clustering structure, see Kaufman & Rousseeuw, 1990; Roszko-Wójtowicz & Grzelak, 2021; Walesiak, 2006).

The classification procedure was conducted employing R package libraries. The investigation made use of specific libraries: ClusterSim for the main analysis (Walesiak & Dudek, 2022), RobustHD for generating silhouette widths (Alphonse, 2022), and factoextra for visualizing the data (Kassambara & Mundt, 2022). The results presented in this research were obtained by applying unitization with a zero minimum as the method for normalizing variables. The calculation of distances between items was achieved using the Generalized Distance Measure.

4 Structure of Final Consumption of Households in the EU

As previously mentioned, households' final energy consumption was 261,771 Mtoe in 2021 (see Fig. 1). This represented 27.9% of total FEC in the EU area. Such a high share placed the household sector in second place after the transport sector (with share equalled 29.2%) and before the industrial sector (25.6%). Thus, these three sectors are responsible for as much as 83% of FEC (see also Rokicki et al., 2022). From 2019 onwards, we have seen considerable fluctuations in total FEC. From a 30-year perspective, these are the largest year-on-year changes. In 2020, compared to 2019, across the economy as a whole, the value of this variable fell by 5.6%. When we look at this variable through the sectors, we observe that the largest decreases were in the transport sector by 12.9% and services by 5.7%, while the household sector remained almost unchanged. This was, of course, due to the massive lockdown of economies as a result of the Covid-19 pandemic. At this time, the IEA (International Energy Agency) experts projected an economic recovery in the next period as a result of the lifting of pandemic limitations, which would contribute to a 4.6% increase in energy demand, which in turn would increase global energy consumption in 2021 (IEA, 2021a). EU total FEC increased by 6.2% and for households by 5.5%. So households started to consume significantly more energy than in the year before the Covid-19 pandemic. When we look at household FEC consumption in the long term, we see that it has oscillated between 234.64 Mtoe and 278.94 Mtoe since 1990. It is currently closer to the upper limit. We see a similar pattern of change during the global financial crisis of 2007, which had a particularly negative impact on the EU economies in 2008. During this period, we initially see a slight decrease in household FEC (2009–2008), followed by a significant increase in this variable of 6.3% the following year.

The energy consumption by product category and the structure of this consumption for the EU area are presented in the Tables 1 and 2. Based on Eurostat data and following the SIEC classification, households use 5 categories of energy products included in the first level (sections): solid fossil fuels, peat and peat products, natural gas, oil and petroleum products (excluding biofuel portion), renewables and biofuels, electricity and heat. Total FEC in households is basically unchanged in 2020 compared to 2019. However, when we look at the individual energy product sections, we see an increase in consumption in the oil and petroleum products (by 4.8%) and solid fossil fuels (by 2.3%) categories and a decrease in consumption in the heat (by

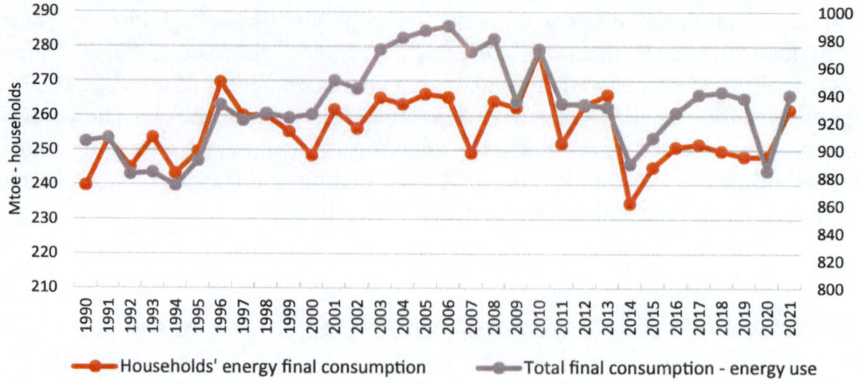

Fig. 1 Total and households' final consumption—energy use in EU in years 1990–2021. 2023a *Source* Own elaboration based on (Eurostat,) data

2.1%) and renewables (by 1.7%) and natural gas (by 1.3%) categories. Given that the winter of 2019/2020 was significantly warmer than the preceding and following season (ECMWF, 2022), and the fact that households use most of their energy for heating, this may have had a significant impact on the reduction in the consumption of product categories that are used for heating purposes. In the following year, the changes were much greater. Consumption in the natural gas, renewables and biofuels and heat categories increased by more than 10% (by 11.6%, 10.2% and 10.8% respectively). Consumption in the electricity category also increased (by 4.9%). In contrast, consumption in the oil and petroleum products category decreased significantly (by as much as 18.1%), as well as solid fossil fuels (4.3%) and peat and peat products (3.8%). This is a significant decrease in categories that are classified as emission sources. Such a tendency is in line with EU climate policy.

Table 2 Shares of energy products by SIEC sections in the households' FEC in the EU in years 1990 and 2019–2021 (in %) and the share of households' FEC in each section to total consumption in the section (sht_i)

SIEC section	1990	2019	2020	2021	sht_i in 2021
Solid fossil fuels	11.3	2.6	2.7	2.5	33.8
Peat and peat products	0.3	0.1	0.1	0.1	55.0
Natural gas	23.5	32.1	31.7	33.5	41.3
Oil and petroleum products	24.3	11.7	12.3	9.5	7.6
Renewables and biofuels	9.6	20.7	20.3	21.2	50.3
Electricity	18.5	24.4	24.7	24.6	30.1
Heat	11.5	8.4	8.2	8.6	47.8
Total	100	100	100	100	27.9

Source Own elaboration based on Eurostat data (Eurostat, 2023a)

It is also worth looking at changes over a longer period. Although household energy consumption has increased by around 9% since 1990, we find significant declines in the categories of solid fossil fuels (76.4%) and peat and peat products (by 76.5%). In contrast, consumption of renewable energy (up 140.9%), natural gas (up 55.4%) and electricity (up 45.2%) has increased significantly. The significant increase in the use of RES in households is due to the development of this sector of energy production, financial support for households to use this source for their own needs (subsidies, refunds, tax deductions, etc.). The increase in gas consumption, on the other hand, is justified by the fact that, although it is a non-renewable source, it is much less polluting than solid fossil fuels, which is in line with EU climate policy. In turn, the increase in electricity consumption occurs through substitution of other categories. As mentioned earlier, EU energy policy actions are moving in this direction, as electricity can be generated from renewable and low-carbon sources. The European Commission says that electricity should supply at least 57% of demand by 2050, and 50–70% for the residential sector (ETIP Wind, 2021). By that time, on the other hand, the European Commission expects renewables to supply at least 81% (ETIP Wind, 2021).

The structure of household's energy final consumption in relation to the SIEC sections is shown in the Table 2. This table also provides information on what share of the energy consumption in the section (category) was attributable to the household sector in 2021 (last column, variable sht_i). These shares were calculated according to the formula: $sht_i = \frac{hec_i}{alec_i} \cdot 100\%$, where hec_i is household's energy final consumption for ith SIEC section (Mtoe) and $alec_i$ is energy final consumption in all sectors for ith SIEC section (Mtoe).

In the years included in the analysis (2019–2021), the structure of household energy consumption by category has not changed significantly. In 2021, natural gas accounted for the largest share of households' final consumption (33.5%). This share was slightly lower in 2019 and 2020 (by 1.4 pp and 1.8 pp respectively, see Table 2). The second most used energy product in the household sector is electricity. Its share in the energy mix was 24.6% in 2021. Despite the significant increase in the value of consumption in this category, the share remained virtually unchanged over the years analysed. This is due to the fact that other categories also recorded significant growth (heat and natural gas). Renewables and biofuels are an important source of energy in the sector analysed, with a share of 21.2%. This share has been increasing steadily over the years as a result of the EU's energy and climate policy targets. When we compare the level of consumption of Renewables and biofuels to that of 1990, we see that this is the product that has gained the greatest increase in share (by as much as 11.6 pp). Analysing the situation between 2019 and 2021, we see that while the share decreased slightly in 2020 compared to 2019 (by 0.4 pp), the following year we recorded an increase of 0.9 pp. This slight decrease in the share in 2020, as mentioned earlier, was rather due to the reduced consumption of energy from this source caused by the warmer winter period. Oil and petroleum products represent 9.5% of the households' final energy consumption. In this section of energy products we observed a significant decrease in shares in 2021 by 2.7 pp compared to 2019 and 2.2 pp compared to 2020. Compared to 1990, it has decreased by as much as 14.8 pp. In 1990, this category represented the largest share of households' FEC. The smallest shares are related to the most carbon-intensive sources: solid fossil fuels (2.5%) and peat and peat products (only 0.1%). While peat and peat products are of marginal importance in households' FEC (the share of this category was only 0.3% in 1990 and 0.1% in 2019–2021), solid fossil fuels have significantly lost ground. The decrease in the share of this category was 8.9 pp in 2021 compared to 1990. This is, of course, a result of the successive EU policy to eliminate the use of sources with high CO_2 emissions.

The shares of each category vary considerably across EU countries. To illustrate this diversity, Fig. 2 shows box plots for the shares of each SIEC section across all EU countries for 2021 data. Natural gas is the most popular energy product in the EU countries. Only in two countries do households not consume this product (Cyprus and Malta), while in the remaining countries, the share of total household final energy consumption ranges from about 0.5% (Finland and Sweden) to 71.2% (the Netherlands) with a median of 20.6%. This variable is characterised by an increased level of variability. Coefficient of variation (for standard deviation) CV_s= 0.81 and quartile coefficient of variation CV_Q= 0,61 ($CV_s = s/\overline{x}$, where $\overline{x}$ - mean of the analysed variable, s - standard deviation od of the analysed variable, (Holmes et al., 2017); $CV_Q = (Q_1 - Q_3)/Q_1 + Q_3)$, where Q_1 - first quartile, Q_3 - third quartile (Botta-Dukát, 2023). This indicates a slightly higher variability of this variable than the moderate variability.

More homogeneous values of household energy product consumption are observed for renewables and biofuels and electricity. The share of renewables and biofuels in final consumption varies from 2.8% (Ireland) to 46.6% (Croatia) with a median of 25.1%

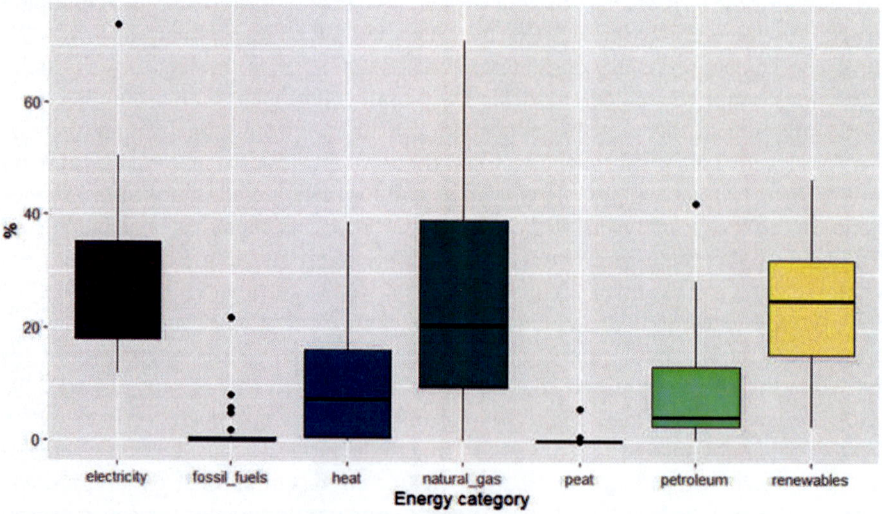

Fig. 2 Boxplots for the shares of each energy product category in the household energy mix in individual EU countries in 2021. *Source* Own elaboration based on (Eurostat, 2023a) data

The coefficients of variation of this variable are $CV_s = 0.47$ and $CV_Q = 0.34$ respectively. In turn, the share of electricity varies from 11.9% (Poland) to 73.6% (Malta) with a median of 20.6%. It is worth noting here that the second country in order with the highest share of this energy product category is Sweden (50.4%). The coefficients of variation are $CV_s = 0.51$ and $CV_Q = 0.32$. Thus, these two variables (shares of renewables and biofuels and electricity in total FEC) are characterised by moderate variability. Heat use ranges from 0% (Ireland, Cyprus, Malta, Louxembourg, Spain, Portugal) to 38.8% (Denmark). In general, it is an unpopular source used by households in southern European countries. In Greece, the share of this category is only 0.8% and in Italy 2.2%. The Benelux countries also make little use of this source. Apart from Luxembourg, in the Netherlands the share is 3.1% and in Belgium 0.2%. The median for this variable equals 7.4%. A high level of volatility is observed for this variable ($CV_s = 1.08$, $CV_Q = 0.94$). The share of oil and petroleum products ranges from 0.3% (in Slovakia) to 42.3% in Ireland, with a median of 4.4%. For this variable, we also observe high variability ($CV_s = 1.07$, $CV_Q = 0.67$). Households use peat and peats products in only three countries: Ireland (with a share of 5.9%), Lithuania (0.8%) and Finland (0.1%). Solid fossil fuels, on the other hand, are used in 20 countries, but in only four of these is this share above 3%: in Ireland (4.9%), in Bulgaria (5.8%), in the Czech Republic (8.1%) and the highest in Poland, (21.9%).

5 Concentration of Household Consumption of Energy Products in EU Countries

The concentration degree of household consumption of individual energy products was measured using the Gini coefficient (see Giorgi & Gigliarano, 2017; Jacobson et al., 2005; Matuszewska-Janica et al., 2021) among may others). The calculated Gini coefficients for individual EU countries and the years 2019–2021 are shown in the Fig. 3.

The Gini coefficient takes values in the range $\langle 0;\ 1\rangle$ and higher values of the Gini coefficient indicate that a country is dominated by specific energy sources used by households. The degree of concentration varied across EU countries from 0.15 to 0.63 in 2019 and from 0.13 to 0.64 in 2021. The highest levels of concentration are observed in the Netherlands and Malta (0.63 and 0.64 in 2021 respectively). This is because more than 70% of the energy used by households comes from a single source. In the case of the Netherlands this is natural gas (71.2% share) and in the case of Malta it is electricity (73.6% share). The next three countries on the list are Sweden (where electricity has a 50.4% share), Louxembourg (with a natural gas share of 50.1%) and Irleand (where oil and petroleum products have the largest share equals 42.3%). The lowest Gini coefficient values were obtained for Poland (0.13 in 2021, 0.15 in 2020 and 0.16 in 2019) and Austria (0.18 in 2019–2021). In Poland, households do not use peat and peat products, and consumption of oil and oil products has a share of 2.8%. In other categories, shares range from 11.9%

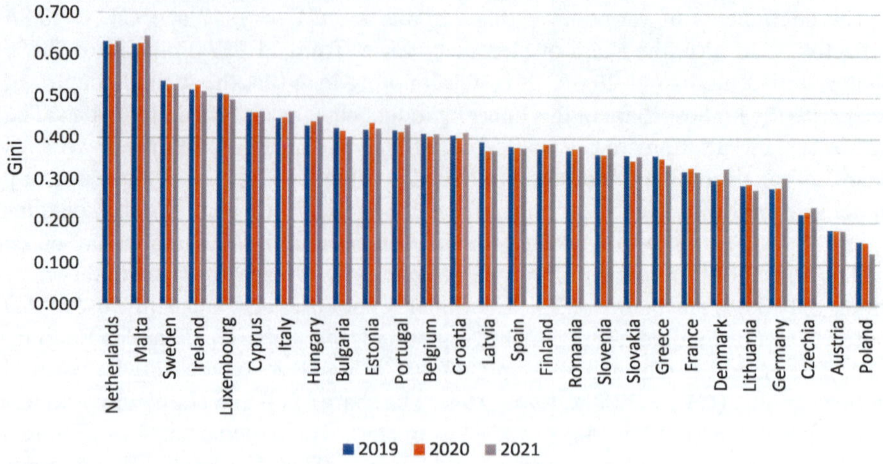

Fig. 3 Concentration (Gini coefficient) of household consumption of energy products (SIEC) in EU countries, years 2019–2021. *Source* Own elaboration based on (Eurostat, 2023a) data

(electricity) to 24.4% (renewables and biofuels).A major problem in Poland is the high consumption of solid fossil fuels, as mentioned earlier. While this reduces the level of concentration, it is not in line with EU energy and climate policy. In the case of Austria, the situation looks much better. Here we see a lack of use of peat and peat products and the share of solid fossil fuels is 0.2. The share of other products ranges from 12.1% (heat) to 30.2% (renewables and biofuels).

6 Clustering of EU Countries by Shares of Energy Carriers in Households' Final Energy Consumption

Clustering of EU countries was conducted employing the k-means algorithm. A silhouette index (SI) was adopted to select the best number of clusters, the values of which are reported in the Table 6 in the Appendix. Various methods were used in the analysis presented here to normalise variables and calculate distances. Various methods were used in the analysis presented here to normalise variables and calculate distances. The best results were obtained applying positional standardisation as the normalisation method and GDM as the distance calculation method. For the 2019 and 2021 data, the best divisions (with the highest SI value) were into two groups, and for 2020 into three groups. In the case of the clustering for 2019 and 2021, there are marginal differences between the SI value for the 2 and 3 group divisions, so to make it easier to compare the three classifications, the results of the 3 group divisions for all the samples analysed are presented. Selected results (group compositions and average values of analysed characteristics within groups) are presented in the Tables 3, 4, and 5 and in the Fig. 4.

Table 3 Clusters averages for the analysed variables—grouping for 2019 data

Cluster k	Countries	fpp_k^{19}	gas_k^{19}	res_k^{19}	$electr_k^{19}$	$heat_k^{19}$
1	Belgium, Germany, Ireland, Luxembourg, Poland	31.42	34.07	11.38	18.27	4.86
2	Austria, Croatia, Czechia, Denmark, Estonia, Hungary, Italy, Latvia, Lithuania, Netherlands, Romania, Slovakia, Slovenia	6.12	30.43	33.53	20.83	17.42
3	Bulgaria, Cyprus, Finland, France, Greece, Malta, Portugal, Spain, Sweden	14.45	8.47	24.06	43.95	9.06

Source Own elaboration based on Eurostat data (Eurostat, 2023a)

Table 4 Clusters averages for the analysed variables—grouping for 2020 data

Cluster	Countries	fpp_k^{20}	gas_k^{20}	res_k^{20}	$electr_k^{20}$	$heat_k^{20}$
1	Belgium, Germany, Ireland, Luxembourg, Poland	33.12	32.98	10.83	18.26	4.81
2	Austria, Croatia, Czechia, Denmark, Estonia, Hungary, Italy, Latvia, Lithuania, Netherlands, Romania, Slovakia, Slovenia	5.93	30.63	33.32	21.32	17.14
3	Bulgaria, Cyprus, Finland, France, Greece, Malta, Portugal, Spain, Sweden	14.51	8.52	24.28	43.89	8.80

Source Own elaboration based on Eurostat data (Eurostat, 2023a)

The least numerous is group 1. This was 5 countries in 2019 and 2020 (Belgium, Germany, Ireland, Luxembourg, Poland), with 4 countries in 2021, as the algorithm did not qualify Germany for this cluster. This cluster is characterised by a very high average share of high-emission energy sources (variable fpp_k^t, where k is cluster number). The average value of this variable in this group was 31.42% in 2019, 33.17% in 2020 and 33.7% in 2021. Among the countries in this cluster, Ireland had

Table 5 Clusters averages for the analysed variables—grouping for 2021 data

Cluster	Countries	fpp_k^{21}	gas_k^{21}	res_k^{21}	$electr_k^{21}$	$heat_k^{21}$
1	Belgium, Ireland, Luxembourg, Poland	33.70	33.01	10.49	18.17	4.63
2	Austria, Croatia, Czechia, Denmark, Estonia, Hungary, Italy, Latvia, Lithuania, Netherlands, Romania, Slovakia, Slovenia, Germany	6.03	32.14	31.91	20.82	16.79
3	Bulgaria, Cyprus, Finland, France, Greece, Malta, Portugal, Spain, Sweden	13.41	8.83	24.33	44.36	9.08

Source Own elaboration based on Eurostat data (Eurostat, 2023a)

the highest share of this type of source, which fluctuated around 54% during the period analysed. For the other countries, it ranged from around 21% in Germany to around 30% in Belgium. In 2021, Germany was not included in this group, as the share of this variable fell from 23% in 2020 to 14.8% in 2021.

Another distinguishing feature of this group is the average low share of renewable sources (res_k^{19}= 11.4%, res_k^{20}=10.8% i res_k^{21}=10.5%). However, when we look at the value of this variable for individual countries, it is worth noting that Poland had a share significantly above the group average (res_{PL}^{19}=2.3%), while for other countries these shares were much lower (res_{DE}^{19}=14.4% for Germany and for the rest do not exceed 9%).

This group is also characterised by the lowest average share of heat ($heat_k^{19}$=4.9%, $heat_k^{20}$=4.8% and $heat_k^{21}$=4.6%). In Ireland and Luxembourg, households do not use this type of energy ($heat_{LU}^t$=$heat_{IE}^t$= 0), while in Belgium it is marginal ($heat_{BE}^t$<0.2%), only in Germany and Poland do values of this variable exceed 6%. However, in the case of Poland, this is quite significant, exceeding 17% in 2019 and 2020 and exceeding 18% in 2021.

For the other two variables, their average values (gas_k^t and $electr_k^t$) are similar to those calculated for cluster 2. The average share of natural gas in the households' energy mix in cluster 1 is about 33% (gas_1^{19}=34%, gas_1^{20}=gas_1^{21}=33%). In 2019 and 2020, in Poland and Ireland, the shares of this source were less than 20% and in the other countries above 37%, in Luxembourg even more than 50% (53.2%). In 2021, these shares increased to around 20% in Ireland and Poland, in Belgium to almost 42%. In Luxembourg, on the other hand, they have decreased to 50.1%. Reducing the share of a single source in the energy mix is in line with the energy security policy, which calls for the diversification of energy sources. The average value of electricity shares in the cluster 1 is approximately 18.2%. This variable within the cluster is characterised by the smallest variability. The values of the variable for individual countries range from 12% in Poland to 23% in Ireland. Whereas Ireland recorded an increase in shares of approximately 1.3 pp in 2021 compared to 2020 and Belgium a decrease of just over 1 pp.

The most numerous cluster cluster 2. For the 2019 and 2020 data, there were 12 countries and in 2021 there were 13. These countries are characterised by the lowest average share of fpp_k^t, the highest average share of res_k^t and the highest average share of $heat_k^t$. For the other two variables, gas_k^t and $electr_k^t$, the cluster averages are similar to those obtained for cluster 1.

The use of fpp_k^t sources in this group ranges from 0.4% in the Netherlands to 14% in Belgium in 2019 and 2020. In 2021, the range was the same, only, the upper limit (14%) was also recorded for Germany. Shares greater than 10% are recorded for Austria and Slovenia (in 2021 as well for Germany). One point is worth noting here. In 2021, the share of this source decreased in Slovenia by more than 2 pp. and in Germany by 8 pp. Such a significant decrease in energy consumption from emission-intensive sources in Germany led the algorithm to assign this country to this cluster in the 2021.

In the case of RES, its share in the household energy mix is greater than 20% (and in the case of 7 countries, greater than 30%). The exception here is the Netherlands, for which the res_{NL}^t was around 6%. Also for Germany, the RES share was well below average, at 15.5%. The largest shares in 2021 were recorded for Croatia (over 46%), as well as Slavonia (almost 45%) and Estonia (over 40%).

Average shares of the $heat_i^t$ variable in this cluster are around 17%. This is the highest value of all the clusters. We recorded the smallest shares of this variable for Italy and the Netherlands (respectively $heat_{IT}^{19}$=$heat_{IT}^{20}$= 2.8%, $heat_{IT}^{21}$= 2.2%, $heat_{NL}^t$= 3.1%) and the largest for Lithuania, Latvia and Estonia (respectively $heat_{LT}^{19}$= 30.8%, $heat_{LT}^{20}$= 29.3%, $heat_{LT}^{21}$= 30.7%, $heat_{LV}^{19}$= 30.9%, $heat_{LV}^{20}$= 31.1%, $heat_{LV}^{21}$= 34.6%, $heat_{EE}^{19}$= 34.3%, $heat_{EE}^{20}$=$heat_{EE}^{21}$= 33.1%).

The average share of electricity in this cluster is approximately 21%. We recorded the smallest shares in Lithuania and Romania (respectively $electr_{LT}^{19}$= 17.3%, $electr_{LT}^{20}$= 18.3%, $electr_{LT}^{21}$= 18%, $electr_{RO}^{19}$= 14.4%, $electr_{RO}^{20}$= 14.6%, $electr_{RO}^{21}$= 14%), and the largest in Croatia and Slovenia (respectively $electr_{HR}^{19}$= 23.8%, $electr_{HR}^{20}$= 23%, $electr_{HR}^{21}$= 23.2%, $electr_{SI}^{19}$= 27.8%, $heat_{SI}^{20}$= 28.6%, $electr_{SI}^{21}$= 28.2%). This variable is characterised by low within-group variability (CV_s^t= 0.2), which is in line with the results for all EU countries (see Sect. 4).

In the case of variable natural gas, its average share in the energy mix in cluster 2 is similar to that of cluster 1 (gas_2^{19}= 30.4%, gas_2^{20}= 30.6% and gas_2^{21}= 32.1%). The increase in the average in 2021 was due to the inclusion of Germany, whose share of natural gas is at (gas_{DE}^{21}= 42.8%). Here we noticed a considerable range of this variable, from around 6% in Estonia to over 50% in Hungary and Italy and even to 71% in the Netherlands.

The third cluster comprises 9 countries (considering all years analyzed). It's important to note that 7 of these countries are either located in Southern Europe or encompass large southern regions (Bulgaria, Cyprus, France, Greece, Malta, Portugal, Spain). The remaining two countries are Finland and Sweden. This cluster

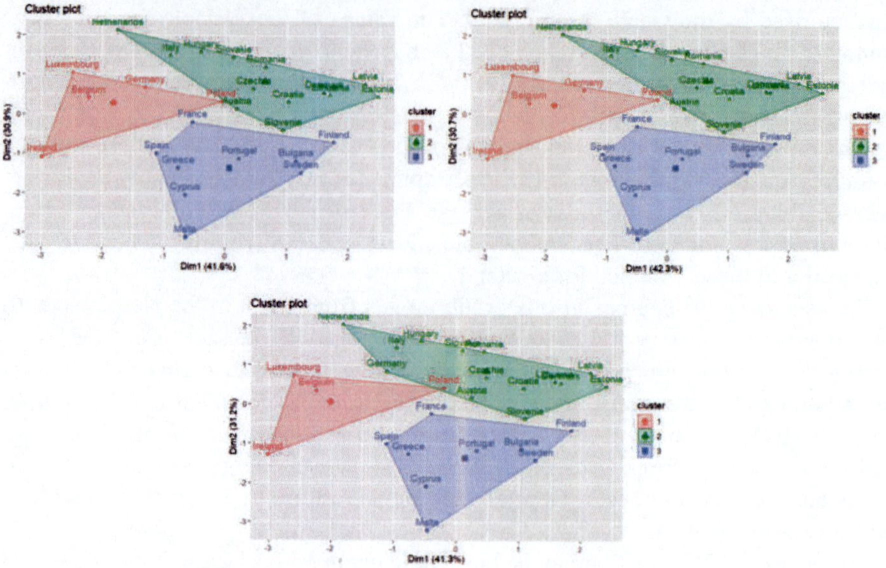

Fig. 4 Visualisation of the obtained divisions. The top left chart shows the division for the 2019 data, the top right chart for the 2020 data and the bottom chart for the 2021 data. *Source* Own elaboration based on (Eurostat, 2023a) data. *Note* Dim1 and Dim2 are artificial variables obtained after employing principal component analysis (PCA). PCA is a statistical procedure applied to reduce dimensionality to facilitate the visualisation of multivariate data. Dim1 and Dim2 represent the two major principal components, which are combinations of the original variables in the dataset (and are ordered in descending order of importance, see Aczel and Sounderpandian (2008)

particularly stands out due to its notably high average share of electricity, reaching 44%. This is more than twice as high as the other two clusters. Notably, two countries stood out within this group: Sweden, with a share exceeding 50%, and Malta, where the share of this energy source ranged from 70.6% in 2019 to 73.6% in 2021 (see Sect. 4).

The second defining characteristic of this group is the low average share of natural gas, registering at 8.5% in 2019 and 2020, and 8.8% in 2021. Natural gas is predominantly utilized by households in France (constituting 28% of the energy mix) and Spain (24%). In contrast, Malta and Cyprus do not use this source at all, while Sweden and Finland use it marginally (less than 0.5%).

The average share of RES is 24%. Within this category, households in Portugal utilize RES the most, accounting for just over 36%, while in Bulgaria and Finland, the share exceeds 30%. Although the differences in the share of this variable in the energy mix are relatively small among other countries, Bulgaria exhibits a notable fluctuation. In 2020, compared to 2019, the share increased by 2.7 pp (from 33.4 to 36.1%) before declining by 4.3 percentage points to 31.8% in 2021.

The average share of the most emission-intensive energy sources (variable fpp_k^t) was approximately 14.5% in 2019 and 2020 before decreasing to 13.4% in 2021. Additionally, a relatively small share is noted for the heat variable, averaging around 9%. Three countries (Cyprus, Malta, and Spain) do not utilize this source, and Portugal and Greece have a marginal share (less than 1%). However, in Sweden and Finland, the usage exceeds 33% and 27%, respectively. This disparity is due to the geographical location of these countries. They are the sole northern European countries categorized in this cluster, and because of their cooler climate, there is a much greater reliance on space heating sources there.

In summary, we note that there were no differences in the clusters identified for the 2019 and 2020 data (maintaining the same cluster compositions, with an adjusted rand index of 1). This indicates that despite some fluctuations in the individual energy source shares within household energy mixes at the country level, these changes were not substantial enough for the k-means algorithm to categorize these countries into different groups.

The only alteration observed in the clustering was for 2021, when Germany moved from cluster 1 to cluster 2. This change might be attributed to a noteworthy decrease in the share of emission-intensive sources (reduced by as much as 8 pp). In this instance, the Rand index for the 2020 and 2021 classifications was 0.9.

These results indicate that the structure of household FEC remained largely unchanged during the Covid-19 pandemic. Although there are observable shifts in the shares of individual sources across countries, these changes are relatively minor. They aren't substantial enough to significantly impact the composition of the resulting clusters. Therefore, the structure of the energy mix can be considered stable and less sensitive to variations in the economic and social environment.

These changes can be observed over an extended period, progressing gradually, and are significantly influenced by the energy policies adopted within the EU (as directed by EU structures). It's important to note an example of variables associated with the energy market that markedly differentiate EU countries, leading to substantial variations in the classifications obtained for individual years. A prime example is electricity prices, specifically their components (Matuszewska-Janica et al., 2023).

7 Conclusion

One of the primary objectives of EU energy policy is to markedly reduce the utilization of emissions-intensive energy sources, aiming ultimately to transition to a zero-carbon economy by 2050. To achieve this goal, several measures are being implemented, significantly impacting the structure of energy consumption across various sectors, including households, which represent the second-largest consumers of final energy.

As mentioned earlier, natural gas is a less emissions-intensive source compared to solid fossil fuels or oil and petroleum products. Therefore, at this stage of the energy transition, it is considered to be a part of the EU's strategy. However, a significant proportion of natural gas is imported. Therefore, considering energy security, its share is likely to decrease.

The analysis presented shows that natural gas currently represents the largest share of households' FEC at 33.5%. As mentioned earlier, natural gas is a less emission-intensive source compared to solid fossil fuels or oil and petroleum products. Therefore, at this stage of the energy transition, it can be considered a part of the EU's strategy. However, a significant portion of it is imported. Thus, with energy security in mind, its share is likely to decrease. The European Commission has outlined that by 2050, a minimum of 50% of energy used in the residential sector should be derived from electricity (ETIP Wind, 2021). Currently, electricity stands as the second-largest source of energy in the FEC, with a share of 24.6%. As of now, the target is significantly distant from being achieved. Encouraging households to adopt renewables and biofuels has long-term effects. Over time, there has been a gradual but consistent increase in the share of these sources in HFEC. Since 1990, this share has risen by 11.6 pp, reaching 21.2%.

It is worth noting that households account for over 50% of the total energy consumption from renewable energy sources (RES) in the EU, making them the primary direct consumers of this energy source.

To analyse the composition of household final energy consumption in each country, a classification process was carried out using the k-means technique. Finally, a division into three clusters was selected for all data sets. The results indicated that the divisions in 2019 and 2020 were the same. Although there were slight differences in the mean values of the analysed variables within the acquired clusters, the composition of the groups remained unchanged. This means that the pandemic period did not have a significant impact on the structure of household energy consumption. In the 2021 group, only one country (Germany) moved clusters, mainly due to a significant decrease in the use of emission-intensive sources.

In summary, changes in the structure of energy consumption are slow but follow a sustainable trend. They depend more on the energy policy pursued in the European Union than on temporary turbulence in the economic and social environment. At present, the main objective of this policy is to increase the share of renewable and emission-intensive resources in energy consumption, covering both direct (employing RES) and indirect consumption (including electricity generated from emission-intensive sources).

Appendix

See Table 6.

Table 6 SI values for individual divisions

Number of clusters	2019	2020	2021
2	0.553*	0.479	0.537*
3	0.542	0.527[a]	0.535
4	0.516	0.515	0.515
5	0.447	0.496	0.399
6	0.444	0.460	0.395
7	0.310	0.535	0.360
8	0.487	0.411	0.469
9	0.355	0.399	0.338
10	0.387	0.367	0.467
11	0.352	0.201	0.230
12	0.226	0.361	0.269

Source Own elaboration based on Eurostat data (Eurostat, 2023a)
Note [a] Highest SI value

References

Aczel, A.D., & Sounderpandian, J. (2008). *Complete business statistics* (7th ed.). McGraw–Hill.

Alphonse, A. (2022). *Package 'robustHD'*.

Biernat-Jarka, A., Trębska, P., & Jarka, S. (2021). The role of renewable energy sources in alleviating energy poverty in households in Poland. *Energies, 14*(10), 2957. https://doi.org/10.3390/en14102957

Bieszk-Stolorz, B., & Dmytrów, K. (2019). Spatial diversity of effectiveness of forms of professional activisation in Poland in years 2008–2014 by poviats. *Oeconomia Copernicana, 10*(1), 113–130. https://doi.org/10.24136/oc.2019.006

Botta-Dukát, Z. (2023). Quartile coefficient of variation is more robust than CV for traits calculated as a ratio. *Scientific Reports, 13*(1), 4671. https://doi.org/10.1038/s41598-023-31711-8

Bouzarovski, S. (2014). Energy poverty in the European Union: Landscapes of vulnerability. *Wires Energy and Environment, 3*(3), 276–289. https://doi.org/10.1002/wene.89

Chen, Y.-H., Li, Y.-Z., Jiang, H., & Huang, Z. (2023). Research on household energy demand patterns, data acquisition and influencing factors: A review. *Sustainable Cities and Society, 99*, 104916. https://doi.org/10.1016/j.scs.2023.104916

Dato, P. (2018). Investment in energy efficiency, adoption of renewable energy and household behavior: Evidence from OECD countries. *The Energy Journal, 39*(3). https://doi.org/10.5547/01956574.39.3.pdat

DESA UN. (2017). *International recommendations for energy statistics (IRES)*, 93.

Dudek, A. (2020). *Silhouette index as clustering evaluation tool* (pp. 19–33).

ECMWF. (2022). *Surface air temperature for February 2022, climate bulletin.*

EEA. (2023). *Primary and final energy consumption in Europe.* 8th EAP. European Environment Agency. Retrieved 18 September 2023 https://www.eea.europa.eu/ims/primary-and-final-energy-consumption

Energypedia. (2023). Energy poverty. *Energypedia.* Retrieved 17 August 2023 https://energypedia.info/wiki/Energy_Poverty

European Commission. (2015). *Communication from the commission to the European parliament, The council, The European economic and social committee, the committee of the regions and the European investment bank. A framework strategy for a resilient energy union with a forward-looking climate change policy.*

European Commission. (2019). *Communication from the commission to the European parliament, The European council, The council, The European economic and social committee and the committee of the regions the European green deal.*

European Commission. (2021). *EU reference scenario 2020. Energy, transport and GHG emissions—trends to 2050.*

European Commission. (2023a). *European green deal: EU agrees stronger legislation to accelerate the rollout of renewable energy.* Retrieved 18 September 2023 https://ec.europa.eu/commission/presscorner/detail/en/IP_23_2061

European Commission. (2023b). *Renewable energy directive.* Retrieved 18 September 2023 https://energy.ec.europa.eu/topics/renewable-energy/renewable-energy-directive-targets-and-rules/renewable-energy-directive_en

European Parliament. (2010). *Directive 2010/31/EU of the European parliament and of the Council of 19 May 2010 on the energy performance of buildings (recast).*

European Parliament. (2018a). *Directive (EU) 2018/844 of the European parliament and of the council of 30 May 2018 amending directive 2010/31/EU on the energy performance of buildings and Directive 2012/27/EU on energy efficiency (text with EEA relevance).*

European Parliament. (2018b). *Directive (EU) 2018/2001 of the European parliament and of the council of 11 December 2018 on the promotion of the use of energy from renewable sources (recast) (text with EEA relevance).*

European Parliament. (2018c). *Directive (EU) 2018/2002 of the European Parliament and of the council of 11 December 2018 amending directive 2012/27/EU on energy efficiency (text with EEA relevance).*

European Parliament. (2018d). *Regulation (EU) 2018/1999 of the European Parliament and of the Council of 11 December 2018 on the Governance of the Energy Union and Climate Action, Amending Regulations (EC) No 663/2009 and (EC) No 715/2009 of the European Parliament and of the Council, Directives 94/22/EC, 98/70/EC, 2009/31/EC, 2009/73/EC, 2010/31/EU, 2012/27/EU and 2013/30/EU of the European Parliament and of the Council, Council Directives 2009/119/EC and (EU) 2015/652 and Repealing Regulation (EU) No 525/2013.*

European Parliament. (2023a). *Directive (EU) 2023/1791 of the European parliament and of the council of 13 September 2023 on energy efficiency and amending regulation (EU) 2023/955 (recast) (text with EEA relevance).*

European Parliament. (2023b). *Parliament adopts new rules to boost energy savings.* Retrieved 13 September 2023 https://www.europarl.europa.eu/news/en/press-room/20230707IPR02421/parliament-adopts-new-rules-to-boost-energy-savings

European Union. (2023). *EU voluntary review on the implementation of the 2030 agenda for sustainable development.* https://doi.org/10.2792/343208

Eurostat. (2023a). *Complete energy balances [NRG_BAL_C].* Retrieved 2 September 2023. https://ec.europa.eu/eurostat/databrowser/view/nrg_bal_c/default/table?lang=en

Eurostat. (2023b). *Population unable to keep home adequately warm by poverty status.* Retrieved 11 September 2023 https://ec.europa.eu/eurostat/databrowser/view/SDG_07_60/default/table?lang=en

Farooqi, M. R., Rahman, A., Ahmad, M. F., & Supriya. (2021). Energy development as a driver of economic growth. In *Energy* (pp. 91–107). Wiley.

Giorgi, G. M., & Gigliarano, C. (2017). The Gini concentration index: A review of the inference literature. *Journal of Economic Surveys, 31*(4), 1130–1148. https://doi.org/10.1111/joes.12185

González-Eguino, M. (2015). Energy poverty: An overview. *Renewable and Sustainable Energy Reviews, 47*, 377–385. https://doi.org/10.1016/j.rser.2015.03.013

Han, J., Kamber, M., & Pei, J. (2012). *Data mining concepts and techniques* (3rd ed.). Morgan Kaufmann Publishers/Elsevier.

Hasheminasab, H., Streimikiene, D., & Pishahang, M. (2023). A novel energy poverty evaluation: Study of the European union countries. *Energy, 264*, 126157. https://doi.org/10.1016/j.energy.2022.126157

Holmes, A., Illowsky, B., & Dean, S. (2017). *Introductory business statistics.* OpenStax.

IEA. (2021a, April). *Economic impacts of Covid-19, global energy review 2021*. International Energy Agency. Retrieved 2 September 2023 https://www.iea.org/reports/global-energy-review-2021/economic-impacts-of-covid-19

IEA. (2021b). *Net zero by 2050. A roadmap for the global energy sector*. International Energy Agency.

Jacobson, A., Milman, A. D., & Kammen, D. M. (2005). Letting the (energy) Gini out of the bottle: Lorenz curves of cumulative electricity consumption and Gini coefficients as metrics of energy distribution and equity. *Energy Policy, 33*(14), 1825–1832. https://doi.org/10.1016/j.enpol.2004.02.017

Jajuga, K., Walesiak, M., & Bąk, A. (2003). On the general distance measure. In *Exploratory data analysis in empirical research* (pp. 104–109). Springer-Verlag.

Jedlińska, K., & Olkuski, T. (2019). The decision-making process in the EU in the field of energy policy. *Polityka Energetyczna–Energy Policy Journal, 22*(1), 5–18. https://doi.org/10.33223/epj/103122

Kassambara, A., & Mundt, F. (2022). *Package 'Factoextra'*.

Kaufman, L., & Rousseeuw, P. J. (1990). *Finding groups in data: An introduction to cluster analysis*. Wiley.

Matuszewska-Janica, A., Żebrowska-Suchodolska, D., Ala-Karvia, U., & Hozer-Koćmiel, M. (2021). Changes in electricity production from renewable energy sources in the European union countries in 2005–2019. *Energies, 14*(19), 6276. https://doi.org/10.3390/en14196276

Matuszewska-Janica, A., Żebrowska-Suchodolska, D., Zalewska, M. E., Ala-Karvia, U., & Hozer-Koćmiel, M. (2023). Differences in the structure of household electricity prices in EU countries. *Energies, 16*(18), 6636. https://doi.org/10.3390/en16186636

McQueen, J. B. (1967). Some methods of classification and analysis of multivariate observations. In L. M. Le Cam & J. Neyman (Eds.) *Proceedings of the Fifth Berkeley Symposium on Mathematical Statistics and Probability*, (Vol. 1, pp. 281–297).

Odum, H. T. (2007). *Environment, power, and society for the twenty-first century*. Columbia University Press.

Pais-Magalhães, V., Moutinho, V., & Robaina, M. (2022). Is an ageing population impacting energy use in the European Union? drivers, lifestyles, and consumption patterns of elderly households. *Energy Research & Social Science, 85*, 102443. https://doi.org/10.1016/j.erss.2021.102443

Pereira, D. S., & Marques, A. C. (2023). How do energy forms impact energy poverty? An analysis of European degrees of urbanisation. *Energy Policy, 173*, 113346. https://doi.org/10.1016/j.enpol.2022.113346

Rita, E., Chizoo, E., & Cyril, U. S. (2021). Sustaining COVID-19 pandemic lockdown era air pollution impact through utilization of more renewable energy resources. *Heliyon, 7*(7), e07455. https://doi.org/10.1016/j.heliyon.2021.e07455

Rokicki, T., Jadczak, R., Kucharski, A., Bórawski, P., Bełdycka-Bórawska, A., Szeberényi, A., & Perkowska, A. (2022). Changes in energy consumption and energy intensity in EU countries as a result of the COVID-19 pandemic by sector and area economy. *Energies, 15*(17), 6243. https://doi.org/10.3390/en15176243

Roszko-Wójtowicz, E., & Grzelak, M. M. (2021). Multi-dimensional analysis of regional investment attractiveness in Poland. *Equilibrium. Quarterly Journal of Economics and Economic Policy, 16*(1), 103–138. https://doi.org/10.24136/eq.2021.004

Simionescu, M., Radulescu, M., Cifuentes-Faura, J., & Balsalobre-Lorente, D. (2023). The role of renewable energy policies in tackling energy poverty in the European union. *Energy Policy, 183*, 113826. https://doi.org/10.1016/j.enpol.2023.113826

Tutak, M., & Brodny, J. (2022). Renewable energy consumption in economic sectors in the EU-27. The impact on economics, environment and conventional energy sources. A 20-year perspective. *Journal of Cleaner Production, 345*, 131076. https://doi.org/10.1016/j.jclepro.2022.131076

United Nations. (2015). *Transforming our world: The 2030 agenda for sustainable development*. Resolution Adopted by the General Assembly on 25 September 2015. https://sdgs.un.org/2030agenda

United Nations. (2018). *Policy brief #27 indicators and data for energy for sustainable development.*

Walesiak, M. (2006). Recommendations for strategies to deal with the classification process of a set of objects. In A. Zeliaś & A. E. Krakow (Eds.), *Spatial-temporal modelling and forecasting of economic phenomena* (pp. 185–203).

Walesiak, M. (2016). Wizualizacja Wyników Porządkowania Liniowego Dla Danych Metrycznych z Wykorzystaniem Skalowania Wielowymiarowego [Visualization of linear ordering results for metric data with the application of multidimensional scaling]. *Ekonometria,* 2. https://doi.org/10.15611/ekt.2016.2.01

Walesiak, M., & Dudek, A. (2022). *ClusterSim package.*

Walesiak, M. (2018). The choice of normalization method and rankings of the set of objects based on composite indicator values. *Statistics in Transition New Series, 19*(4), 693–710. https://doi.org/10.21307/stattrans-2018-036

Wang, Y., & Lin, B. (2022). Can energy poverty be alleviated by targeting the low income? Constructing a multidimensional energy poverty index in China. *Applied Energy, 321*, 119374. https://doi.org/10.1016/j.apenergy.2022.119374

ETIP Wind. (2021). *Getting fit for 55 and set for 2050 electrifying Europe with wind energy, European technology and innovation platform on wind energy.*

Yadav, J., & Sharma, M. (2013). A review of k-mean algorithm. *International Journal of Engineering Trends and Technology, 4*(7), 2972–2976.

Dr Aleksandra Matuszewska-Janica graduated with a master's degree in Computer Science and Econometrics from the Faculty of Economics and Sociology at the University of Lodz. She received her Ph.D. degree in Economic Sciences from the Nicolaus Copernicus University in Torun. In 2007 she joined the Warsaw University of Life Sciences (WULS) where she currently works in the Department of Econometrics and Statistics, part of the Institute of Economics and Finance. She is also the coordinator of the English-language studies at the WULS, majoring in Informatics and Econometrics with a specialization in Big Data Analytics.

Her current main research activities are related to two fields. The first concerns labour market. Her scientific interests include analyses of regional differences among European labour markets and changes over time in the labour market. Current studies under way are focused on women's labour market activity and women's compensation for contract work. The second research area focuses on the situation of households in the energy market.

She has authored or co-authored more than 100 scientific presentations at conferences and seminars and more than 60 peer-reviewed papers and two monographs. She has been involved in numerous research projects. She has completed two international research internships in recent years. The first internship was at the Ruralia Institute at the University of Helsinki with the topic *Economic determinants of women's labour market participation decisions in selected European Union countries. An econometric analysis.* During her second internship at the University of Castilla la Mancha in Toledo, she worked on the project's theme *Statistical and econometric methods for spatial analysis of the labour market data.*

Application of New Technologies to Sustainability

Integrating Blockchain Technology in Supply Chain Risk Management for Sustainable Development

Fahim ul Amin, Qingkai Ji, Wasim ul Amin, and Azka Amin

1 Introduction

The numerous industries have embraced blockchain technology, frequently hailed as a revolutionary force. It offers solutions that guarantee efficiency, security, and transparency. The area of supply chain risk management is where one of its most significant effects can be seen. Modern supply chains, characterized by their extensive global reach and complex interdependencies, necessitate innovations that deal with various problems, from ensuring that products are ethically sourced to real-time tracking. This book delves deeply into the intersection of supply chain management and Blockchain, critically analyzing its prospects, practical applications, and stakeholder viewpoints (Chang et al., 2020). We seek to understand the transformative potential of Blockchain, its practical advantages, the roadblocks in its way, and the broader implications for businesses and society through a series of structured explorations. In the following sections, readers will be guided through in-depth case studies, perspectives from business executives and sustainability advocates, and forward-looking analyses of changing trends and impending difficulties. As we set out on this analytical journey, it is crucial to realize that adopting Blockchain in supply chains represents more than just a technological development; it also represents a paradigm shift in how global trade functions, handle risks and looks to the future. Blockchain's appeal goes beyond its technological prowess (Esmaeilian et al., 2020). At its core, it stands for a wider movement towards openness, moral behaviour, and

F. ul Amin (✉) · Q. Ji
International Business School, Hainan University, Haikou 570228, Hainan, China
e-mail: fahimulamin@gmail.com

W. ul Amin
School of Economics and Management, North China Electric Power University, Beijing, China

A. Amin
International Business School, Hainan University, Haikou 570228, Hainan, China

M. d. C. Valls Martínez and J. M. Santos-Jaén (eds.), *Environmentally Sustainable Production*, https://doi.org/10.1007/978-3-031-52656-5_11

increased accountability in international business. The leaders, academics, and innovators have all been drawn to its transformative potential, which has sparked discussions, research, and pilot projects in various industries. The supply chains become more complicated, businesses and consumers seek assurance and clarity regarding ethical standards and product origins. The rising demand, the nexus of supply chain management and Blockchain is even more crucial. Our thorough investigation aims to fill in knowledge gaps by understanding how Blockchain is changing the fundamental frameworks of commerce (Zaoui & Souissi, 2020). By doing this, we hope to give organizations, decision-makers, and enthusiasts a road map for navigating this quickly changing environment.

1.1 Triple Bottom Line

The Triple Bottom Line (TBL), which challenges conventional corporate success measures that frequently only consider profit, emerged as a revolutionary framework for sustainable business development. The TBL necessitates a broader viewpoint on how businesses operate because it is based on three fundamental dimensions: economic, social, and environmental (Ertz & Boily, 2019). While profit is unquestionably important from an economic standpoint, the TBL philosophy encourages companies to consider their broader economic impacts, such as local job creation and community investment. The critics have questioned the depth of this economic perspective, pointing out that some businesses may publicize the creation of local jobs without ensuring that those jobs pay a living wage. TBL emphasizes the value of ethical business practises, employee welfare, and community involvement from a social perspective, arguing that a company's core business activities should have a positive social impact (Dwivedi & Paul, 2022). The distinction between sincere social responsibility and PR campaigns that boost brand recognition is becoming hazier. Although beneficial, many corporate social responsibility (CSR) campaigns are frequently seen as superficial brand-building initiatives rather than initiatives for real societal change. TBL challenges companies to examine their ecological footprint, including resource consumption, waste management, and general environmental stewardship. However, this environmental commitment is frequently questioned, with "greenwashing" claims that some businesses deceptively highlight small green initiatives while obscuring their larger environmental transgressions. Stakeholders in the modern business environment, from investors to customers, are keenly aware of a company's overall performance.

The market's reputation and ethical consumption trends are inextricably linked to TBL observance. The core value of TBL is its capacity to operationalize the more general objectives of sustainable development, guaranteeing business success without adversely affecting society or the environment. The "sustainability" is still a moving target with many possible meanings (Keesara et al., 2020). While TBL offers a basic framework, its genuine, thorough integration into corporate cultures represents the real challenge. Without sincere commitment and open methodologies, TBL risks losing its transformative power and becoming another business fad. Expanding on the Triple Bottom Line's (TBL) complexities reveals new nuances and complexities, especially in our increasingly interconnected global business ecosystem. For instance, as supply chains become more interconnected globally, stakeholders on different continents can feel the effects of a company's decisions. TBL must therefore consider a company's global impact and localized effects (Haq & Ali, 2018). The impact of the dimension on the economy goes beyond creating new jobs right away and touches on global economic equity. The TBL promotes economic wellness; critics contend it fails to address the complex issues associated with global economic disparities. Not only should equitable wealth distribution be the aim, but also wealth creation. The social pillar of TBL explores the complex web of cultural respect and understanding, particularly for businesses operating across diverse regions, while highlighting community engagement and ethical practices. More is needed for businesses to adhere to ethical labour practises as defined by Western standards; they must also be aware of and respectful of regional cultural differences. Failure to do so may result in unintentional sociocultural sensitivities, undermining the social pillar's fundamental tenet (Fig. 1).

The developing digital platforms and social media have increased corporate accountability. The amplified effects of the interconnected digital age can be seen in how quickly a small social responsibility slip can turn into a global controversy (To, 2020). The conversation about the environment has expanded beyond carbon footprints and waste disposal. The climate change picks up speed; businesses are increasingly being questioned about their contributions to global ecological crises,

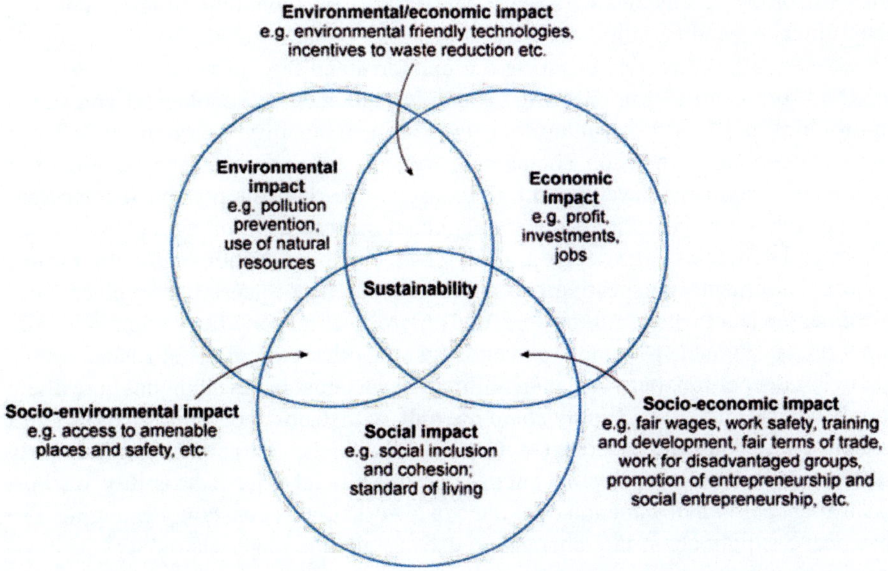

Fig. 1 The three impact components of sustainability and their interconnections

such as water scarcity and biodiversity loss. The environmental pillar of TBL requires proactive actions rather than merely reactive ones. Critics stress that businesses must change from merely reducing harm to actively aiding in ecological restoration to be truly sustainable. Technology has a significant impact on how TBL will develop in the future. The potential for monitoring, reporting, and optimizing TBL components has increased dramatically with the development of technologies like artificial intelligence, Blockchain, and big data analytics (Kyriazis, 2020). The challenge of data privacy and the moral application of technology comes along with this potential and broadens the scope of TBL. Although it is a fundamental concept, the Triple Bottom Line is a dynamic framework that adapts to the changing socioeconomic and environmental conditions worldwide. It is a continuous adaptation, reflection, and improvement process rather than a static benchmark. The TBL offers more than just a direction as businesses face unprecedented global challenges; it also acts as a compass, directing businesses towards overall prosperity. The realizing its transformative potential requires a deep-seated commitment, not just a superficial allegiance.

1.2 Supply Chain Risk Management: A Necessity for Sustainable Development

The Supply Chain Risk Management (SCRM) is a complex procedure focusing on locating, evaluating, and minimizing potential supply chain disruptions. The SCRM aims to fortify supply chains against vulnerabilities, including supplier failures, geopolitical unpredictability, and natural disasters (Bhaskar et al., 2021). Its significance goes beyond merely ensuring the uninterrupted flow of goods and services. SCRM's role is crucial in ensuring economic resilience and ecological and social equilibrium in a world that is turning towards sustainability. The delicate balance between economic expansion, social inclusion, and environmental preservation is at the core of sustainable development. In this nexus, resilient supply chains are essential. The recurrent disruptions in a supply chain can cause production to stop, costs to rise, and deliveries to be delayed, all of which endanger economic growth. From a socio-environmental perspective, companies that purchase materials from places with questionable labour standards or degraded environmental conditions unintentionally support unsustainable behaviour and endanger the other two pillars of sustainability. There is a deeper link between adaptability and sustainable development in resilient supply chains. A resilient supply chain not only withstands shocks but also responds to them by evolving and adapting (McCarthy et al., 2022). This adaptability is directly related to sustainable development at its core. It is crucial to have the ability to adjust to shifting socio-environmental dynamics in a world that is constantly changing. For instance, a supply chain that changes to use sustainable materials reduces its environmental impact while protecting itself from potential regulatory restrictions on non-sustainable resources. However, the problems with conventional SCRM models can occasionally make it difficult to achieve sustainable development. In the past, SCRM prioritized cost-effectiveness. The procurement process frequently favours the least expensive options, which, while they may be economical in the short term, come with many risks. These choices occasionally drew their supply from nations with unstable geopolitical conditions or suppliers with questionable labour practises. Even though they are economically alluring, these cost-driven models frequently put sustainability at risk and are susceptible to disruptions.

The reactivity of traditional SCRM is a significant criticism as well. Conventional models frequently have response structures rather than preventive mitigation strategies. In the long run, this defensive stance is more expensive and disruptive (Yaqoob et al., 2021). Additionally, the lack of transparency in traditional supply chains makes tracking the origin of goods challenging, masking unsustainable practices and challenging them to change. Traditional SCRM frequently works in isolation, integrating other business functions infrequently. This siloed strategy overlooks the comprehensive viewpoint required for sustainability. The SCRM might overlook important environmental risks or fail to take advantage of sustainable innovations without integrating with sustainability teams. Although SCRM is crucial in sustainable development, traditional approaches have drawbacks (Pettit et al., 2019). The SCRM model needs to be rethought for the future, with value-driven resilience taking precedence over merely cost-efficiency. It demands proactive, integrated, and open SCRM models that support the more general objectives of sustainable development and economic imperatives. Supply Chain Risk Management is essential for sustainable development and not just from a logistical standpoint. The paradigm shift is necessary from a constrained, cost-centric perspective to an expansive, sustainability-driven one to fully realize it's potential (Fig. 2).

The SCRM be able to effectively connect business demands today with sustainability in the future. Further extending the evolving SCRM and sustainability landscape, it is critical to comprehend the macro-environmental pressures on supply chains. On the one hand, globalization has made accessing a wide range of suppliers and affordable solutions easier. Still, on the other hand, it has made supply chains more susceptible to disruptions from distant regions of the world. Because a geopolitical crisis in one place can impact multiple continents, SCRM must adopt a global perspective (Raimundo & Rosário, 2021). Data analytics and the development of digital technologies present both opportunities and difficulties. While technology can enable real-time monitoring and provide insights into supply chain vulnerabilities, it also creates new risk vectors like cyber threats and data breaches. Because of this, modern SCRM must include cybersecurity safeguards to prevent digital supply chain integrations from becoming a weakness. One must recognize the rising consumer awareness of sustainability. Today's Consumers are informed and frequently prefer brands that share their moral and environmental principles. It is crucial for SCRM to proactively integrate with sustainability goals because any irregularities or disruptions linked to unsustainable practices can harm the brand's reputation. Considering these details, it becomes clear that SCRM's role goes beyond operational effectiveness (Murray, 2018). Its domain now includes geopolitical awareness, technological

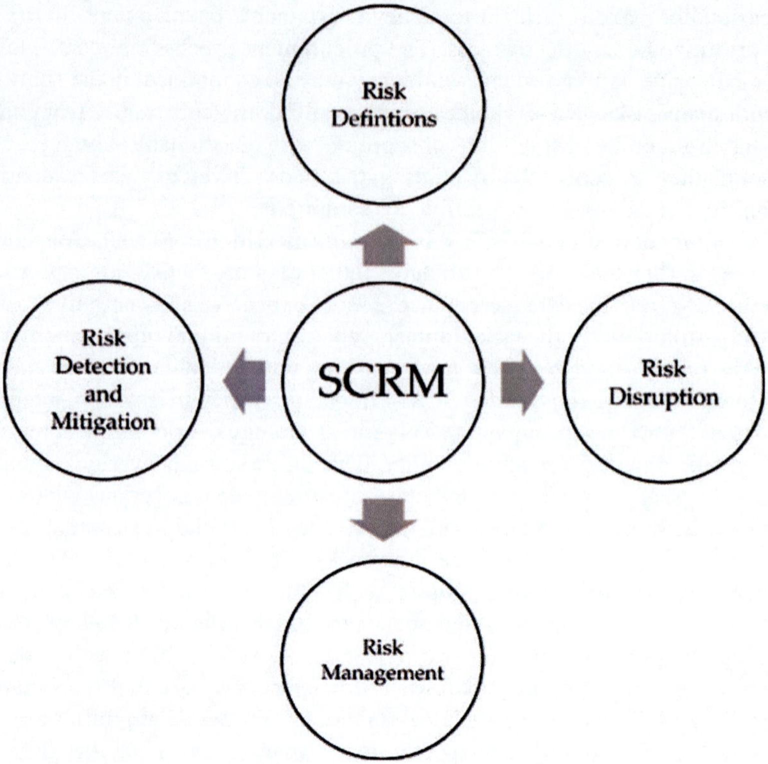

Fig. 2 SCRM (https://www.mdpi.com/2227-9091/9/1/16)

foresight, and a deep-seated alignment with shifting societal values, reinforcing its crucial role in sustainable development in our age of global connectivity.

2 Blockchain Technology

See Fig. 3.

2.1 Origins and Principles of Blockchain Technology

The Blockchain has its roots in cryptocurrencies and is frequently hailed as a revolutionary advancement in digital ledger technology. Its fundamental idea was to develop a decentralized system where transactions could be verified without the need for central authorities like banks, according to Satoshi Nakamoto's seminal

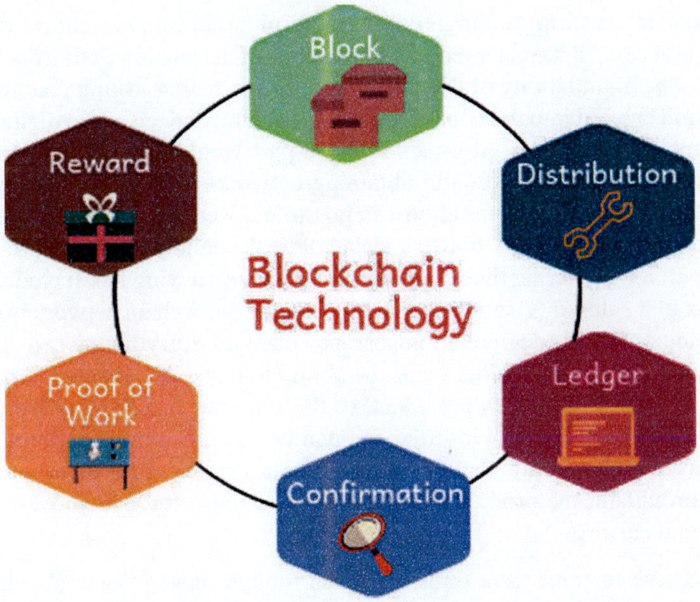

Fig. 3 Blockchain Technology (https://www.tekshapers.com/blog/Introduction-to-Blockchain-Technology-:-A-New-Technology)

2008 Bitcoin whitepaper. The due to its underlying mechanics, Blockchain's potential resonated in several sectors outside digital currencies (Roth et al., 2022). A blockchain is fundamentally made up of linked data blocks. After validation, each block is added chronologically, forming a nearly impossible chain to break without agreement.

Critique: Although the decentralized nature of Blockchain presents a challenge to conventional centralized systems and offers a fresh approach to data storage and verification, it also raises questions. Decentralization can result in inefficiencies and scalability problems, which may only be suitable for some applications even though it reduces single points of failure.

2.2 *Key Features Beneficial to Supply Chains*

There are many benefits to using blockchain technology, but some are particularly revolutionary for supply chains (Cai et al., 2021). The traditional supply chain databases frequently operate as potential bottlenecks or points of failure because a single party typically controls them. However, with its decentralized approach, Blockchain ushers in a paradigm shift. The control is distributed across a huge network of nodes, ensuring no single entity has absolute control. In addition to democratizing data access, this greatly reduces the dangers of data manipulation or unilateral tampering. The immutability of Blockchain is another distinguishing feature. A data or transaction entered into the Blockchain is permanently added to the digital ledger. Any attempt to change this information in the past would call for changing every block after that and, more difficult, obtaining network-wide consensus (Sylvester, 2019). Data tampering becomes almost impossible, given the computational weight that would be required. This bolsters stakeholders' confidence in the accuracy of the information by assuring them that the data they are viewing has stayed the same since it was first entered. Viewing transactions on the blockchain network is possible for all participants. This degree of transparency changes everything when applied to supply chains. Every stakeholder gains the ability to track a product's journey, from the initial producers, logistics providers, to the ultimate end users (Alketbi et al., 2018). As a result, they can close the frequently enormous information gaps that plague conventional supply chains and promote a greater sense of accountability and trust throughout the product lifecycle. They can also track its movements and authenticate its claims.

Critique: These features have trade-offs even though they are undoubtedly transformative. Blockchain's transparency could be a double-edged sword, exposing potentially sensitive corporate data. The ensuring data integrity, immutability also means that incorrect entries are permanently logged, calling for additional layers of verification.

2.3 Successful Case Studies of Blockchain in Other Sectors

The Blockchain technology has the potential to completely transform many industries, and some notable successes have been seen outside of supply chains. Cryptocurrencies were the Blockchain's first and most well-known use in finance. Even though Bitcoin continues to serve as the model for decentralized digital currencies, the financial industry is still exploring other uses for blockchain technology (Caputo et al., 2019). Blockchain has been used by financial institutions worldwide to reinvent cross-border payments. These transactions used to take days to settle because they had to go through several intermediaries. Blockchain significantly decreases this latency, with transaction times falling to just minutes, increasing efficiency and cutting costs. In the healthcare industry, the potential of Blockchain has been utilized for the vitally important task of storing patient data. The Sensitive patient data is abundant in the healthcare sector and must be accessible and secure. The Blockchain offers a solution that enables seamless data sharing between healthcare providers. This decentralized ledger system facilitates controlled access while guaranteeing the data's accuracy. Thus, timely and accurate access to patient histories can optimize care delivery while protecting patient privacy for doctors and other medical professionals (Tao et al., 2022). The real estate industry, historically associated with bureaucracy and paperwork, has also been impacted by Blockchain. Property transactions, particularly title transfers, can be complicated, prone to delays, and fraught with fraud risks. Blockchain offers a streamlined approach, bringing speed and transparency to real estate transactions. Property titles are recorded on a blockchain, reducing the risk of erroneous claims. The transparency of Blockchain speeds up procedures and significantly reduces the red tape that frequently complicates real estate transactions. The fundamental principles of blockchain decentralization, transparency, and immutability are highlighted in these industries, highlighting its adaptability and potential to transform established operational paradigms.

Critique: Although promising blockchain applications have been in these fields, it is important to remember that not all have been implemented successfully (Carson et al., 2018). The advantages have occasionally been overshadowed by scalability issues, energy consumption (especially in proof-of-work systems), and regulatory ambiguities. The Blockchain is still in its infancy; many alleged use cases are experimental and need time and iterations to mature.

3 Advantages of Integrating Blockchain in Supply Chain Risk Management

3.1 Traceability

The adoption of blockchain technology in modern supply chain management has emphasized the need for traceability, bringing together operational effectiveness and consumer-driven ethics. Blockchain's ability to enable real-time tracking is advantageous and essential in an age defined by instantaneous responses and a pervasive desire for immediacy. The Blockchain permanently records every transactional shift and logistical movement, giving stakeholders a previously unheard-of ability to track the movement of goods (Chaudhary, 2019). This is more than just a test of logistical prowess; it also acts as a safety net against unanticipated disruptions, ensuring the supply chain is responsive and resilient. The threat of fake goods and worries about questionable sourcing methods loom large in global trade. The meticulous documentation of Blockchain provides a powerful remedy. The technology significantly reduces the entry of counterfeit goods into the supply chain by meticulously validating and documenting each product's provenance and subsequent journey. The capacity to guarantee and demonstrate ethically sourced products becomes a commercial necessity as consumer paradigms shift towards a more conscious and ethically attuned purchasing ethos (Angrish et al., 2018). In this setting, blockchain traceability transforms into a strategic asset that helps businesses meet changing customer demands. Extolling the virtues of Blockchain while ignoring its complexities and difficulties would be negligent. The intersection of blockchain and other emerging technologies, such as the Internet of Things (IoT) and Artificial Intelligence (AI), presents a significant challenge. Although Blockchain is a bulwark of data integrity, the fineness and accuracy of this data are frequently dependent on the successful integration of these supporting technologies. This leads to a related worry about the confidentiality of data. The adage "inaccuracies once inscribed, forever remain" becomes especially relevant given the immutable nature of blockchain technology (Aggarwal et al., 2019). Any deviation or error is irreversible once it has been ingrained, highlighting the critical necessity of stringent data validation procedures to guarantee that the traceability it offers is not only transparent but also flawlessly accurate.

3.2 Transparency

The introduction of blockchain technology has given the idea of transparency new life in contemporary business dynamics, weaving a story that balances accountability and tactical discretion. Blockchain's transparency produces a complex range of benefits that call for careful consideration in this period, marked by changing consumer expectations and increased scrutiny. The public and permissioned ledger options that Blockchain offers demonstrate how flexible its transparency is. This adaptable strategy recognizes that transparency is only sometimes required in all circumstances (Wang & Su, 2020). The public ledgers encourage an environment of unrestricted disclosure because of their open accessibility, making it possible for anyone to examine supply chain procedures. Permissioned ledgers, on the other hand, extend controlled transparency while limiting access to only the right stakeholders. This customized transparency responds pragmatically to different business needs, acknowledging that while some supply chain elements might call for universal visibility, others might require a more specialized viewpoint. Transparent ledgers on the Blockchain have become effective resources for building trust in a time when brand integrity and transparency are paramount.

An unambiguous display of supply chain practices is necessary because modern consumers tend to support transparent brands driven by ethics. The businesses can promote their ethical and effective practises in an environment supported by Blockchain's indelible record-keeping, enabling them to back up their claims with verifiable data (Hartley et al., 2022). Investors who seek to align with organizations that uphold values of credibility and integrity are deeply moved by this newfound transparency, which naturally elevates trust among consumers. The adoption of complete transparency necessitates caution, just like any other innovation. The eagerness to project an unwaveringly transparent front could unintentionally reveal confidential pricing information, proprietary methodologies, or delicate business strategies. Such information could risk the competitive advantage businesses work so hard to protect. A strategic tightrope walk is required to strike the right balance between openness and preserving competitive differentiators. The introducing Blockchain into the transparency space creates a complex interplay between strategic safety and accountability. The dual facets of public and permissioned ledgers aptly capture the nuanced approach necessary in the modern business landscape (Zhu et al., 2022). While crucial to building trust and maintaining authenticity, transparency must be used carefully because long-term success depends on a delicate balance between openness and strategic protection.

3.3 Resilience

A new era of resilience is ushered in with the introduction of blockchain technology into the complex world of supply chains, strengthened by decentralized architecture and cryptographic integrity. The Blockchain's characteristics offer a variety of benefits in a world dotted with flaws and disruptions, but they also bring up subtle issues that demand careful thought. The Blockchain's ability to reduce single points of failure is one of its most effective characteristics for boosting supply chain resilience (Schlecht et al., 2021). The dangerous clutch of centralized databases or crucial logistical nodes frequently traps traditional supply chains. As a result, the system becomes fragile and vulnerable to catastrophic disruptions. The decentralized nature of Blockchain, control and data are dispersed across a network of nodes, creating a complex web of redundancy. This complex interaction prevents a disturbance at one node from starting a catastrophic domino effect, giving rise to a newfound agility essential for navigating contemporary commerce's ups and downs. The Blockchain architecture's cryptographic underpinnings usher in a heightened security mantle that fortifies supply chains against the growing threat of data breaches or manipulations. The immutability of blockchains implies that once data is entered, it cannot be changed without authorization. The consensus-based data alteration mechanism creates an impassable barrier to unauthorized access. This high level of security protects the confidentiality of sensitive data and fosters trust among stakeholders, creating the ideal environment for resilient supply chains (Capece et al., 2020). Like any transformative innovation, Blockchain has its share of complexities that must be worked out in order for it to be resilient. Decentralization strengthens against single points of failure but also creates complexity, especially in complex, large-scale supply chains with diverse stakeholders. Decentralized node orchestration requires careful coordination and is frequently fraught with difficulties that require skilled management.

The blockchain security is commendable, it is important to recognize that it is not impervious. Unknown blockchain variations might contain hidden flaws that could be used against them. Therefore, carefully evaluating the selected blockchain version becomes essential to guaranteeing its enhanced resilience (Voinea, 2019). The introduction of Blockchain into the web of supply chains heralds a paradigm shift in favour of strengthened resilience. It ushers in an era where disruptions are met with agility and sensitive information is shielded with unprecedented robustness thanks to the decentralization it promotes and the cryptographic security it enfolds. This heightened resilience, though, is not without its complexities and limitations. The

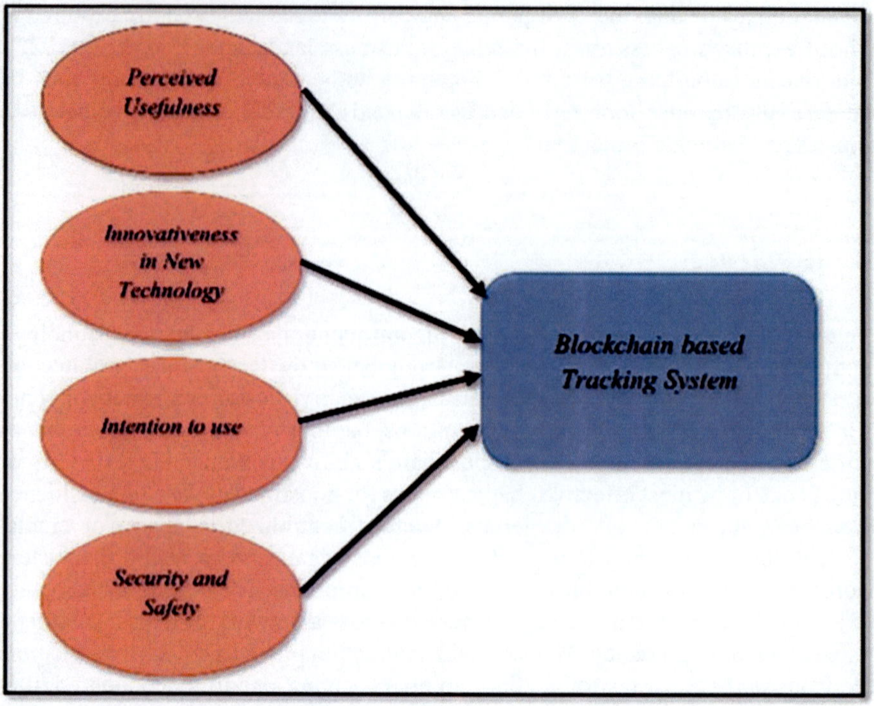

Fig. 4 Blockchain based tracking system (https://www.mdpi.com/2071-1050/15/8/6519)

hallmark of shrewd implementation emerges as Blockchain navigates the constantly changing landscape of supply chain dynamics. This is accomplished by carefully balancing its benefits with an astute assessment of its drawbacks (Fig. 4).

3.4 Efficiency

It has demanded a nuanced assessment of its implications and challenges. Still, the adoption of blockchain technology into supply chain operations has ushered in an era of improved efficiency characterized by streamlined transactions and cost reductions (Danial, 2023). The ability of Blockchain to speed up transactions and reduce the red tape that frequently characterizes traditional processes is a key component of its efficiency. The Blockchain's architecture enables nearly instantaneous transactions where time is of the essence, doing away with the complicated paperwork and protracted validation times that characterized earlier methodologies. This effectiveness is further enhanced by introducing smart contracts, which can automate and orchestrate processes based on predetermined criteria. By combining the efficiency of smart contracts with the speed of Blockchain, traditional slowness is transformed into an accelerated pace, improving resource use and response times (Wright, 2021). The ability of Blockchain to eliminate the need for intermediaries redefines the financial landscape of supply chains. Because intermediaries are required for validation in traditional supply chains, each layer raises costs and complexity.

Blockchain, built on decentralization and consensus, naturally gets around the need for many of these intermediaries. This seismic shift reduces overhead costs, leading to tangible cost savings. The result is a supply chain ecosystem that is leaner, more agile and allocates resources more wisely, giving it a competitive operational edge (Kiu et al., 2022). Smart considerations go along with the blockchain integration's commendable efficiency gains. The initial integration of Blockchain can be capital-intensive both financially and in terms of time commitments. The upfront investment in infrastructure, technology, and training demands a long-term perspective that weighs the efficiency dividends received compared to the initial investment. The elimination of human involvement does not follow from the elimination of intermediaries. As new positions, like blockchain managers, auditors, or technicians, become necessary, there must be a complex rebalancing of how human resources are allocated. In order to achieve sustainable efficiency, it becomes strategically imperative to balance these new demands with current knowledge (Dobrovnik et al., 2018). The introduction of Blockchain into supply chain operations has revealed a new level of operational efficiency that boosts transactional speed and cost savings. However, as with any truly transformative paradigm shift, moderation is key. Careful examination of the initial costs and emerging human resource needs is necessary to ensure that long-term efficiency gains truly outweigh the initial investments. The dynamic realignment of resources and expertise demonstrates the multifaceted nature of the efficiency transformation brought about by Blockchain, emphasizing that its integration is an evolution rather than a revolution.

4 Implementing Blockchain for a Sustainable Supply Chain

The Blockchain technology can usher in a new era of sustainability, transparency, and efficiency when integrated into supply chains. There are many obstacles, complexities, and strategic considerations from conceptualization to successful implementation (Peterson, 2018). Businesses must thoroughly understand the procedures, factors to consider, and difficulties if they want to fully realize the transformative potential of Blockchain in creating a sustainable supply chain.

4.1 *Steps for Businesses to Integrate Blockchain in Their Supply Chains*

The journey starts with identifying the most pertinent and significant use cases for Blockchain within the supply chain (Vatiero, 2018). Understanding pain points, weaknesses, and areas where traceability, transparency, or efficiency could be improved by blockchain technology are necessary for this.

Design a Blockchain Network: The architecture of the blockchain network must be carefully planned once the use cases are established. The consensus mechanism, required level of transparency, and blockchain type (public or private) must all be decided.

Choose the Right Blockchain Platform: The abundance of blockchain platforms available necessitates a thorough assessment. This choice should be made based on scalability, security features, and integration potential.

Data Collection and Integration: It is essential to collect pertinent data from every phase of the supply chain. Standardization protocols are needed when integrating data from various sources to ensure seamless interoperability.

Develop Smart Contracts: These self-executing contracts automate procedures and make sure that predetermined rules are followed. Smart contracts have the potential to increase transparency while streamlining supply chain operations.

Pilot Implementation: Performing a pilot implementation is essential before scaling up. This stage provides information on real-world difficulties, the efficiency of the selected architecture, and potential areas for improvement.

Scale Up Gradually: Gradual scaling after the pilot phase enables businesses to fine-tune the implementation, address unforeseen issues, and evaluate the system's scalability and performance under varying loads (Fig. 5).

4.2 *Navigating Scalability, Interoperability, and Data Privacy in Blockchain Integration*

Three factors—scalability, interoperability, and data privacy—play a crucial role in determining the success and sustainability of the endeavour when integrating blockchain technology into supply chains (Capece et al., 2020). As supply chains expand in scope, scalability becomes an urgent concern. The blockchain network must seamlessly handle the increased load without compromising performance as business operations grow and transaction volumes soar. A strong blockchain framework will be able to support expanding demands while maintaining rapid transaction processing. This calls for carefully assessing blockchain platforms that can scale

Fig. 5 Significance of blockchain and IoT integration in logistics

without sacrificing efficiency. In order to avoid potential bottlenecks and guarantee seamless continuity,

It is essential to ensure that the technology is in line with the changing needs of the supply chain. The foundation for harmonious integration is interoperability. Modern supply chains are made up of a variety of emerging and ageing technologies. In this situation, it is a strategic necessity for the blockchain network to communicate seamlessly with existing systems, suppliers, and partners (Madine et al., 2021). The switch to Blockchain needs to be orchestrated like a symphony, with the technology blending in with the existing orchestration. It is crucial to strike a balance between the novel blockchain possibilities and the existing infrastructure to prevent disruptions, hiccups, or inefficiencies. When delicate supply chain information is involved, the importance of data privacy is highlighted. It is crucial to protect private information from unauthorized access. Blockchain's immutability guarantees data integrity, but data privacy requires extra security measures. Strong encryption methods and precise access controls keep sensitive data secure from prying eyes (Michael et al., 2018). The keys to unlock particular data elements should only be in the hands of authorized entities. This dual layer of security, which provides privacy through encryption and integrity through Blockchain, fortifies trust throughout the supply chain ecosystem by creating a strong barrier against data breaches. Dealing with the trifecta of scalability, interoperability, and data privacy becomes a litmus test for the effectiveness of the transformation in the challenging environment of integrating Blockchain into supply chains. A sustainable, effective, and resilient blockchain-driven supply chain is built on carefully considering these factors, using strategic judgement, and adhering to changing best practices.

4.3 Navigating the Challenges of Implementing Blockchain and Crafting Solutions

The enterprises must navigate a maze of obstacles when integrating Blockchain into supply chains because doing so could impede progress. No matter how formidable, each obstacle offers a chance for strategic growth and adaptation, necessitating creative solutions to open the door for a successful transformation (Boulos et al., 2018). The initial investment presents a significant barrier because integrating Blockchain requires up-front infrastructure, technology, and training investments. The businesses must take a cautious stance in order to reduce this financial stress. Businesses must conduct a thorough cost–benefit analysis to determine whether the long-term efficiency gains associated with blockchain integration outweigh the upfront costs. This analytical examination demonstrates to stakeholders the value of the investment and directs resources to the most promising areas of blockchain implementation. Integration Complexity is a major worry, especially in established supply chains. The complex interplay of systems and processes can make it difficult to integrate Blockchain effectively. A phased approach represents a practical solution. Businesses can gradually reduce interruptions and manage complexity by implementing a gradual integration strategy (Novo, 2018). This enables time for process improvement, feedback incorporation, and refinement within the blockchain framework, resulting in a more seamless transition. A concern that frequently hangs over blockchain implementation is regulatory uncertainty. The constantly changing legal environment can create uncertainty and possibly halt progress. The proactive collaboration with legal experts is essential to overcome this obstacle. The establishing a partnership that constantly keeps track of regulatory shifts and modifies the blockchain strategy in accordance ensures compliance with the law, which promotes compliance and lowers the risk of disruptions.

There needs to be more Blockchain talent in many industries. Businesses should fund training programmes for their current workforce and develop internal expertise to close this gap. Working with specialized companies or consultants familiar with blockchain technology can offer momentary assistance during the knowledge transfer (Tseng et al., 2018). This proactive strategy addresses the talent gap and establishes the company as a knowledge hub. Despite the inherent cryptographic security of Blockchain, security concerns present an important challenge. Organizations should implement a multifaceted security strategy to strengthen the network. Regular audits and vulnerability assessments can help proactively find flaws, enabling prompt correction. A culture of security awareness should permeate the organization, and strict security protocols should be established. As businesses think about how Blockchain will affect the environment, the necessity of sustainability becomes apparent. Even though Blockchain can potentially improve sustainability, energy use is still a concern. Eco-friendly blockchain platforms and energy-efficient consensus mechanisms are emerging as solutions (Andoni et al., 2019). Organizations can successfully combine technological innovation with environmental responsibility by carefully choosing technologies that support sustainability objectives. Challenges stop being obstacles in the blockchain implementation tapestry and start acting as stimuli for innovation. Each challenge calls for organizations to use their strategic knowledge, adaptability, and dedication to best practices that are constantly changing. The businesses can pave the way for a sustainable, effective, and resilient blockchain-driven supply chain ready to navigate the constantly changing currents of modern commerce by deftly navigating these challenges and creating customized solutions.

The Blockchain to build sustainable supply chains is a complex task that calls for an all-encompassing strategy (Al-Megren et al., 2018). The businesses can use Blockchain to improve their operational efficiency and contribute to the global imperative of sustainability by carefully adhering to a structured implementation process, considering scalability, interoperability, and data privacy, and proactively addressing the challenges. It is crucial to approach this integration critically because, despite the transformative potential of Blockchain, its success depends on shrewd planning, ongoing improvement, and flexibility in the face of shifting technological conditions (Fig. 6).

5 Real-World Case Studies of Blockchain in Sustainable Supply Chains

The convergence of blockchain technology and sustainable supply chains is not just a theoretical idea but a real-world reality that forward-thinking businesses worldwide are embracing. These real-world case studies shed light on success stories, lessons learned, and the avoidance of potential pitfalls, providing insights into the transformative potential of Blockchain (Radanović & Likić, 2018). The Blockchain technology's successful integration into sustainable supply chains shows how well innovation can address contemporary problems. These case studies highlight the practical application of Blockchain and show how it transcends purely theoretical discussion to become an agent of transformation. We learn important lessons about

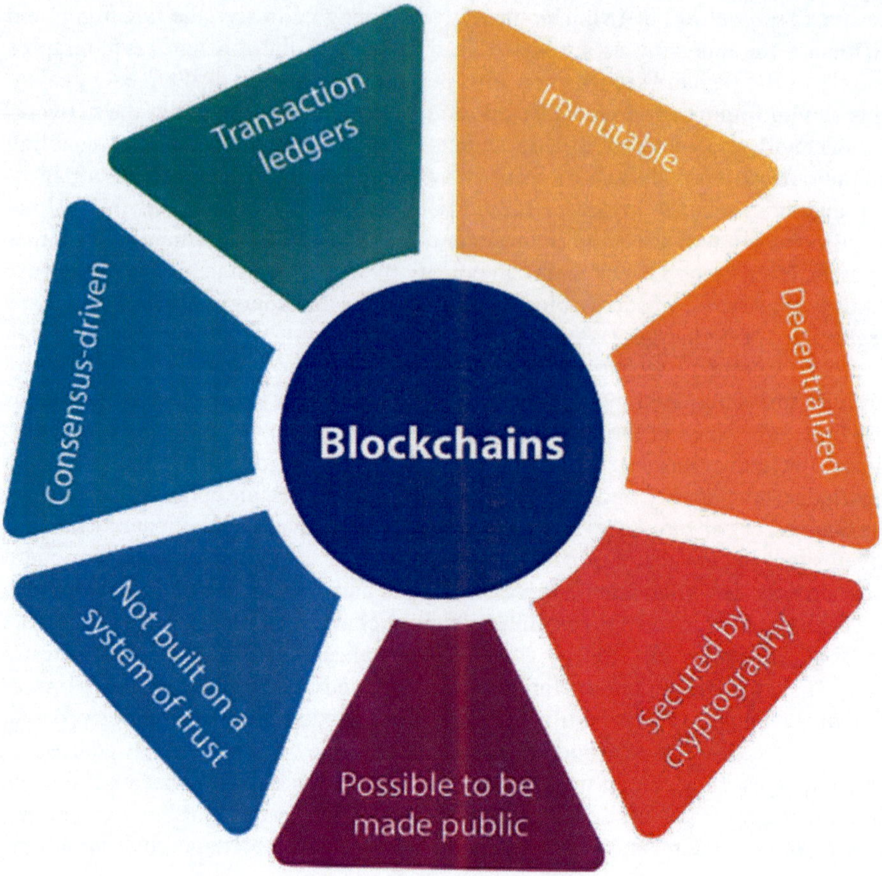

Fig. 6 Blockchains (https://www.softwebsolutions.com/resources/implementing-blockchain-in-business.html)

how businesses use blockchain technology to reshape supply chains, improve transparency, and spark ethical behaviour by digging deeper into these narratives (Erol et al., 2022). These case studies serve as evidence that Blockchain is more than just a trendy term; it is a powerful force that has the potential to transform entire industries, increase consumer confidence, and contribute to a more sustainable future.

6 Walmart and the Mangoes from Mexico

Retail behemoth Walmart understood the critical need for transparency and traceability in its international supply chains. They tested a blockchain-based method of tracking mangoes from Mexican farms to American stores in partnership with

IBM. Walmart achieved unmatched transparency by tracking each step of the supply chain's journey, from cultivation and harvest to transportation and storage (Alkaraan et al., 2023). This sped up recalls in the event of contamination and allowed customers to scan QR codes to learn about the journey of the mangoes they were buying. The fact that this pilot was so successful and spread to other products supports the idea that blockchain technology's transparency and traceability capabilities are not just theoretical and incredibly useful.

The mango initiative by Walmart is a prime example of how blockchain technology can fundamentally alter supply chain dynamics. Before this pilot, tracing a product's path to its origin could take days or weeks, depending on how complex the product was and how many intermediaries were involved (Hartley et al., 2022). Walmart cut this time down to just 2.2 s using Blockchain. This has significant implications for consumer confidence and food safety, making it more than an impressive technological achievement. Rapid traceability makes it possible to quickly identify and remove potentially affected batches from shelves in case of contamination or a health scare, reducing risk to customers and the brand's reputation. This openness encourages consumers and the products they buy to connect on a new level. A consumer now has an unrivalled perspective into a product's lifecycle thanks to the ability to scan a code and learn where their food comes from, who handled it, and under what conditions it was grown. Customers want assurance that their purchases are ethical and sustainable, so this satisfies their curiosity and fits with a larger shift towards ethical consumption (Capece et al., 2020). The mango project set a standard by demonstrating that Blockchain is a front-facing asset that enhances customer experience and trust, not just a backend tool for operational efficiency.

7 De Beers and Ethical Diamonds

Conflict or blood diamonds have long been a concern for the diamond industry. De Beers, a major producer of diamonds globally, has taken a stand against this moral conundrum. They introduced Tracr, a blockchain platform that verifies ethical sourcing and eliminates the possibility of shady origins by tracking the provenance of diamonds from the mine to the market (Aggarwal et al., 2019). This programme strengthened De Beers' dedication to ethical sourcing, winning over customers looking for reassurance about the reliability of their diamond purchases. The success of Tracr demonstrates that blockchain technology is not just a technological advance but also an ethical one that resonates strongly with consumers who are demanding more ethical products. The diamond industry revolutionized when De Beers started using the Tracr platform. This blockchain-based project tackled a deeply ingrained ethical issue that has dogged the diamond trade for decades. It went beyond simply tracking a diamond's journey. Conflict diamond issues, fueled by regional wars and leading to human rights violations, could no longer be disregarded or dealt with in cursory (Zhu et al., 2022). Consumers today are more savvy and discriminating, seeking clarity in their diamonds and the principles guiding their purchase.

De Beers developed an ecosystem where every diamond's history is unquestionably and openly documented by incorporating blockchain technology. Each process step, from its extraction from the earth to its refinement and eventual sale, is recorded on an unchangeable ledger. This initiative establishes the diamonds' ethical history and creates a compelling story about them. A consumer buying a diamond certified by Tracr is not just getting a gem; they also get a story about how it was obtained ethically and traded responsibly. De Beers' action has wider ramifications for the market for luxury goods (Voinea, 2019). It illustrates how provenance and authenticity can develop into important differentiators when supported by technological innovation. Tracr serves as a beacon for De Beers and the diamond industry, demonstrating how, with the right application, technology can support both business and ethical requirements.

8 Lessons Learned and Pitfalls Overcome

These success stories are motivating, but they come with difficulties. For blockchain implementations to be successful, all supply chain participants must work together. Comprehensive adoption depends on getting everyone on board with the technology's transformative potential, from producers to consumers. Scalability was identified as a danger (Wright, 2021). The technology's scalability was tested as Walmart expanded its blockchain initiative to include more products. The difficulty lay in accommodating higher transaction volumes and preventing a decline in network performance. This required careful platform selection and ongoing optimization to handle an increasing data load while maintaining efficiency. Regulatory landscapes can be challenging to navigate. The difficulty of complying with the various regional regulations governing the diamond industry was faced by De Beers' Tracr.

It took more than technological prowess to overcome this; it also required collaboration with regulatory bodies and legal know-how to ensure the blockchain initiative remained compliant without being hindered from reaching its full potential. The integrity of the data was crucial. It is said that "garbage in, garbage out." Transparency and traceability are rendered useless by inaccurate or fraudulent data, even with the most advanced blockchain system (Peterson, 2018). To ensure that data inputs were precise and verifiable from the start, Walmart and De Beers had to invest significantly in thorough training programmes and quality assurance procedures. Understanding that Blockchain is not a stand-alone solution and works best when combined with other technologies was another lesson learned. For Walmart's products, IoT devices were essential in capturing real-time data. The final challenge that emerged was consumer education, which was subtle but important. The term "blockchain" may be popular in the tech community, but the average consumer may not know its nuances (Dobrovnik et al., 2018). Both businesses launched extensive awareness campaigns to educate consumers about blockchain-based products' benefits, building greater trust and credibility for their respective brands.

These actual case studies serve as evidence that incorporating blockchain technology in sustainable supply chains is not merely a theoretical ideal but a practicable reality. Companies like Walmart and De Beers are paving the way for a future in which trade is characterized by accountability and trust by promoting transparency, traceability, and ethical consumption. The takeaways from their experiences emphasize the value of cooperation, scalability, and regulatory navigation (Al-Megren et al., 2018). These examples highlight that blockchain technology can revolutionize various industries, alter consumer expectations, and pave the way for a more ethical and sustainable future.

9 Stakeholder Perspectives

9.1 Company Executives and Managers

The business operations and technological advancement are converging for company executives. For them, Blockchain offers greater security, efficiency, and transparency. The Blockchain's decentralized structure can significantly reduce fraud, human error, and the need for intermediaries (McCarthy et al., 2022). The difficulties of high initial investment burden these advantages, a difficult integration process, and the requirement for organizational-wide training. In the fiercely competitive market of today, standing out is essential. With the help of Blockchain, businesses can provide customers with unmatched traceability and transparency, which could turn into key selling points. The quick and secure transactions can boost operational effectiveness, improving service delivery and growing market share.

9.2 Investors and Shareholders

A company's strategic use of Blockchain communicates innovation and adaptability to investors. A company's use of blockchain technology can help it stand out from the competition and become a more appealing investment as companies come under increasing scrutiny for their ethical business practices. They increased operational effectiveness can result in improved financial performance and higher returns on investment (Yaqoob et al., 2021). Although the benefits of Blockchain are clear, there are risks as well. The regulatory environment is unstable, and the technology is still in its early stages. For shareholders, the balance is in separating businesses that use Blockchain as a trendy buzzword from those that use it for competitive advantage.

9.3 *Students and Academics*

Academic research has much to gain from the convergence of supply chains and blockchain technology. The students can explore how Blockchain affects global trade dynamics, ethical sourcing, and how it interacts with other cutting-edge technologies. Such research can spur innovation and present new viewpoints to the business community. The business curricula must change as Blockchain becomes more widely used (Pettit et al., 2019). The future business leaders must focus on this technology because of its strategic ramifications and potential to transform global trade.

9.4 *Consultants and Advisors*

The theoretical promise of Blockchain must be transformed into real-world business results, and consultants are essential for this. They can provide expertise in staff training, ensuring seamless integration with existing systems, and choosing the best blockchain platform. Their strategic and technological contributions help businesses align blockchain initiatives with corporate objectives (Raimundo & Rosário, 2021). With the growth of Blockchain, a new consulting market is opening up. Beyond simple integration, consultants can provide services like blockchain audits, educational courses, and custom blockchain solutions to meet a particular sector's demands. This creates opportunities for new business models and sources of income.

9.5 *Sustainability Activists and Advocates*

The transparency of blockchain technology may be advantageous for sustainability. Companies can unequivocally demonstrate their commitment to the environment by using ethical sourcing practices for raw materials and eco-friendly manufacturing practices. This can encourage businesses to act more responsibly because blockchain technology ensures they cannot just "greenwash" their image but live up to their commitments. The Blockchain creates opportunities for cooperation between corporations and advocacy groups (Murray, 2018). In order to ensure real sustainability, advocacy groups can work with businesses to set up blockchain parameters. Together, they can create industry norms and best practices, utilizing Blockchain for a more moral and sustainable business environment (Fig. 7).

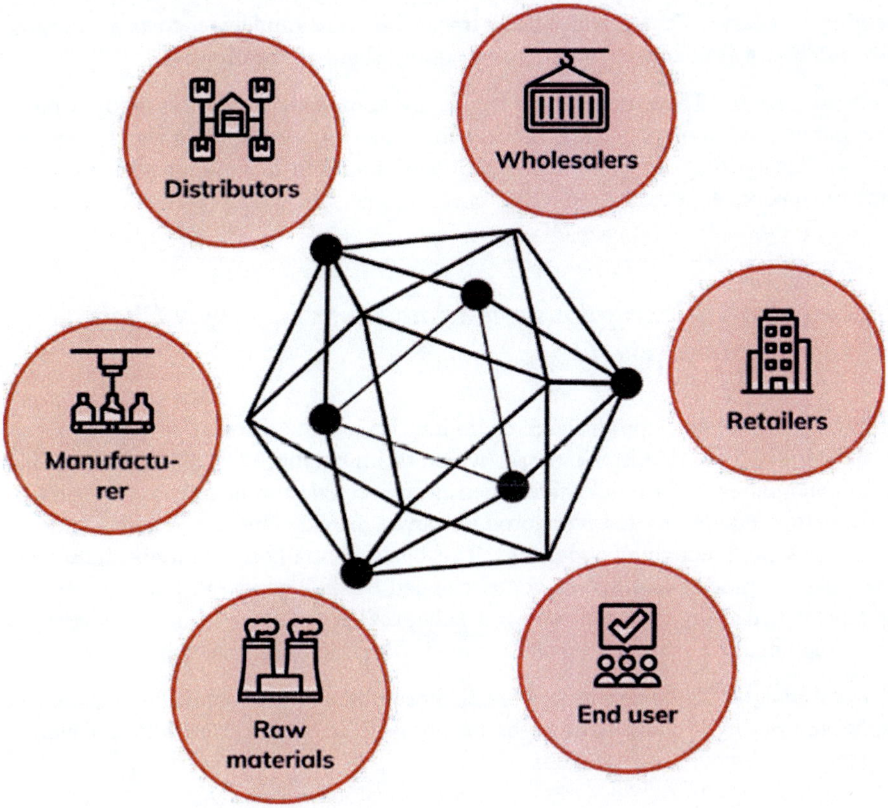

Fig. 7 Blockchain in supply chain industry

10 Future Prospects and Challenges

10.1 Evolving Trends in Blockchain Technology

We are transitioning from simple transactional usage to more sophisticated and nuanced applications as blockchain technology develops. Advanced smart contracts are becoming more complex and can automatically carry out a variety of conditions. Greater automation, accuracy, and efficiency in supply chains are predicted as a result of this evolution. End-to-end digitalized supply chains are being made possible by the convergence of Blockchain with technologies like IoT, AI, and big data analytics (Roth et al., 2022). Real-time monitoring, demand forecasting using predictive analytics, and more proactive risk management are all possible outcomes of this interaction. Major Blockchains have scalability issues, so layer-2 solutions—additional frameworks or protocols built on top of an existing blockchain—are gaining popularity. They promise faster transaction times and lower costs, increasing the viability of Blockchain for extensive supply chain applications.

Critical Insight: These developing trends are complex despite presenting a positive picture. As more complex blockchains emerge, user adoption and comprehension issues may arise. The integration of multiple technologies necessitates domain-specific expertise.

10.2 Potential Disruptions and Innovations in Supply Chain Risk Management

Real-time audits of supply chain operations can be performed using advanced analytics and blockchain technology, which is immutable (Cai et al., 2021). This means anomalies can be quickly identified and corrected, significantly lowering risks. A decentralized identity for every product can be provided by Blockchain, ensuring authenticity and reducing counterfeits. This can potentially revolutionize industries like pharmaceuticals and luxury goods plagued by fake goods. Physical assets can be represented as digital tokens on a blockchain. This can make ownership transfers, asset tracking, and even new business models like fractional ownership easier.

Critical Insight: These innovations are disruptive but require significant adjustments to the current supply chain. Businesses must be flexible, willing to abandon old habits, and open to new working methods.

10.3 The Role of Governments, International Bodies, and Regulations

The standardized protocols and best practices are necessary as blockchain adoption rises. International organizations can be crucial in developing these standards, ensuring that cross-border supply chains reap the full benefits of blockchain technology. Regarding blockchain regulations, many areas continue to operate in a grey area. Blockchain in supply chains and the technology itself requires clarification from the government (Sylvester, 2019). This clarity can encourage quicker adoption and lessen legal ambiguity for companies. Governments and international organizations can mediate to promote cooperation between various supply chain stakeholders. Cohesive blockchain implementations can result from collaborative efforts, benefiting entire industries rather than individual businesses.

Critical Insight: Government involvement has both benefits and drawbacks. While their participation can help Blockchain in supply chains become standardized and legitimate, excessive regulation may stifle innovation. A balance must be struck. Optimism is clear when predicting the future of Blockchain in supply chain risk management. The technology's inherent characteristics fit contemporary supply chains' requirements well. The path ahead, though, is with obstacles (Carson et al., 2018). Because technology, business practices, and regulations are intertwined, a comprehensive strategy is necessary to fully realize the potential of Blockchain while navigating its difficulties.

11 Conclusion

The integrating blockchain technology and supply chain management is no longer just a novel concept in an increasingly globalized world; it is necessary for ethical, transparent, and sustainable business practices. Our thorough investigation has made it clear that the promise of Blockchain extends beyond just transactional efficiency; it offers a paradigm shift in how businesses function, engages stakeholders, and support global sustainability. However, problems still exist with all innovations. Huge obstacles include scalability, regulatory environments, and the need for cross-sector cooperation. However, the discussed success stories and proactive solutions indicate real momentum. The fusion of supply chains and Blockchain will undoubtedly serve as a cornerstone for next-generation commerce as industries develop. In order to take advantage of the opportunities this convergence presents, leaders, academics, and practitioners must remain flexible. At the same time, they must make sure that the benefits are comprehensive and extend beyond boardrooms to the final consumers and the environment. The journey of Blockchain in supply chains, in which technology and ethics come together for a better, more accountable future, embodies the evolution of business in the twenty-first century. This investigation also highlights the responsibility of all stakeholders, from business executives to sustainability proponents. Although Blockchain provides a powerful toolkit to transform supply chains, its potential is only realized when stakeholders work together and contribute their distinct perspectives and expertise. As more industries test out blockchain technology, the lessons learned can be shared, improving best practices and reducing pitfalls. There are complexities in the way that must be navigated. Complexity is increased by the quick development of technology, shifting geopolitical environments, and the unpredictability of the world economy. The main narrative, though, is still upbeat. The business world is on the verge of an era where transparency, efficiency, and sustainability are ideals and actualized realities thanks to knowledge, collaboration, and a dedication to innovation.

References

Aggarwal, A., Gupta, I., Garg, N., & Goel, A. (2019, August). Deep learning approach to determine the impact of socio economic factors on bitcoin price prediction. In *2019 Twelfth International Conference on Contemporary Computing (IC3)* (pp. 1–5). IEEE.

Alkaraan, F., Elmarzouky, M., Hussainey, K., & Venkatesh, V. G. (2023). Sustainable strategic investment decision-making practices in UK companies: The influence of governance mechanisms on synergy between industry 4.0 and circular economy. *Technological Forecasting and Social Change, 187*, 122187.

Alketbi, A., Nasir, Q., & Talib, M. A. (2018, February). Blockchain for government services—Use cases, security benefits and challenges. In *2018 15th Learning and Technology Conference (L&T)* (pp. 112–119). IEEE.

Al-Megren, S., Alsalamah, S., Altoaimy, L., Alsalamah, H., Soltanisehat, L., & Almutairi, E. (2018, July). Blockchain use cases in digital sectors: A review of the literature. In *2018 IEEE International Conference on Internet of Things (iThings) and IEEE Green Computing and Communications (GreenCom) and IEEE Cyber, Physical and Social Computing (CPSCom) and IEEE Smart Data (SmartData)* (pp. 1417–1424). IEEE.

Andoni, M., Robu, V., Flynn, D., Abram, S., Geach, D., Jenkins, D., McCallum, P., & Peacock, A. (2019). Blockchain technology in the energy sector: A systematic review of challenges and opportunities. *Renewable and Sustainable Energy Reviews, 100*, 143–174.

Angrish, A., Craver, B., Hasan, M., & Starly, B. (2018). A case study for blockchain in manufacturing:"FabRec": A prototype for peer-to-peer network of manufacturing nodes. *Procedia Manufacturing, 26*, 1180–1192.

Bhaskar, P., Tiwari, C. K., & Joshi, A. (2021). Blockchain in education management: Present and future applications. *Interactive Technology and Smart Education, 18*(1), 1–17.

Boulos, M. N. K., Wilson, J. T., & Clauson, K. A. (2018). Geospatial blockchain: promises, challenges, and scenarios in health and healthcare. International journal of health geographics, 17.

Cai, Y. J., Choi, T. M., & Zhang, J. (2021). Platform supported supply chain operations in the blockchain era: Supply contracting and moral hazards. *Decision Sciences, 52*(4), 866–892.

Capece, G., Levialdi Ghiron, N., & Pasquale, F. (2020). Blockchain technology: Redefining trust for digital certificates. *Sustainability, 12*(21), 8952.

Caputo, F., Cillo, V., Candelo, E., & Liu, Y. (2019). Innovating through digital revolution: The role of soft skills and Big Data in increasing firm performance. *Management Decision, 57*(8), 2032–2051.

Carson, B., Romanelli, G., Walsh, P., & Zhumaev, A. (2018). Blockchain beyond the hype: What is the strategic business value. *McKinsey & Company, 1*, 1–13.

Chang, Y., Iakovou, E., & Shi, W. (2020). Blockchain in global supply chains and cross border trade: A critical synthesis of the state-of-the-art, challenges and opportunities. *International Journal of Production Research, 58*(7), 2082–2099.

Chaudhary, D. (2019). The decentralization, devolution and local governance practices in Nepal: The emerging challenges and concerns. *Journal of Political Science, 19*, 43–64.

Danial, K. (2023). *Cryptocurrency investing for dummies*. John Wiley & Sons.

Dobrovnik, M., Herold, D. M., Fürst, E., & Kummer, S. (2018). Blockchain for and in Logistics: What to Adopt and Where to Start. *Logistics, 2*(3), 18.

Dwivedi, A., & Paul, S. K. (2022). A framework for digital supply chains in the era of circular economy: Implications on environmental sustainability. *Business Strategy and the Environment, 31*(4), 1249–1274.

Erol, I., Neuhofer, I. O., Dogru, T., Oztel, A., Searcy, C., & Yorulmaz, A. C. (2022). Improving sustainability in the tourism industry through blockchain technology: Challenges and opportunities. *Tourism Management, 93*, 104628.

Ertz, M., & Boily, É. (2019). The rise of the digital economy: Thoughts on blockchain technology and cryptocurrencies for the collaborative economy. *International Journal of Innovation Studies, 3*(4), 84–93.

Esmaeilian, B., Sarkis, J., Lewis, K., & Behdad, S. (2020). Blockchain for the future of sustainable supply chain management in Industry 4.0. *Resources, Conservation and Recycling, 163*, 105064.

Gao, W., Hatcher, W. G., & Yu, W. (2018, July). A survey of blockchain: Techniques, applications, and challenges. In *2018 27th international conference on computer communication and networks (ICCCN)* (pp. 1–11). IEEE.

Haq, H. B., & Ali, S. T. (2018). Navigating the Cryptocurrency Landscape: An Islamic perspective. arXiv preprint arXiv:1811.05935

Hartley, J. L., Sawaya, W., & Dobrzykowski, D. (2022). Exploring blockchain adoption intentions in the supply chain: Perspectives from innovation diffusion and institutional theory. *International Journal of Physical Distribution & Logistics Management, 52*(2), 190–211.

Hennelly, P. A., Srai, J. S., Graham, G., & Fosso Wamba, S. (2020). Rethinking supply chains in the age of digitalization. *Production Planning & Control, 31*(2–3), 93–95.

Keesara, S., Jonas, A., & Schulman, K. (2020). Covid-19 and health care's digital revolution. *New England Journal of Medicine, 382*(23), e82.

Kiu, M. S., Chia, F. C., & Wong, P. F. (2022). Exploring the potentials of blockchain application in construction industry: A systematic review. *International Journal of Construction Management, 22*(15), 2931–2940.

Kyriazis, N. A. (2020). Herding behaviour in digital currency markets: An integrated survey and empirical estimation. *Heliyon, 6*(8).

Madine, M., Salah, K., Jayaraman, R., Al-Hammadi, Y., Arshad, J., & Yaqoob, I. (2021). Appxchain: Application-level interoperability for blockchain networks. *IEEE Access, 9*, 87777–87791.

McCarthy, M. W., Naggie, S., Boulware, D. R., Lindsell, C. J., Stewart, T. G., Felker, G. M., & Accelerating COVID-19 Therapeutic Interventions and Vaccines (ACTIV)-6 Study Group. (2022). Fluvoxamine for outpatient treatment of COVID-19: A decentralized, placebo-controlled, randomized, platform clinical trial. medRxiv, 2022-10.

Michael, J., Cohn, A. L. A. N., & Butcher, J. R. (2018). Blockchain technology. *The Journal, 1*(7), 1–11.

Murray, M. (2018). An examination of the blockchain, the emerging technology that promises to transform digital tracking of assets. In *AMCIS*.

Novo, O. (2018). Scalable access management in IoT using blockchain: A performance evaluation. *IEEE Internet of Things Journal, 6*(3), 4694–4701.

Peterson, M. (2018). Blockchain and the future of financial services. *The Journal of Wealth Management, 21*(1), 124–131.

Pettit, T. J., Croxton, K. L., & Fiksel, J. (2019). The evolution of resilience in supply chain management: A retrospective on ensuring supply chain resilience. *Journal of Business Logistics, 40*(1), 56–65.

Radanović, I., & Likić, R. (2018). Opportunities for use of blockchain technology in medicine. *Applied Health Economics and Health Policy, 16*, 583–590.

Raimundo, R., & Rosário, A. (2021). Blockchain system in the higher education. *European Journal of Investigation in Health, Psychology and Education, 11*(1), 276–293.

Roth, T., Utz, M., Baumgarte, F., Rieger, A., Sedlmeir, J., & Strüker, J. (2022). Electricity powered by blockchain: A review with a European perspective. *Applied Energy, 325*, 119799.

Schlecht, L., Schneider, S., & Buchwald, A. (2021). The prospective value creation potential of blockchain in business models: A Delphi study. *Technological Forecasting and Social Change, 166*, 120601.

Sylvester, G. (2019). *E-agriculture in action: Blockchain for agriculture, opportunities and challenges*. FAO.

Tao, R., Su, C. W., Naqvi, B., & Rizvi, S. K. A. (2022). Can Fintech development pave the way for a transition towards low-carbon economy: A global perspective. *Technological Forecasting and Social Change, 174*, 121278.

To, K. (2020). *Reimagining the future of aviation with blockchain*. Stern School of Business New York.

Tseng, J. H., Liao, Y. C., Chong, B., & Liao, S. W. (2018). Governance on the drug supply chain via gcoin blockchain. *International Journal of Environmental Research and Public Health, 15*(6), 1055.

Vatiero, M. (2018). Smart contracts and transaction costs. *Dipartimento Di Economia e Management Discussion Papers, 238*, 1–18.

Voinea, D. V. (2019). Blockchain for journalism-potential use cases. *Social Sciences and Education Research Review, 6*(2), 244–256.

Wang, Q., & Su, M. (2020). Integrating blockchain technology into the energy sector—From theory of blockchain to research and application of energy blockchain. *Computer Science Review, 37*, 100275.

Wright, A. J. (2021). The rise of decentralized autonomous organizations: Opportunities and challenges.

Yaqoob, I., Salah, K., Jayaraman, R., & Al-Hammadi, Y. (2021). Blockchain for healthcare data management: Opportunities, challenges, and future recommendations. *Neural Computing and Applications*, 1–16.

Zaoui, F., & Souissi, N. (2020). Roadmap for digital transformation: A literature review. *Procedia Computer Science, 175*, 621–628.

Zhu, Q., Bai, C., & Sarkis, J. (2022). Blockchain technology and supply chains: The paradox of the atheoretical research discourse. *Transportation Research Part E: Logistics and Transportation Review, 164*, 102824.

Dr. Fahim ul Amin is a dedicated postdoctoral researcher specializing in the dynamic Supply Chain Risk Management (SCRM) field. Currently he is affiliated with International Business School of Hainan University. Dr. Fahim ul Amin earned his doctoral degree in [Logistics Management] from Chang'an University, where his passion for supply chain dynamics and risk management flourished. His academic journey reflects a commitment to excellence and a drive to contribute impactful solutions to real-world challenges.

Dr. Qingkai Ji is a distinguished professor at the International Business School of Hainan University, where he brings a wealth of expertise to the intersection of Supply Chain Management and Blockchain technology.

Dr. Wasim Ul Amin is a dedicated and ambitious Ph.D. candidate currently pursuing advanced studies at North China Electric Power University in Beijing.

Dr. Azka Amin is currently serving as postdoctoral researcher at International Business School of Hainan University. She did Ph.D. from Shandong University.

The Metaverse as a Solution to Depopulation: Creation of Interconnected Macro-villages Through Digital Twins and Transportation. The Theory of Population Leverage

Álvaro Bueno-Ferrer and Jaime de Pablo Valenciano

1 Introduction

According to data from the Institute of Statistics and Cartography (2018), it is anticipated that the most significant increases in population will occur in medium-sized localities, which are home to between 10,000 and 100,000 inhabitants, and where almost half of the inhabitants of Andalusia already live. It is estimated that, by 2040, these municipalities will see an additional increase of 161,000 residents. On the contrary, those less than 10,000 inhabitants will be emptied progressively, including the majority at risk of depopulation by the European Union at different levels of severity. However, it is a common problem not only in Andalusia, but also in Spain and other countries around the world.

In the twenty-first century, Spain's population has grown in the last 20 years by 15% in terms of its inhabitants. However, the process of depopulation has been accentuated, increasing the pace in the last decade. More specifically, 61.98% have lost population since 2001 and in the last decade, the figure rises to 75.71% of the total number of small villages (Fourth Vice Presidency and Ministry for Ecological Transformation and the Demographic Challenge, 2020). Hence the importance of finding new forms of action to fight a problem that affects 3 out of 4 municipalities in Spain.

Á. Bueno-Ferrer (✉) · J. de Pablo Valenciano
Department of Economics and Business, University of Almeria, Carretera de Sacramento s/n, 04120 Almeria, Spain
e-mail: alvaro.bueno.ferrer@gmail.com

J. de Pablo Valenciano
e-mail: jdpablo@ual.es

M. d. C. Valls Martínez and J. M. Santos-Jaén (eds.), *Environmentally Sustainable Production*, https://doi.org/10.1007/978-3-031-52656-5_12

So far, metaverse and depopulation have been lines of research not connected to each other, which is an innovation with potential for implementation. The greatest approximation is the article by Allam et al. (2022), in which it is stated that the metaverses conceived as a coexisting digital universe, which presents new ways of living and working in virtual urban environments, offering a different path from the smart cities of the future. It also identifies the emerging products and services inherent to it, examining how these could contribute to smart cities through their virtual presence. This focuses especially on environmental, economic and social factors, in addition to sustainability goals. However, it has not delved into aspects of depopulation and repopulation, so this study is a pioneering exploration into this gap in international literature.

This study addresses concepts such as the shift from sustainable local development to sustainable digital development in small interconnected villages through the metaverse and the transport sector. With this, the progress and growth of small villages is addressed, to improve their competitiveness against larger towns and cities.

Due to the innovative nature of this research, there are no existing theories that directly relate to it. Therefore, the so-called theory of population leverage is created and justified. This defends that if the villages are interconnected by uniting their inhabitants and services, the competitiveness of the grouping of small villages is improved to cause a greater increase in their inhabitants by becoming a more attractive option to live equipped with a greater number of services. In other words, it is the strategic process of using the significant increase of companies and people in a region in an interconnected digital twin as an advantage to bring about greater economic and social development in vulnerable areas making them comparable with large cities in terms of provision of services and population in the digital twin. All of this, without losing the unique identity of each community. The population exodus from small communities and municipalities is becoming an increasing challenge for many areas, and urgent action is required to counter it (González-Bustos & de Andalucía, 2022).

The metaverse, through digital twins, serves as a digital extension of communities. It provides a greater abundance of services available using augmented reality and virtual reality. The scarcity of services and resources in these areas can foster demographic decline, and it is crucial to seek innovative strategies to address the fragility of the most marginalized regions (Rodríguez, 2022). It also increases the number of products available, which through metaheuristics and efficiency in transport, interconnect peoples in the real world to get their products to them without the need to travel. This has a positive environmental impact, as reducing the need to travel reduces the carbon footprint. In turn, it affects social and environmental sustainability.

The main objective of this research is to explore and validate the new theory of population leverage. It is argued that by interconnecting small communities through the implementation of digital twins in the metaverse, it is possible to enhance their repopulation, constituting a valuable resource in the fight against depopulation. It is proposed to analyze the applicability and benefits of this theory in 596 villages of Andalusia with up to 10,000 inhabitants, seeking to build a new path towards social and business progress. And thus, expanding their possibilities beyond the geographical borders that usually limit them. The research will look for evidence of how the inclusion of the metaverse can boost population growth and, in turn, decrease pollution by eliminating the need for physical travel for the acquisition of goods and services.

The metaverse is a great challenge for the future, with the present being the time to plan it. It is seen as a strategy of sustainable digital repopulation, in which new technologies acquire a fundamental role in the redefinition and reconfiguration of small villages, being applicable to other similar areas of Spain and the world.

It should be noted that this study aligns with the framework of the European policies of economic, social and territorial cohesion for the period 2021–2027, in which specific strategies are being considered to address depopulation in sparsely populated regions. Specifically, with the development of transport infrastructures, the search for territorial balance and the investment in digital connectivity supported by the Just Transition Fund (European Parliament, 2023).

2 Theoretical Framework

The metaverse is a virtual space that leverages technologies such as virtual reality, augmented reality, and artificial intelligence to build an alternative universe that coexists with reality (Campos-Sánchez et al., 2022). It incorporates technologies such as virtual reality, augmented reality, blockchain, artificial intelligence, the Internet of Things, and telecommunications to forge an advanced tool (Bale et al., 2022). It offers users a personalized and immersive experience, driving companies to build their own metaverse and transform their customers' experiences. This digital world can be linked to reality using a digital twin, allowing visits to any store, company, or place instantly without physically moving. The connection between both worlds, physical and virtual, is an opportunity for the development of the world as we know it, as it brings together the economy, interaction, health, work and other factors in a new digital environment (Hernán-Guerrero, 2023). These aspects gain special relevance when viewed as a strategy to group small villages together, aiming to strengthen and increase their appeal. Among these aspects for the fight against depopulation, it is necessary to pay attention to artificial intelligence, the new possibilities of the internet and the construction of the metaverse.

Also, the digital twin mitigates the need to travel in villages far from large cities to access services. Thus, the new immersive three-dimensional internet, known as the metaverse, could eliminate one of the main weaknesses: distance. This article illustrates how, as the distance to the capital increases, the number of inhabitants per population decreases. This can be directly linked to the tendency to concentrate investments and highly educated people in cities can worsen the differences between urban and rural areas, and accelerate depopulation in the latter (Rosas-Villar, 2023). To avoid this, it is important to put in place policies that promote a more equitable distribution of resources and talent among all regions.

Although artificial intelligence required decades of research and investment, its true impact emerged in 2023, sparking a global revolution. In fact, companies with potential to implement these advancements but fail to do so might find themselves lagging behind competitors in the coming years. This simile is applicable to the development of the metaverse. This concept was first introduced by Stephenson (1992), where he envisioned a cyberspace connected to the real world as a part of fiction. From this moment arose the first research led by Budiansky (1995) in which he explained how the construction of this metaverse could be taking advantage of the beginnings of the internet and its future development.

After more than two decades of research, in the second decade of the twenty-first century, it is when large corporations such as Meta (new name of Facebook after its commitment to the metaverse), Apple, Microsoft, Telefónica, Nvidia and Alphabet work on the development of this cyberspace (McKinsey & Company, 2022). The present bet of these large companies is artificial intelligence for its transformative power in the short term. However, the metaverse is understood as a medium-long term bet because it is still in the creation phase. Both technologies complement each other, serving as tools for social transformation. While big brands aspire to be market leaders, the most significant market breakthrough is anticipated when regional and local companies start to integrate into the built metaverse (Hollensen et al., 2022). This part of the equation is where village enterprises enter as a new form of digitization for villages.

The development of a digital twin remains in line with the so-called smart villages, which are rural communities that adopt novel approaches based on technology to increase their resilience and optimize their social, economic and environmental circumstances (European Commission, 2021). These innovative solutions not only promote the improvement of the quality of life of the inhabitants, but also enhance the self-sufficiency and sustainability of these rural areas, aligning with the objectives of sustainable development in the long term. The metaverse can transform services, especially transport. By leveraging it, consumers can instantly access any part of the physical world within the digital twin, eliminating the need for physical travel. This would help reduce the carbon footprint and make the world more sustainable.

Research indicates that the metaverse can enhance sustainability. Empirical data revealed that, through a metaverse approach, companies saw their environmental profits grow by up to 19.6% per month in 2021. These figures suggest that digitalization from the metaverse perspective has a greater potential to favor carbon neutrality and green and sustainable development (Cao et al., 2023). In relation to transport, the metaverse can facilitate route simulations, helping transport metaheuristics through artificial intelligence to reduce pollutant emissions and transport costs. Thus, it would not only influence the transport of products purchased in the metaverse but could also enhance public transport by making it more efficient and eco-friendlier (Deveci et al., 2023).

In addition, this new way of promoting small villages would allow services and industries to be expanded, enhancing their attractiveness through the development of a social and economic community. This approach might be likened to the European Union's strategy, aiming to make villages competitive with major cities and towns in terms of repopulation. All this, without the loss of local identity and eliminating the barrier of distance in the acquisition of products and services.

This is not just an economic revolution; it also offers significant benefits to consumers. It also addresses issues like the lack of leisure and socialization. It provides possibilities such as attending concerts, sports, art, virtual tourism, training, health, entertainment and live events in a virtual environment. These options have garnered significant interest, with 66% of respondents expressing enthusiasm. It also redefines commerce by introducing innovative ways of buying and selling online, eliciting positive emotions in 64% of surveyed users (McKinsey & Company, 2022). In fact, from an anthropological perspective, rural depopulation affects the leisure habits of both young people and adults in small villages, indicating a need for new resources to fulfill their requirements (Campos-Díez, 2023).

At the same time, it connects villages, people and companies with opportunities for personal and professional development. This suggests that the metaverse can serve as a platform for remote work and collaboration, enhancing interaction and cooperation among individuals in a virtual setting. Beyond allowing instant communication free of geographical restrictions, it also provides the opportunity to establish new ways of working and collaborating (Rojas-Concepción & Guerra-Chagime, 2022). This digital revolution could lead to the formation of international work teams, the holding of meetings and conferences in immersive contexts and real-time cooperation on tasks and projects, regardless of where the participants are physically located.

It is estimated that the metaverse could have an economic impact in the range of 28–53 billion euros by 2035, and in a broader context, it could add between 259 and 489 billion euros per year to the Gross Domestic Product of the European Union by 2035, which would correspond to between 1.3 and 2.4% of GDP (Target, 2023). In addition, a solid digital infrastructure is reported in the country (Spain), being the seventh among the EU member states and ranking third in terms of connectivity within the European Union (European Commission, 2023a). Moreover, workers in Spain possess essential digital skills that surpass the EU average (Meta, 2023). This opens up a range of growth possibilities, being able to decentralize work remotely and helping interconnected smart villages connect with digital workers in Spain and Europe, strengthening their position. To do this, it starts from immersive digital development towards local development as part of the innovation and commitment of small villages. For this reason, it is about minimizing the weaknesses of villages of up to 10,000 inhabitants using an opportunity and strength as previously mentioned.

In Spain, factors such as the migration of rural population to urban areas, the increase in the average age of residents and the scarcity of economic opportunities in rural areas have been the cause of rural depopulation (González, 2022). The lack of economic progress has affected young people to seek job opportunities beyond their villages, since each of them has particular characteristics within their business fabric in which many of them do not fit as professional profiles. Within this scenario, strategies such as rural tourism have been proposed to reactivate and promote development in rural areas with poor socioeconomic conditions. Rural tourism has proven to be a relevant economic instrument for these territories and a fundamental pillar in regional development, although it is disparate in terms of what each municipality can offer (Sánchez & Alba-Sánchez, 2018).

Therefore, to foster growth, each small village's unique growth opportunities should be evaluated and leveraged, such as the commitment to the metaverse in terms of new employment opportunities that satisfy young people and greater possibilities of access to services beyond the basic ones that retain and attract new population. The exodus of young people and highly qualified professionals from rural areas to the main cities can lead to a decrease in education levels and cause a new cycle of depopulation in rural regions and small cities in which present and future human capital disappears (González-Leonardo et al., 2019).

All these initiatives prioritize sustainability. The creation of green jobs in cities dedicated to sustainable practices can effectively combat climate change while generating economic growth opportunities in both urban and rural settings (Rosas-Villar, 2023).

3 Methodology

The theory of population leverage is inspired by financial leverage, although with the focus of the fight against depopulation. The term financial leverage, according to Borja-Peñaranda et al. (2022), refers to the debts assumed by organizations for the purpose of carrying out activities or projects with the aim of increasing profitability or fulfilling obligations that cannot be covered solely with available own resources. Van-Horne and Wachowics (2010) point out that leverage involves using fixed expenses or costs in order to increase (or leverage) profitability. This helps to have a greater amount of capital with which to invest through the use of debt.

This parallel means that while financial leverage uses external debt to increase its production and profitability, population leverage uses the external resources of other interconnected communities to increase their productive, economic and social potential. Through the metaverse, this allows people to reside in each small village without the need to move, thereby extending each village to the collective sum of all interconnected villages. For this, the main axes are the development of the virtual world to empower small communities in a conglomerated virtual city through the use of virtual reality, augmented reality and extended reality.

This type of leverage can be very beneficial for small villages, which due to their location or number of inhabitants, face the challenge of attracting and retaining people to their villages. Thus, leveraging web 4.0 and the immersive capabilities of this three-dimensional internet can address one of Spain's primary challenges in the twenty-first century: depopulation. Thus, it seeks to improve its competitiveness and attractiveness, especially for younger generations, through the creation of an interconnected macro-city inhabited by thousands or even millions of people. In this way, small communities will be able to benefit from the typical characteristics of large cities.

Therefore, the theory of population leverage is proposed, whose focus is on expanding the number of inhabitants of small villages through a strategy of unification of communities in the metaverse to boost the growth of territories without the loss of identity of each small village and attract a greater number of inhabitants, as cities do. For this, it is important that a joint strategy of the member communities is created for its success, and that each village can enhance its positioning and user experience to prosper effectively. In this framework, a more expansive vision of what constitutes a "village" is suggested, allowing these places to become larger and more dynamic communities thanks to digitalization and connectivity. Interconnectedness across the metaverse allows people to live in small villages, but be part of much larger communities, benefiting from the resources and opportunities they offer.

Efficient and sustainable transport, through the application of metaheuristics, unites the physical spaces of each small village, allowing each inhabitant of each community to access in seconds any place or establishment of any of the member municipalities and obtain any service as he would if he went in person. From this standpoint, transport serves as the axis of connection in the physical realm, while the metaverse acts as the virtual extension of each village. Some of the guidelines for it to succeed is to understand that a community strategy needs each community to be able to think beyond themselves, because fragmentation would limit their growth. And, on the other hand, the digital twin must be developed, so that the user experience is as realistic as possible, being able to include aspects such as the ≪ gamification of life ≫.

This theory is supported by the 2030 objectives of the strategic program of the Digital Decade, which states that virtual universes will change the way we interact, presenting both possibilities and challenges that must be managed (European Commission, 2023c).

The formulation of population leverage is described as follows:

- In terms of the number of inhabitants and leisure possibilities:

$$\text{Effective population } (E_p) = \text{Physical population } (P_p) \\ + \text{Virtual population } (V_p)$$

- E_p *is understood as the total population of the community in the* metaverse through the union of small villages, in which the physical population is added to the rest of the small villages of the community that extends each village.
- P_p *is* understood as the number of inhabitants of the particular village that is part of the conglomerate community of villages.
- V_p is understood as the number of inhabitants that extends each village by unifying in the metaverse.

This approach enables the effective population of small villages to mirror that of larger cities, bridging distance barriers via transport services.

So, Population leverage (P_{le}) could be defined as:

$$\begin{aligned} &\textit{Population leverage } (P_{le}) \\ &= \textit{Effective population } (E_p) / \textit{Physical population } (P_p) \end{aligned}$$

Thus, if there's a larger virtual population with the same amount of physical population, it results in greater population leverage, which aids growth. However, achieving this requires the operation of artificial intelligence systems with algorithms that optimize community-wide decisions and employ effective big data strategies. This ensures that as the population grows, population leverage increases accordingly. So, it is influenced by a correct management so that this leverage grows.

- In the section on the number of services:

$$\begin{aligned} &\textit{Services density in the metaverse community } (SD_{mc}) \\ &= \textit{Number of services in the metaverse community } (NS_{mc}) \\ &/ \textit{Effective population } (E_p) \end{aligned}$$

- SD_{mc} measures the level of services offered to each physical village in the community by the number of inhabitants. This data should be considered as an aspect for business efficiency and growth strategies.
- NS_{mc} *is the* sum of all available services, with the exception of the hospitality sector, since it is the only sector that governs face to face. However, tourism and hospitality are promoted within the member small villages, using cross-selling strategies, so that in each sale discount coupons can be offered with a term of X days to promote and enhance local tourism.
- E_p it is important to calculate SD_{mc} with the E_p to estimate the number of services in the metaverse for each person in the village, to constitute the SD_{mc} of the interconnected macro-city, comparable to that of large cities.

$$\begin{aligned} &\textit{Leverage of services } (L_s) \\ &= \textit{Number of services in the metaverse community } (MS_{mc}) \\ &/ \textit{Physical population } (P_p) \end{aligned}$$

$$\begin{aligned} &\textit{Total Population Leverage } (TP_{le}) \\ &= \textit{Population Leverage } (P_{le}) * \textit{Leverage of services } (L_s) \end{aligned}$$

In this way, the potential of the implementation of virtual worlds in the metaverse is measured at the level of constitution of the population number, taking advantage of the benefits of large cities in small villages. And, also, expanding the number of services to favor the level of attraction and retention of inhabitants.

To measure the growth of each village's inclusion in the interconnected smart metaverse:

$$\textit{Effective population growth potential } (\Delta E_p) = \textit{Virtual population } (V_p) - \textit{Physical population } (P_p)$$

$$\textit{Growth potential of total services } (\Delta T_s) = \text{Number of services in the metaverse community } (NS_{mc}) - \textit{Number of services in each village } (NS_{ev})$$

In addition, comparisons can be made between the number of inhabitants of the community with that of the cities in terms of number of inhabitants and services. This theory serves as a strategy for potential growth and sustainable digital development. Since, in addition to fostering the growth of small villages through their amalgamation, the digital twin and the transport system reduces the number of trips thus contributing to the reduction of the carbon footprint.

The reduction in carbon footprint results from the ability of each transport vehicle to carry a multitude of products from various residents of each village, so that through efficient transport routes, displacements of the inhabitants are saved for the purchase of products or purchase of services in the digital twin. For this reason, it can be estimated that the use of transport services has the potential to reduce the carbon footprint by reducing the number of trips made by the inhabitants of the area.

Finally, it should be noted that the sectors that would be directly enhanced include:

- Agriculture, livestock, forestry and fisheries.
- Industry, energy, water and waste management.
- Construction.
- Commerce.
- Artificial intelligence and virtualization of the world connected in real time.
- Transport and storage.
- Information and communications.
- Banking and insurance.
- Public administration, education and health.
- Real estate, professional, auxiliary, artistic and other services.

While indirectly:

- Hospitality and tourism (Figs. 1, 2, 3, 4, 5, 6, 7 and 8).

Additionally, using the CNAE 2009 code, data from 10,258 potential companies from the studied villages were analyzed to identify sectors relevant to the metaverse. Companies within the hospitality sector were excluded, as they require face-to-face interactions for their services. In this manner, an 'X-ray' of the business fabric affected by the metaverse can be obtained, as shown in Fig. 9. Finally, density plots are used from quantitative variables of the number of firms and number of firms per capita to see which sectors increase in per capita density as the population grows. In this way, the variations that an increase in the real population can cause from the

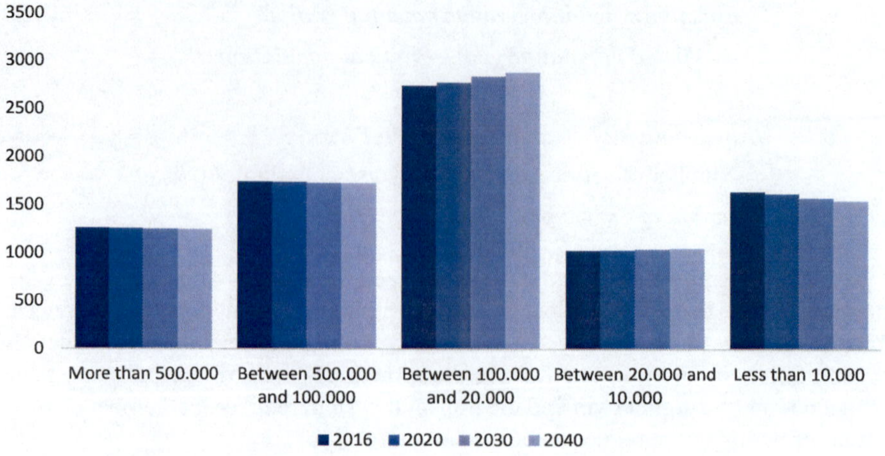

Fig. 1 Andalusian population by municipality size, growth or decrease estimate. *Source* Own elaboration based on the Institute of Statistics and Cartography (2018)

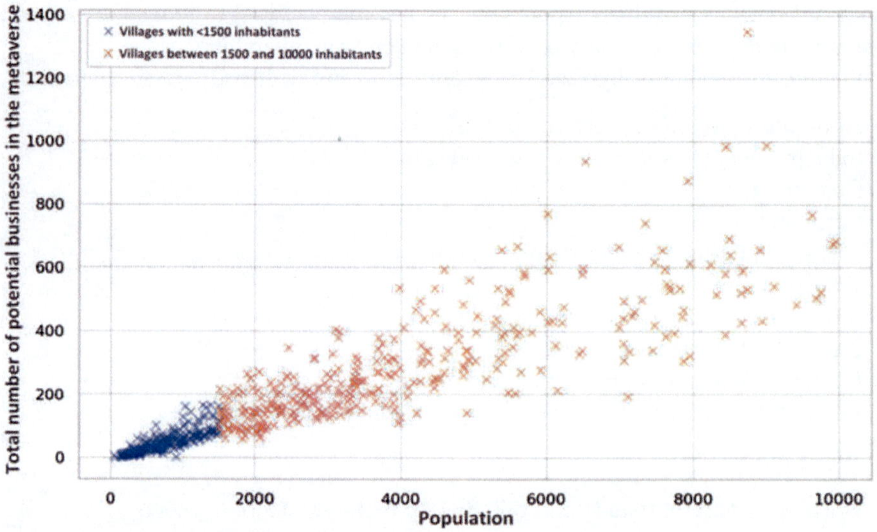

Fig. 2 **Relationship between the number** of inhabitants and the number of companies with potential. Segmented into less than 1500 inhabitants and between 1500 and 10,000 inhabitants. *Source* Authors, based on SIMA data (2021)

population leverage strategy based on each sector are identified and can be appreciated. Density plots visually represent the distribution of a continuous variable. Unlike histograms, which segment the data into separate ranges and calculate the number of observations in each of these ranges, density plots build a continuous representation

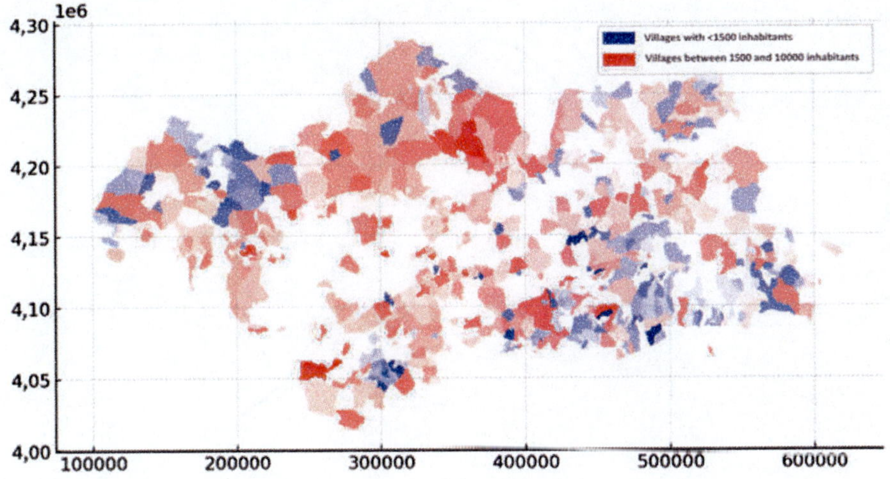

Fig. 3 Companies with potential by territory within each segment. *Source* Authors

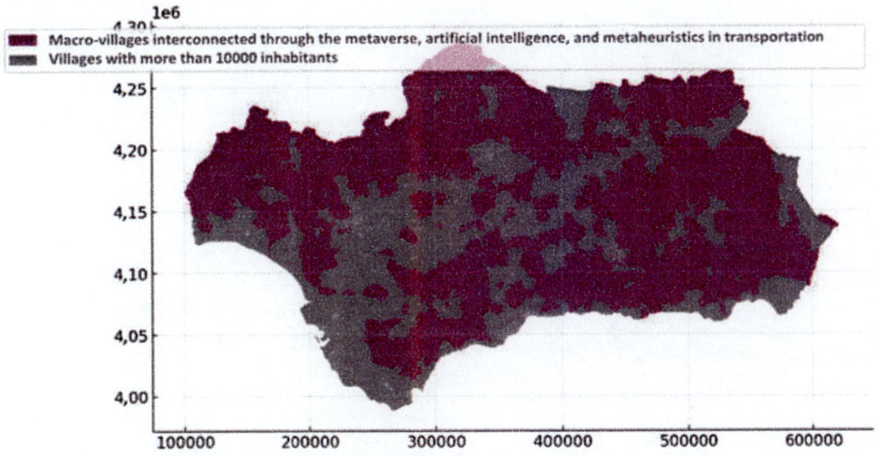

Fig. 4 **Territorial map of the interconnected villages in the** metaverse of the villages ≤ 10,000 inhabitants in Andalusia. *Source* Authors

of probability across the data, creating a smooth function that describes the density of the different values (Gibran-Juárez, 2018).

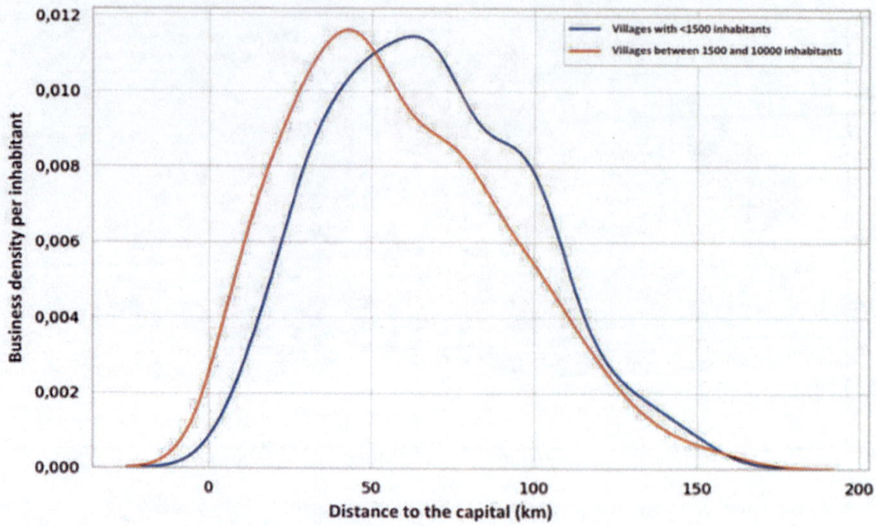

Fig. 5 Density of companies per inhabitant as a function of distance to the capital. *Source* Authors

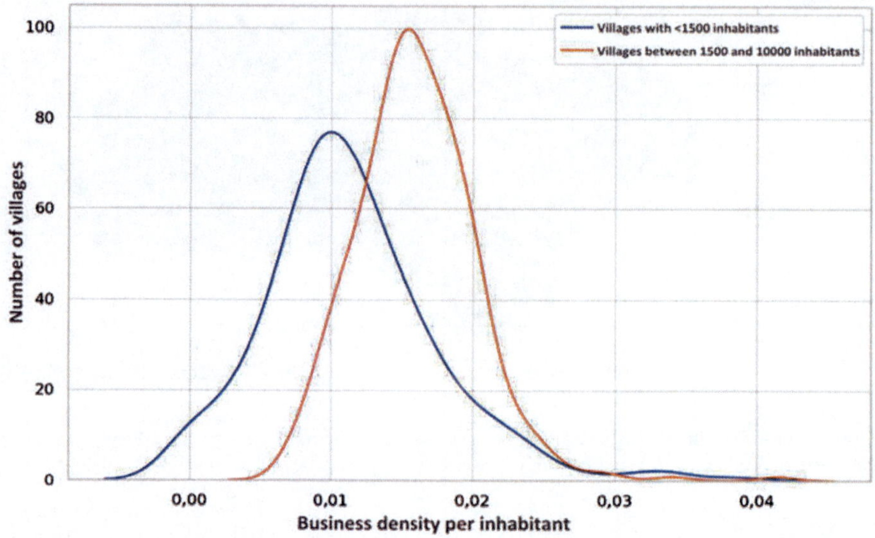

Fig. 6 Density of commercial companies per number of inhabitants in Andalusia. *Source* Authors

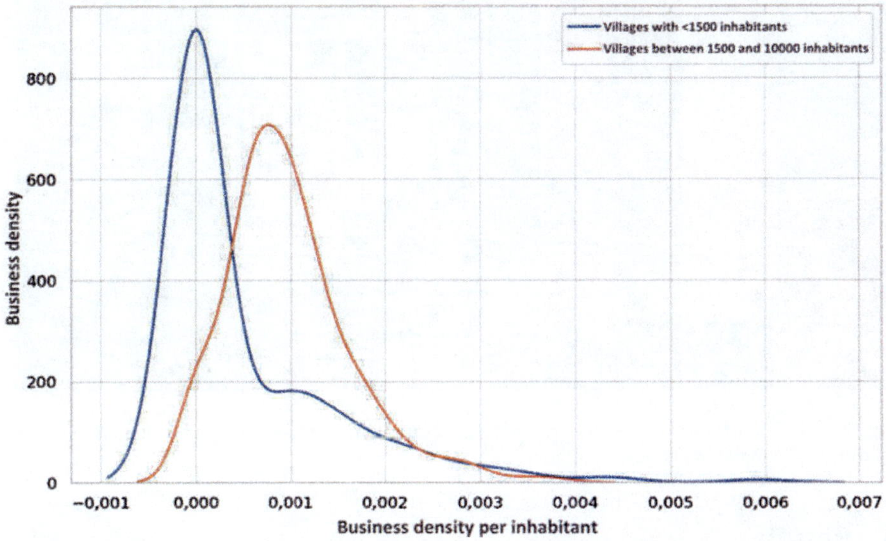

Fig. 7 Density of banking and insurance companies by number of inhabitants in Andalusia. *Source* Authors

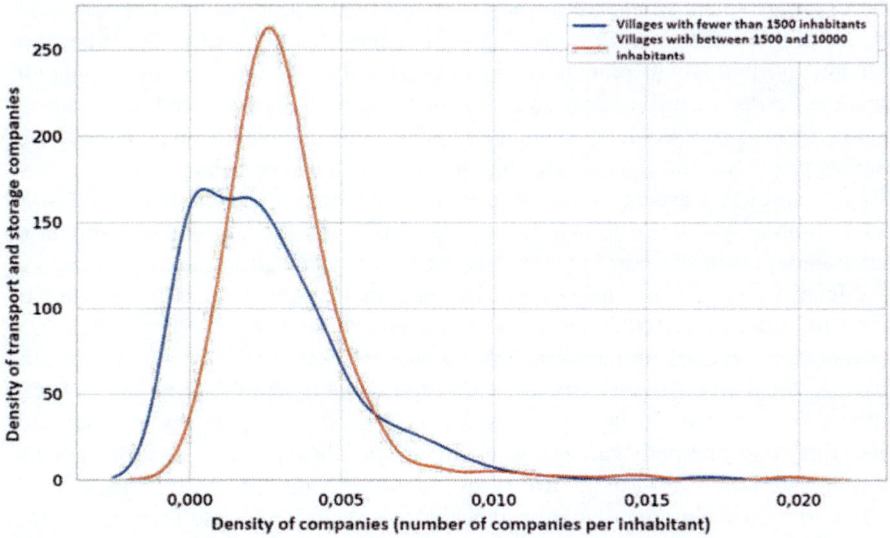

Fig. 8 Density of transport and storage companies by number of inhabitants in Andalusia. *Source* Authors

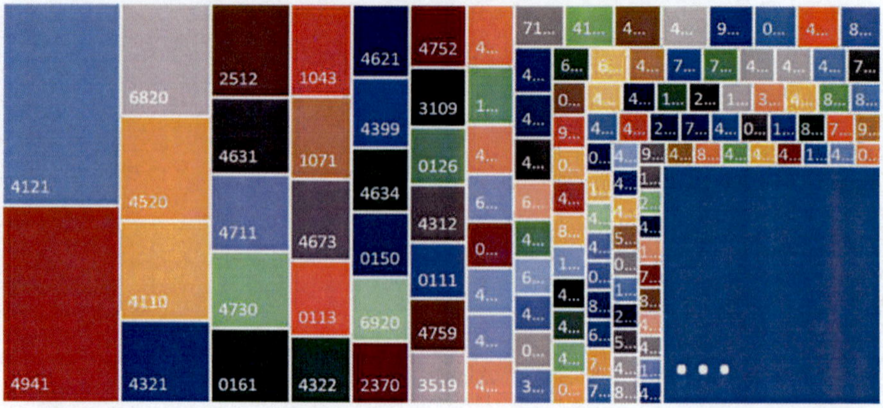

Fig. 9 CNAE 2009 codes main by presence in sectors with metaverse potential in small villages up to 10,000 inhabitants. *Source* Authors

4 Results

How Can the Metaverse Address the Depopulation Issue in Andalusia?
The union of small villages becomes essential in the fight against depopulation, since in this way they can take advantage of systems of joint growth. This is because individually, their growth potential is lower because their resources for the population are more limited as well as opportunities. In addition, the metaverse has the ability to attract young people who seek to develop and settle in territories where rental and housing prices are not in an overheated market as is the case of large cities.

Another relevant aspect, as indicated by the projections of the Institute of Statistics and Cartography (2018), villages of up to 10,000 inhabitants would continue to lose population in the coming years, hence the need to group them to carry out an effective struggle that helps them increase their population as in the case of populations between 10,001 and 100,000 inhabitants. In this way, the risk of severe and moderate depopulation of small villages could be reduced through joint strategies that help them emerge, since these data do not contemplate the grouping of small villages in Andalusia in the metaverse to enhance their population and services. Therefore, this grouping strategy is based on the cluster criterion, following the disaggregation by size of municipality, as indicated by the IEC itself in Table 1.

For the fight against depopulation, it becomes essential to choose a territory, in this case Andalusia, and carry out the population leverage strategy based on official data, since as can be seen, small villages of up to 10,000 inhabitants are less competitive than villages of between 10,001 and 100,000 inhabitants. From this starting point, growth is enhanced through the use of virtual worlds, being one of the pillars of the new society driven by virtual worlds, whose potential is defended by the European Commission. Thus, it represents a commitment to the future, focused on the fight against depopulation that represents an innovation in the way of understanding virtual worlds. This potential of the metaverse and digital twin will need to be harnessed

Table 1 Population projection by subregional areas of Andalusia 2016–2040

Type	2016	2020	2030	2040	Trend since 2020
Disaggregation by territorial domains					
Coast	3262.5	3310.7	3396.1	3452.8	Growth
Sierra Morena-Los Pedroches	232.3	225.3	212.5	203.6	Decrease
Sierras y Betic Valleys	1400.6	1386.3	1.357	1338.1	Decrease
Guadalquivir Valley	3508.4	3499.1	3470.6	3451.1	Decrease
Disaggregation by type of territorial unit					
Mid-sized interior cities	1979.2	1.957	1908.4	1871	Decrease
Mid-sized coastal cities	1338.2	1363.8	1423.8	1459.9	Growth
Regional centers	4688.5	4711.7	4744.7	4770.3	Growth
Rural centers	397.8	383.8	359.3	344.5	Decrease
Disaggregation by municipality size					
More than 500,000	1257.8	1256.9	1250.4	1247.5	Decrease
Between 500,000 and 100,000	1740.4	1737.8	1728.3	1722.8	Decrease
Between 100,000 and 20,000	2740.9	2776.7	2839.8	2879.6	Growth
Between 20,000 and 10,000	1028.2	1033.7	1043.3	1050.8	Growth
Less than 10,000	1636.4	1616.3	1574.3	1544.8	Decrease
Disaggregation by capital status					
Provincial capitals	2390.9	2379.6	2.349	2328.5	Decrease
Rest of the municipalities	6012.8	6041.7	6087.1	6117.1	Growth

In thousands
Source Institute of Statistics and Cartography (2018)

as one of the challenges that the EU may need to address in social, technological, economic and political dimensions (European Commission, 2023b).

The segment expected to decrease the most comprises towns with up to 10,000 inhabitants, followed by towns of between 500,000 and 100,001 inhabitants. On the other hand, the one with the most growth potential, not including virtual worlds in the estimate, is the one with a population of between 100,000 and 10,001 inhabitants, followed by towns of between 20,000 and 10,001 inhabitants. Finally, although cities with more than 500,000 inhabitants are decreasing in population, this drop is negligible. Therefore, if smaller towns with up to 10,000 inhabitants offer the services of large cities (more than 500,000 inhabitants) and the advantages of villages are maintained, their power of attraction could increase as happens in territories of between 100,000 and 10,001 inhabitants, promoting a social and technological revolution with the aim of repopulating the most weakened areas.

For this purpose, 596 villages in Andalusia, each with up to 10,000 inhabitants, are grouped together to generate an interconnection between villages in a way that promotes sustainable digital development using SIMA data (2021) (Table 2):

Table 2 Number of inhabitants, companies by sectors and companies with direct potential in the metaverse of small villages in Andalusia

Inhabitants	1550.344
Agriculture, livestock, forestry and fishing companies	37,813
Companies industry, energy, water and waste management	7,519
Construction companies	11,156
Commerce companies	24,740
Transport and storage companies	5,017
Information and communications companies	709
Banking and insurance companies	1,547
Companies public administration, education and health	4,416
Companies real estate, professional, auxiliary, artistic and other services	16,231
Total companies with potential	109,148

Data 2021
Source Own elaboration based on SIMA (2021)

This would imply that the population could reach up to 1550,344 inhabitants if the 2021 data persisted and all individuals in small villages leveraged the opportunities of the metaverse, thus accessing a total of companies with potential of 109,148 companies. When calculating the average number of potential companies, it's determined that in small villages of up to 10,000 inhabitants, there is an average of 183 companies per municipality without the metaverse. Thus, using the metaverse and digital twins, each municipality can increase its access to services, ranging from 80.97 to 1550.344 times its current capacity. This would be a substantial change capable of revolutionizing the concept of small villages. For comparison, based on the SIMA data (2021), there would be a network of businesses and services expanding throughout the villages in the metaverse, of 109,148 companies, being higher than the sum of the capitals of Granada and Huelva together, 99,295 total companies.

As the number of inhabitants grows, there is a linear growth in the number of companies, which reflects a greater dispersion in terms of the number of companies as their size increases. Municipalities with low population behave in terms of number of companies very homogeneously in quantity. Each village needs to grow, but the individual strategies of each municipality don't leverage the trend of collective growth. Because when the number of services increases, the population potentially grows. And if the population increases, the number of services can continue to increase, in a positive trend line in terms of repopulation, to take advantage of this multiplicative potential taking advantage of the inertia and the economies of agglomeration of large cities. Thus, the concept of virtual agglomeration economies is introduced into the metaverse through augmented reality and digital twins. Currently, agglomeration economies are known to describe the advantage that companies experience when grouping in the same area. By being together, they can enjoy a larger set of workers, share facilities and resources, and mutually benefit from innovative ideas and advances (Cabrera-Moya, 2021).

Small villages with fewer than 1500 inhabitants are less competitive and are home to a large number of villages at risk of severe depopulation. For this reason, villages with less than 1500 inhabitants are distinguished from those with a population between 1500 and 10,000 inhabitants, since the latter is the majority of villages at moderate risk. This differentiation helps visualize them separately, even though the overarching strategy is collective. In addition, it allows us to appreciate that the villages of less than 1500 inhabitants that need a greater effort to repopulate their areas are mainly on the northern border of Andalusia, and in the eastern part of the region. In smaller villages, there are fewer basic services, so it is a key aspect to take into account to implement previous policies for access to fiber optic networks, improvement of transport to avoid isolation and inefficiencies in transport. This could offer a wider and more diversified range of services to residents throughout the community, which would otherwise not be possible due to size constraints.

It should be noted that this double segmentation is justified, since they are the ones with the highest risk of depopulation and through the sum of small villages at risk of severe and moderate depopulation, they could strengthen their presence in a joint market in the virtual plane that connects with the physical world. For villages with less than 1500 inhabitants, the average number of inhabitants is 698, with a minimum of 50 and a maximum of 1499. The average number of enterprises is 49, with a minimum of 1 and a maximum of 172.

For villages with 1500 to 10,000 inhabitants, the average number of inhabitants is 4104, with a minimum of 1506 and a maximum of 9951. The average number of enterprises is 316, with a minimum of 62 and a maximum of 1379.

In this way, it is argued that the peoples at a higher risk of depopulation could benefit from those at lower risk, and in turn all the peoples could be nourished by the services of the community, to compose a socioeconomic union with the aim of prospering together. To achieve this, we need to visualize potential companies by municipality, where a more intense color indicates a higher number.

Figure 4 shows the map with Andalusia interconnected in the metaverse as an extension of each small village, in which it can be seen that most of the territories with more than 10,000 inhabitants are from the coastal area and the areas near the Guadalquivir River. On the other hand, there are two areas remarkably different from s such as those of the north and northwest of Andalusia; and on the other hand, those of the south and east of the region. A peculiarity is that areas with populations of less than 1500 inhabitants are concentrated in the southeast of the region. It is essential to contemplate on the map of Andalusia the territories of the interconnected peoples in the metaverse to create efficient transport systems and nuclei of union. In addition, the metaverse could be used to improve the planning and design of transport infrastructure, as well as optimize environmental and economic sustainability. The verse could play an essential role in reducing emissions from areas such as transport, industrial production and power generation (Allam et al., 2022).

The number of companies created by the number of inhabitants is an index used to measure the density of commercial enterprises in the region (Andalusian Institute of Statistics, 2011). Based on the data on the density of companies per inhabitants, it can be seen in Fig. 5, a change in trend at kilometer 50 away from the capital. This type of analysis can be useful for understanding how a small village's geographic location (in terms of its proximity to the capital) can influence the local economy, such as the number of businesses in different sectors. Therefore, it is divided into two groups, of less than 1500 inhabitants and between 1500 and 10,000 inhabitants to see if there are differences. There is a higher density of enterprises per inhabitant in the case of small villages with less than 1500 inhabitants up to the distance of 50 km. However, from this kilometer the villages of the second group, acquire a substantial advantage in density of companies per inhabitant until the position of approximately 120 km, where the difference as the distance grows closer considerably. The density of firms per inhabitant and the distance to the capital can be explained in a multifactorial way with a complex interplay of economic, geographical, social and historical factors. However, some of the possible explanations may be due to aspects such as:

- Competition: In smaller villages (less than 1500 inhabitants) and close to the capital, there may be less competition and more market opportunities, leading to a higher density of companies in micro business ecosystems. Proximity to the capital can provide access to resources, employment, markets and services.
- Agglomeration economies: In larger towns (between 1500 and 10,000 inhabitants), but still at a reasonable distance from the capital (50–120 km), economies of scale could facilitate the growth of more companies. As the distance from the capital increases, economies of scale converge between these villages and the capital does not become a negative incentive for population density.

Therefore, by creating a metaverse that unifies small villages and offers them services, we can eliminate distance barriers, favoring that there are no negative slopes in terms of the density of companies per inhabitant from 0 km to approximately 50 km, since the capitals absorb the entrepreneurial potential of small villages. Nor in the case of increasing the distance by more than 50–60 km, the competitiveness of companies is reduced based on population density, due to factors such as the power of influence of the territories of > 10,000 inhabitants that are growing due to the preferences of residing in these localities and the growth forecasts until 2040. In this way, it seeks to improve the competitiveness of small villages through economies of scale and creating a possible solution to a change in trend by being perceived as villages of a size less than (or equal) to 10,000 inhabitants as an intelligent macro-city united by transport, the use of digital twins and the interconnection by transport services for the demographic and economic growth of these localities. Making the distance to the capital not a factor that limits its growth. Neither would the size of the population, since there would be an extension of each village giving a new dimension and opportunity to the inhabitants of each village. By unifying forces between small villages and removing the barriers of distance, displacements could be made in the metaverse. In order to reduce, in turn, the carbon footprint, by reducing the number of trips while optimizing the well-being of the population through new technologies.

Of all sectors, the one that has the greatest influence on growth power as population increases is trade, as shown in Fig. 6. The density of trading enterprises by number of inhabitants increases with population growth. Therefore, if the effective population is increased in the metaverse, this growth trend may favor expansion into new markets through digitalization, which could result in an increase in the number of companies (Table 3).

As villages with fewer than 1,500 inhabitants grow, they may generally experience a variation in their percentage of companies by sector. In this case, by taking advantage of population leverage, these small villages could witness a notable variation in the growth/decrease of other sectors. In the case of the agriculture, livestock, forestry, and fisheries sector, the percentage could decrease by an average of 6.7% (of the total number of companies). The same applies to the industry, energy, water and waste management sector, which could reduce its presence by 0.3%. As the total population of the villages grows, there are transformations in the total number of companies, and it should be noted that a reduction in percentage does not imply a loss of companies by sectors, rather a greater growth in certain sectors.

On the other hand, there is growth in the total of construction companies (0.2%), transport and storage (0.5%), hospitality (2.8%), commerce (4.3%) and real estate, professional, auxiliary, artistic and other services (4.6%). This means that as the total population of each village grows, new employment opportunities arise. So, by making a union of small villages interconnected by the metaverse and the transport sector, the populations are unified to feed on the benefits of each village. Thus, natural population growth can be generated from the growth of the effective population, which would allow competing with urban areas of more than 10,000 inhabitants, including cities, by offering better living conditions. Thus, it is possible to fight against depopulation in the most disadvantaged areas, since the paradigm is combated that the decrease in the population of rural areas is related to the lack of job opportunities and migration to urban areas in search of better living conditions (Aranzana & Santero-Sánchez, 2021).

Table 3 Variation of the main percentages of presence of companies by sectors

Sector	< 1500 Inhabitants (%)	Between 1500 and 10,000 inhabitants (%)	Variation in municipal population growth (%)
Agriculture, livestock, forestry and fisheries	38	31.3	– 6.7
Industry, energy, water and waste management	6.6	6.3	– 0.3
Construction	9.3	9.5	+ 0.2
Transport and storage	3.8	4.3	+ 0.5
Hospitality industry	10.1	7.3	+ 2.8
Commerce	17.1	21.4	+ 4.3
Real estate, professional, auxiliary, artistic and other services	9.6	14.2	+ 4.6

Characteristics by type of village
Source Authors

In most cases, the density of enterprises in a sector by number of inhabitants is similar in both groups, except in the aforementioned sector of commerce and banking and insurance, and transport and storage. In these three sectors, there is an increase in the density of companies per inhabitant.

In addition to the increased density of transport companies, the exogenous factor of the metaverse and new digitization would, in turn, enhance the development of the transport sector, being responsible for distributing the physical products purchased in the digital twin. The optimization of transport could lead to a lower emission of polluting gases, but the required technological infrastructure could have an environmental impact. Therefore, the net balance of the environmental impact will depend on how the system is implemented and managed. In other words, the reduction of emissions of polluting gases thanks to the metaverse must be greater than the increase in the environmental impact of the technology.

This approach would promote sustainable economic, social, and environmental development. In addition, by strengthening sectors and providing a new dimension to small villages, a series of additional challenges must be faced, such as investment, security, environmental impact and cultural change. To achieve this, energy consumption must come from renewable energy sources and, without the disturbance of local ecosystems and the excessive consumption of natural resources.

The CNAE 2009 codes are analyzed in terms of 4 digits to make visible the main types of companies found in municipalities with up to 10,000 inhabitants. Specifically, those with the greatest presence in number of companies are:

- **Construction and Maintenance**:

  4121: Construction of residential buildings
  4520: Maintenance and repair of motor vehicles
  4110: Real estate development
  4399: Other specialized construction activities N.E.C.
  4312: Land preparation.

- **Transport and Fuel**:

  4941: Carriage of goods by road
  4730: Retail sale of motor fuel in specialized stores.

- **Real Estate and Related Services**:

  6820: Self-employed rental of real estate
  4321: Electrical installations
  4322: Plumbing, heating and air conditioning system installations.

- **Manufacturing and Industry**:

  2512: Manufacture of metalwork
  1043: Manufacture of olive oil
  1071: Manufacture of bread and fresh bakery and pastry products
  2370: Cutting, carving and finishing of stone
  3109: Manufacture of other furniture.

- **Wholesale and Retail Trade**:

  4631: Wholesale of fruit and vegetables
  4711: Retail trade in non-specialized stores
  4673: Wholesale of wood, building materials and sanitary appliances
  4621: Wholesale of cereals, raw tobacco, seeds and feeding stuffs
  4634: Wholesale of beverages
  4752: Retail sale of hardware, paint and glass in specialized stores
  4759: Retail sale of furniture, lighting fixtures and other household goods in specialized stores.

- **Agriculture and Livestock**:

  0161: Activities in support of agriculture
  0113: Cultivation of vegetables, roots and tubers
  0150: Agricultural production combined with livestock production
  0126: Cultivation of oleaginous fruits
  0111: Cultivation of cereals (except rice), legumes and oilseeds.

- **Energy and Professional Services**:

  3519: Production of other types of electrical energy
  6920: Accounting, bookkeeping, auditing and tax advisory activities.

It should be noted that from a sample of 10,258 companies, it is in second place, the transport of goods by road with 6.67%. What is of special interest for the unification of the peoples and shows us the power of the economic impact that it can have in the small villages, which would be enhanced together with the rest of the sectors with potential. Collaboration between these different sectors, facilitated by technology, could be key to addressing common challenges such as geographical isolation, combating depopulation, environmental management and promoting local economic development.

Regardless, by uniting these communities through virtual/augmented reality and interaction with this new form of the internet, there is potential for growth without losing the unique identity of each community. It would happen as it appears in Table 4 from having an average of 2601.24 inhabitants per village to being 1550.344 potential people. This results in community and opportunities multiplying by 596 on average. And the average number of companies goes from 183.13 to 109,148, basing estimates on 2021 data. This information serves to highlight its potential.

In addition, it should be noted that the impact it will have on each village and province will be different. Proportionately, it is small villages that experience the greatest multiplier of increase in population and services. This means that the higher the leverage values, the lower the number of population or services. This indicator shows those provinces with higher (or lower) average population. And, also the provinces with the largest number of services (companies) from which the metaverse is nourished.

Table 4 Potential growth of effective (virtual) population and services through the metaverse in small villages of up to 10,000 inhabitants, in Andalusia

Stats	Potential of inhabitants of effective population (with metaverse)	Number of inhabitants by population (no metaverse)	Potential of companies in joint metaverse	Number of companies per village (no metaverse)
Mean	1550.344	2601.24	109,148	183.13

Source Authors

In any case, the average potential of each province for the inclusion of the metaverse in towns with up to 10,000 inhabitants can be visualized by variables (Table 5).

Population leverage can be understood as the factor by which the virtual population of the community multiplies the actual population of municipalities. Therefore, higher values will indicate the potential for growth of the virtual population. This also helps to detect provinces with villages of less than 10,001 inhabitants smaller on average. The highest values are the province of Almeria with a potential average multiplicative factor of the population of 3215.98; Huelva with 2298.32; followed by Granada with 1970.57 and then Malaga with 1779.45. These 4 provinces are above the average value. For this reason, they are the ones that experience a greater change of virtual population in the community. Below the average values, from lowest to highest are Seville with 627.89; Cadiz with 664.56; Córdoba with a value of 923.44 and then Jaen with 1014.04.

Service leverage refers to the factor by which each small village can multiply its number of companies and services with potential in the metaverse and digital twin. So, the fewer companies with potential, the greater the leverage and perception of the number of services offered in each village. The order is the same, indicating that the

Table 5 Metaverse potential in small villages by Andalusian provinces

Province	Population leverage	Service leverage	Total population leverage	Population growth potential
Almeria	3215.97	226.41	1588,498.00	107,392.70
Cadiz	664.55	46.78	74,890.13	104,563.50
Cordoba	923.44	65.01	126,868.90	105,717.30
Granada	1970.56	138.73	526,083.60	107,234.40
Huelva	2298.32	161.80	1542,340.00	106,984.10
Jaen	1014.04	71.39	129,018.00	106,314.10
Malaga	1779.44	125.27	512,850.90	106,875.90
Seville	627.88	44.20	68,911.86	104,706.10
Average total	1739.06	122.43	628,445.50	106,546.80

Source Authors

amount of population indicates the number of companies (or vice versa) in average terms.

If you multiply the population leverage and the services leverage, you get the total population leverage. In this way, the potential for improvement of depopulation is measured and the current deficiencies are glimpsed in two dimensions: the lack of population and the number of companies, services and opportunities. In this regard, Almeria and Huelva stand out as the provinces most affected by depopulation, showing the greatest variation in effective population in average terms.

In this way, from the effective population, an economic agglomeration is created capable of competing with large cities in which the most benefited according to the data are the small villages, without forgetting that all grow considerably. The average effective population growth potential is between 104,563.50 and 107,392.70. So, an economy of the agglomeration would be created creating a virtual macro-city capable of dealing with towns and cities, eliminating geographical and economic isolation. This would improve the attractiveness of small villages in search of a repopulation with increasing economic returns as in the case of a large city, improving the opportunities and leisure of its inhabitants. The aim is to increase the population of each municipality, based on the leverage produced by the effective population and the increase in services in the community. Developing a new way of understanding depopulation and creating strengths of union and growth (Figs. 10 and 11).

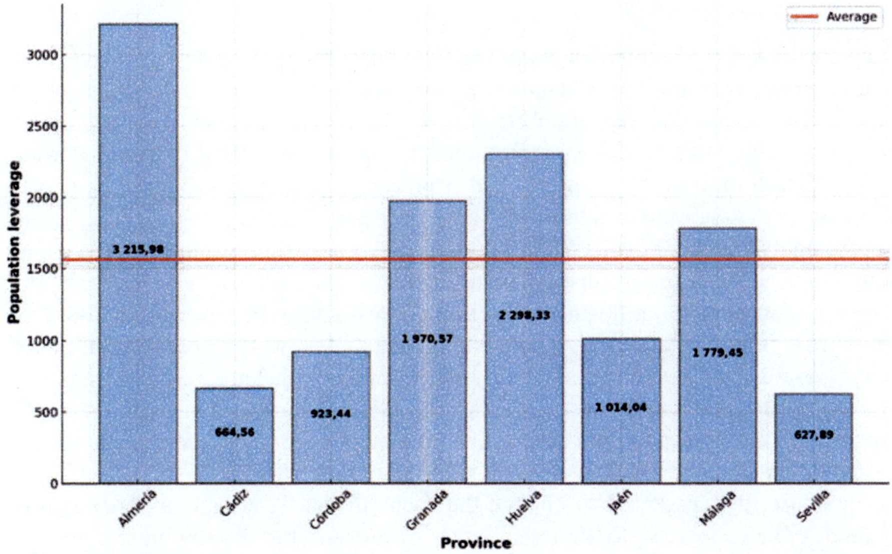

Fig. 10 Levels of population leverage by province, in Andalusia. *Source* Authors

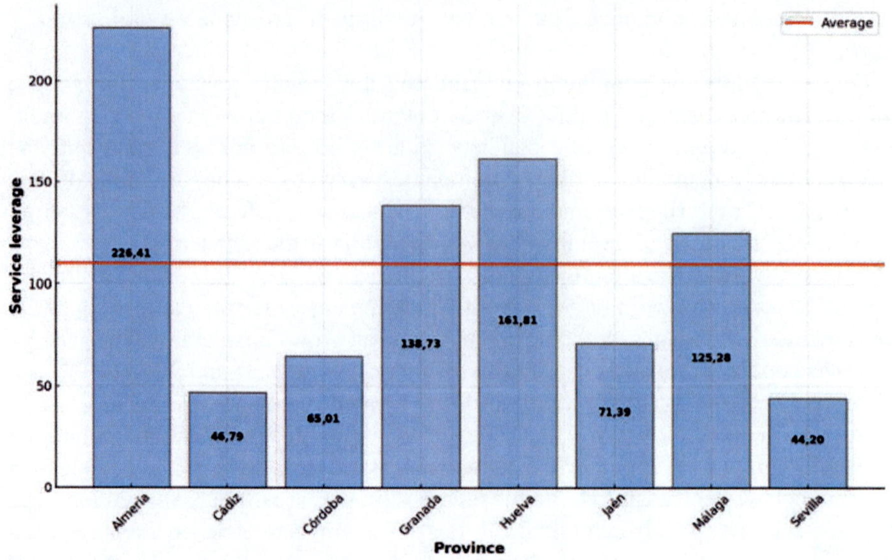

Fig. 11 Levels of leverage of services by province, in Andalusia. *Source* Authors

5 Discussion

From a historical perspective, it should be noted that depopulation is a consequence of a complex combination of factors, many of which can only be influenced at best partially and indirectly by public policies. It is a gap between rural and urban living conditions that has been accentuated in Spain since 1950 (Vicente-Pinilla, 2020). Therefore, it must be approached from a strategy that maximizes the power of influence of small villages to make them competitive in the face of repopulation. According to Karcagi-Kovaits et al. (2009) the key factors contributing to depopulation in rural areas of the European Union include inadequate living conditions, low wages, a reduction in tax revenues and limited availability of services. In addition, the growth in living standards and the emergence of new workplaces also influence this phenomenon. These factors remain significant a decade later, now accompanied by even greater population loss. In the last decade, 785 localities in Andalusia have experienced a decrease in their population, and 82 of them face a serious danger of depopulation (Institute of Statistics and Cartography, 2018).

Therefore, it is essential to change the focus in the fight against depopulation in small villages to attract new inhabitants. At present, the development approach focuses on the rural aspect. This sector, as its productivity increases, requires fewer staff and often offers lower pay than other jobs. It can be observed that as a population increases from less than 1500 inhabitants to between 1500 and 10,000 inhabitants, the total number of companies decreases on average by 6.7%. On the contrary, job opportunities increase in sectors such as construction, with an increase of 0.2%; transport and storage, with an increase of 0.5%; hospitality, with a growth of 2.8%;

commerce, with an increase of 4.3%; and real estate, professional, auxiliary, artistic and other services, with an increase of 4.6%. This provides a broader diversity of opportunities. This aligns with the professional report McKinsey & Company (2022), in which the metaverse could transform several sectors: from communication and technology to advanced industry; from financial services and insurance to retail and food. It can also influence areas such as energy, health and the public sector, as well as tourism and logistics, revolutionizing how we interact, produce and consume in these areas, among others.

The development of a metaverse that unifies small villages of up to 10,000 inhabitants, allows to carry out a Benefit of the Doubt methodology, with an index built from factors of four dimensions: economic, demographic, social and environmental. This is applied in different proposals for sustainable development in rural municipalities as in the case of López-Penabad et al. (2022) with the region of Galicia. However, a new paradigm is also added, such as the inclusion of the metaverse and digital twins, aspects never before addressed in terms of depopulation. Through the use of virtual and extended reality, coupled with the rise of artificial intelligence, a community can be established that enhances the potential of small villages for sustainable digital development. The 4 pillars of the initiative and that align with the proposal of the "Road to the Digital Decade" of the European Commission (2023a) and are:

- Economic development: The construction of a metaverse could help the population access a greater number of services and leisure opportunities to retain and attract the population. This would help level the playing field for businesses located in small villages, giving them the possibility to compete in living conditions with large towns and cities. To achieve this, a digital twin would be developed to interconnect communities through transportation and the evolved form of the internet. On the other hand, it could encourage innovation and entrepreneurship in rural areas, attracting young entrepreneurs and generating employment. This section is linked to the European Commission's goal (2023a) of the digital transformation of companies and the digitalization of public services. Also, with the promotion of collective competitiveness and resilience in the global context.
- Social development: The implementation of the metaverse would contribute to promoting the digital transformation of companies and the need to address the digital literacy of the population. It could also offer access to health and wellness services that would otherwise be limited in rural areas, such as telemedicine, virtual therapy, and wellness programs. In this way, a greater number of public and private services could be accessed by eliminating space barriers and promoting the rise of economic sectors with the engine of transport and technological development. It could serve as a marketing tool to promote tourism in small villages by offering virtual experiences that entice real visitors. As well as cross-selling strategies with promotional coupons for consumption in any of the villages of the metaverse in person. It is linked to the European Commission's goal (2023a) of the digital transformation of companies and the digitalization of public services.
- Environmental development: The creation of a smart digital macro-village through a digital twin by extended reality could reduce the need to travel, which could

have environmental benefits. In addition, it could encourage sustainable practices in rural areas. It is linked to the European Commission's goal (2023a) of transformation towards secure and sustainable digital infrastructures.

All these concepts align with the European Commission's (2021) ideas and initiatives aimed at revitalizing rural services through digital and social innovation.

- Demographic development: Through the strategy of population leverage, inspired by financial leverage, it seeks to increase the effective population of the community, through the virtual population. In this way, the number of services is increased, employment possibilities are increased, technological development is promoted and innovation is linked beyond the agricultural aspect. It is necessary to promote network learning that motivates joint actions, innovative and adapted to the context of the territory so that the inhabitants can return to small villages of up to 10,000 inhabitants (Andreu-Abela & Andreu-Pérez, 2017). This boosts the appeal of small villages. By joining together, an average village population of 2601 can effectively rival larger towns and cities with a combined strength of 1550.344 inhabitants. It is linked to the European Commission's (2023a) goal of the population with digital skills and highly qualified digital professionals.

A real-life digital twin uses physical, augmented, and virtual reality to offer a metaphorical representation of real existence for its members in a virtual environment (Cali et al., 2022). The metaverse has the capacity to prolong the social and labor transformations that have already been accelerated by the pandemic, and to generate new forms of interaction improving inclusion in the future (Hutson, 2022). This transformation can dismantle the barriers of geographical isolation, as physical presence is no longer required to access company services and leisure activities. Carrying out a strategy that minimizes weaknesses and strengthens interconnection to be perceived as a macro-city of more than one million inhabitants is one of the keys to the fight against depopulation in the future. To this end, it is currently working on strategies such as offering tax and economic incentives to attract new residents, as well as investing in digital infrastructures to enable teleworking and attract remote workers (European Commission, 2023b). In this way, a strategy could be generated that modifies the forecasts of the Institute of Statistics and Cartography (2018) in favor of towns with up to 10,000 inhabitants; improving the competitiveness of villages at risk of depopulation. Simultaneously, this strategy promotes social, economic, demographic, and environmental development, leveraging population dynamics to enhance inhabitants' quality of life.

6 Conclusion

The implementation of the metaverse and digital twins in combating depopulation introduces a new paradigm in socioeconomic organization, leveraging these technologies as strategic allies. This offers a fresh perspective on conceptualizing smart villages, interconnected and augmented via digital twins. This approach can be extended to any territory that meets certain conditions, such as having a good transport network and high-speed internet access. In this study, we have specifically analyzed the case of Andalusia, considering its economic, demographic and sustainable digital development implications.

The objective is to explore how clustering and virtual interconnection could help combat depopulation from the theory of population leverage. The approach is rooted in fostering economies of agglomeration by eliminating distance barriers through transportation. Consequently, the community's effective population grows, along with the number of services, making small villages more appealing. This attracts young individuals seeking interesting job opportunities. This creates an interconnected macro-city capable of facing large cities in number of inhabitants and services, through the metaverse.

The fight against moderate and severe depopulation, using innovative tools like the metaverse, aims to foster the growth of villages with up to 10,000 inhabitants to make them more competitive. Estimations from the Institute of Statistics and Cartography (2018) suggest that municipalities with populations of up to 10,000 inhabitants will continue experiencing a decline in resident numbers in the forthcoming years. This trend emphasizes the urgency of implementing measures to increase the population in these areas, thus mitigating the continued danger of depopulation.

These technologies enable the creation of virtual worlds that can transform the growth dynamics of small villages, offering new opportunities and overcoming geographical and physical barriers. Geographical and demographic isolation is also avoided, while the business sectors are strengthened with the fundamental axis of transport and technology. The metaverse emerges as a tool to attract young people and digital nomads seeking development opportunities in areas where costs, such as housing, are lower.

Population leverage, involving the virtual multiplication of municipal populations, emerges as a central concept in this study. The union of 596 localities in Andalusia, each with a population of no more than 10,000 inhabitants, is aimed at fostering an advance in digital sustainability. The combined total of inhabitants would reach 1550.344, and there would be 109,148 companies with growth opportunities from virtual worlds. This would imply a considerable transformation, reinventing the concept of small villages and even surpassing the business structure of important city capitals. The proposed strategy emphasizes grouping villages together to capitalize on collective growth systems because their individual growth potential is more constrained.

Patterns like the influence of capital proximity and population size on firm density per capita have been recognized. However, the metaverse can eliminate these distance-related impediments. Furthermore, it potentially reduces the carbon footprint by minimizing physical movements and optimizing transport networks for sustainability.

The data indicates that as the population increases, the density of commercial enterprises also rises. This relationship implies that a rise in the effective population within the metaverse might foster expansion into fresh markets via digitalization, facilitating company growth through virtual agglomeration economies. Upon deeper analysis, we observe a change in the business landscape as villages with fewer than 1500 inhabitants grow, reaching the threshold of 1500 to 10,000 inhabitants. Concretely:

- Reduction in agriculture, forestry and farming: A 6.7% decrease in the percentage of these enterprises could reflect a change in the local economy as the population grows.
- Growth in construction, transport, marketing, trade and real estate: These sectors are experiencing a growth in the percentage of total firms, indicating new employment opportunities and greater economic diversification beyond the agricultural sector. This enhances the allure for younger individuals.

On the other hand, there are differences in the density of enterprises per capita in the banking and insurance, and transport and storage sectors between villages with up to 1500 inhabitants and those between 1500 and 10,000. Therefore, the implementation of digital twins has the potential to increase the importance of transport and increase its density while boosting the rest of the economic sectors. Road freight transport, in particular, stands out as a critical sector in Andalusia, with 6.67% of the total number of companies in the sample.

The metaverse presents significant growth potential in two key dimensions: population and services.

- In terms of population leverage, the metaverse can act as a virtual multiplier, transforming an average of 2601.24 inhabitants per village into a virtual community of 1550.344 people. In this way, it is sought to leverage the population by increasing the real population from the effective population of the community.
- Regarding service leverage, the average number of companies could increase from 183.13 to 109,148, underlining the expansionary potential of the metaverse.

In addition, the distribution of this potential varies between the provinces of Andalusia. Almeria, Huelva, Granada and Malaga have the highest values of population leverage, which indicates a significant potential for growth of the virtual population. Conversely, Seville, Cádiz, Córdoba, and Jaen, despite showing significant potential, have values below average in terms of leverage.

This study contributes to providing new solutions to one of the main challenges of the European Union and the world, such as the problem of depopulation. The study provides a pioneering contribution in this field and introduces the so-called population leverage theory. It also contributes to improving environmental sustainability by reducing the number of population displacements and boosting the road transport sector.

The study's limitations stem from its projective nature, as this digital twin does not yet exist. Therefore, the potential for the future has been measured, based on analysis, theoretical models and elimination of factors that affect depopulation through the metaverse and the construction of an interconnected macro-city that represents an extension of small villages. This means that it has not been possible to empirically analyze how the initiative affects population growth year after year. However, this lack of future historical data does not diminish the importance of the study. Rather, it underscores the innovation and anticipation of work, by exploring uncharted territory, presenting a vision and potential solutions to current and future problems of population loss.

As for the future development of the subject, other affected territories in Spain and other countries could be analyzed. As well as, implementing the theory of population leverage once the metaverse is extended and a community of small villages is created to be conceived as a digital city in the metaverse in the future. This initiative unifies two fundamental aspects not previously connected by the European Commission, such as the creation of virtual worlds and the fight against depopulation, especially in areas such as Andalusia.

References

Allam, Z., Sharifi, A., Bibri, S. E., Jones, D. R., & Krogstie, J. (2022). The metaverse as a virtual form of smart cities: Opportunities and challenges for environmental, economic, and social sustainability in urban futures. *Smart Cities, 5*(3), 771–801. https://doi.org/10.3390/smartcities5030040

Andalusian Institute of Statistics. (2011). *The transformation of Andalusia 1990–2010*. Retrieved August 13, 2023, from https://www.juntadeandalucia.es/institutodeestadisticaycartografia/InformacionEstadistica/RevistaAniversario-en.pdf

Andreu-Abela, J., & Andreu-Pérez, A. R. (2017). Youth depopulation in rural Andalusia: Factors of return. *Journal of Youth Studies, 122*, 49–68. https://www.injuve.es/sites/default/files/adjuntos/2019/11/injuve_122.pdf

Aranzana, C., & Santero-Sánchez, R. (2021). Public policies and female entrepreneurship in rural areas. A case analysis: Alto Bernesga Biosphere Reserve (León). *Applied Economics Studies, 39*(3). https://doi.org/10.25115/eea.v39i3.3875

Bale, A. S., Ghorpade, N., Hashim, M. F., Vaishnav, J., & Almaspoor, Z. (2022). A comprehensive study on metaverse and its impacts on humans. *Advances in Human-Computer Interaction, 2022*, 1–11. https://doi.org/10.1155/2022/3247060

Borja-Peñaranda, L. D., Villa-Guanoquiza, O. P., & Armijos-Cordero, J. C. (2022). Financial leverage and profitability of the manufacturing industry of the canton in Cuenca, Ecuador. *Quipukamayoc, 30*(62), 47–55. https://doi.org/10.15381/quipu.v30i62.22103

Budiansky, S. (1995). How to build metaverse. *New Scientist, 148*(1999), 32–35.

Cabrera-Moya, D. R. R. (2021). Economies of agglomeration, mass public transport systems type BRT (Bus Rapid Transit) and decisions of geographical location of companies. *City. Revista Brasileira de Gestão Urbana*. https://doi.org/10.1590/2175-3369.013.e20190217

Cali, U., Kuzlu, M., Karaarslan, E., & Jovanovic, V. M. (2022). Opportunities and challenges in metaverse for industry 4.0 and beyond applications. In *IEEE 1st Lobal Emerging Technology Blockchain Forum: Blockchain & Beyond, Igetblockchain*. https://doi.org/10.1109/igetblockchain56591.2022.10087104

Campos-Díez, M. A. (2023). Dance, disco and boxes: Evolution in the modes of nightlife of the young people of Alcoba de los Montes (Ciudad Real). *Aries*. https://doi.org/10.11156/aries/2023.ar0001400

Campos-Sánchez, C. M., Guillén-León, L. A., Acosta-Yanes, R. C., & Gil-Oloriz, M. A. (2022). Metaverse: The future of medicine in a virtual world. *Metaverse Basic and Applied Research, 1*, 4. https://doi.org/10.56294/mr20224

Cao, W., Cai, Z., Yao, X., & Chen, L. (2023). Digital transformation to help carbon neutrality and green sustainable development based on the metaverse. *Sustainability, 15*(9), 7132. https://doi.org/10.3390/su15097132

Deveci, M., Mishra, A. R., Gokasar, I., Rani, P., Pamučar, D., & Özcan, E. (2023). A decision support system for assessing and prioritizing sustainable urban transportation in metaverse. *IEEE Transactions on Fuzzy Systems, 31*(2), 475–484. https://doi.org/10.1109/tfuzz.2022.3190613

Ecology and the Demographic Challenge. (2020). The demographic challenge and depopulation in figures. In La Moncloa. Retrieved July 14, 2023, from https://www.lamoncloa.gob.es/presidente/actividades/Documents/2020/280220-despoblacion-en-cifras.pdf

European Commission. (2021). *Smart villages.* Retrieved July 17, 2023, from https://ec.europa.eu/enrd/enrd-thematic-work/smart-and-competitive-rural-areas/smart-villages_es.html

European Commission. (2023a). *Digital economy and society index 2022.* European Commission. Retrieved July 17, 2023, from https://ec.europa.eu/newsroom/dae/redirection/document/88764

European Commission. (2023b). *Next generation virtual worlds. Societal, technological, economic and policy challenges for the EU.* Publications Office of the European Union. https://doi.org/10.2760/51579

European Commission. (2023c). *Towards the next technological transition: The Commission presents the EU's strategy to lead the web 4.0 and virtual worlds.* Retrieved July 18, 2023, from https://ec.europa.eu/commission/presscorner/detail/es/ip_23_3718

Gibran-Juárez, L. (2018). *Practical handbook of basic statistics for research.* Kresearch Group eBooks. https://doi.org/10.24944/isbn.978-1-945721-24-3

González, A. H. (2022). Demographic vulnerability and distance to large cities: Consequences on the adult-young population in Castilla y León. *Geographical Studies, 82*(291). https://doi.org/10.3989/estgeogr.202198.098

González-Bustos, M. A., & de Andalucía, J. (2022). Sustainable development of the territory and rural world. The impact of Law 7/2021, of December 1, on the promotion of the sustainability of the territory in Andalusia. *Administration of Andalusia, 114*, 125–155. https://doi.org/10.46735/raap.n114.1367

González-Leonardo, M., López-Gay, A., & Recaño-Valverde, J. (2019). Educational decapitalization and second wave of depopulation. *Political Science*, 1–4. https://doi.org/10.46710/ced.pd.esp.16

Hernán-Guerrero, L. (2023). Crypto-assets as a means of payment to obtain a new intelligent world. Cryptocurrencies and the metaverse. *Episteme. Journal of Socioterritorial Studies, 13*(1), 26–37. https://doi.org/10.15332/27113833.8320

Hutson, J. H. (2022). Social virtual reality: Neurodivergence and inclusivity in the metaverse. *Societies, 12*(4), 102. https://doi.org/10.3390/soc12040102

Institute of Statistics and Cartography. (2018). *Population projection by subregional areas 2016–2040.* Junta De Andalucía. Retrieved July 14, 2023, from https://ws089.juntadeandalucia.es/institutodeestadisticaycartografia/blog/2018/05/proyeccion-subregionales-2016-2040/

Karcagi-Kovaits, A., Odor, K., & Kuti, I. (2009). The problem of rural exodus in the national sustainable development strategies and national rural development plans of EU members. *Rural Development, 4*(1). https://portalcris.vdu.lt/server/api/core/bitstreams/8f524893-227c-421c-9b2c-10df05756265/content

López-Penabad, M. C., Casal, A. I., & Rey-Ares, L. (2022). Proposal for a sustainable development index for rural municipalities. *Journal of Cleaner Production, 357*, 131876. https://doi.org/10.1016/j.jclepro.2022.131876

McKinsey & Company. (2022). *Value creation in the metaverse. The real business of the virtual world.* McKinsey & Company. Retrieved July 17, 2023, from https://www.mckinsey.com/~/media/mckinsey/business%20functions/marketing%20and%20sales/our%20insights/value%20creation%20in%20the%20metaverse/Value-creation-in-the-metaverse.pdf

Meta. (2023, May 9). The economic impact of the metaverse in Spain could reach 53,000 million euros in 2035. *Information About Meta.* Retrieved July 17, 2023, from https://about.fb.com/es/news/2023/05/el-impacto-economico-del-metaverso-en-espana-podria-llegar-a-los-53-000-millones-de-euros-en-2035/

Parlamento Europeo. (2023). *Regional and cohesion policy. Economic, social and territorial cohesion.* Retrieved July 15, 2023, from https://www.europarl.europa.eu/factsheets/es/sheet/93/la-cohesion-economica-social-y-territorial

Rodríguez, J. M. (2022). Review of "The social economy as a lever for sustainability in rural territories". *Journal of Labour and Social Security Legal Studies, 4*, 401–406. https://doi.org/10.24310/rejlss.vi4.13622

Rojas-Concepción, A. A., & Guerra-Chagime, R. (2022). World Metaverse Index (WMI): A necessary tool to evaluate the implementation of the metaverse and its impact at a global level. *Metaverse Basic and Applied Research, 1*, 5. https://doi.org/10.56294/mr20225

Rosas-Villar, I. (2023). Sustainable cities and green jobs: Disruptive solutions to climate change. *Revista Digital Universitaria, 24*(2). https://doi.org/10.22201/cuaieed.16076079e.2023.24.2.10

Sánchez, A., & Alba-Sánchez, F. (2018). Impact of rural tourism on employment in Spain: An approach at provincial level. *Rural Development Notebooks, 15*(82), 1–18. https://doi.org/10.11144/javeriana.cdr15-82.itre

SIMA. (2021). *Andalusia village by village. Statistics by sector.* Retrieved July 19, 2023, from https://www.juntadeandalucia.es/institutodeestadisticaycartografia/sima/info.htm?f=h01

Stephenson, N. (1992). *Snow Crash* [Penguin]. William Gibson. http://malsyned.net/tmp/snowcrash.pdf

Van-Horne, J. C., & Wachowics, J. M. (2010). *Fundamentals of financial management* [Pearson Education]. *Pearson Educación eBooks* (13th ed.). http://104.207.147.154:8080/handle/54000/1142

Vicente-Pinilla, F. C. (2020). The true story of rural Spain's depopulation and how it can help us improve our policies. *Working documents of the Spanish Association of Economic History, 20.* https://dialnet.unirioja.es/servlet/articulo?codigo=7560659

Álvaro Bueno-Ferrer is a Doctoral student in Applied Economics at the University of Almería. He holds a Master's degree in Accounting and Corporate Finance from the University of Almería; a Master's in Teacher Training (specializing in economics, business, and commerce) from the University of Almería; a Master's in Digital Marketing from ThePowerMBA; and a Bachelor's degree in Marketing and Market Research. He is a grant recipient from the Department of Economics and Business and a researcher for the Research Group SEJ147—Almeria Group of Applied Economics at the University of Almería. His areas of interest include new technologies, metaverse, efficiency, financial risks, business schoolsmetaheuristics, real estate, and sustainable social transformation. He worked at Real Estate Business School as the director of marketing, advertising, and business data analysis.

Jaime de Pablo Valenciano with a degree in Economic and Business Sciences from the University of Alcalá de Henares and a Ph.D. from the University of Almería, began his teaching career in 1991 at the University of Granada. In 1993, he joined the emerging University of Almería, becoming its first Ph.D. holder. By 1997, he was already a tenured professor. He currently teaches at the Faculty of Economics and Business and the Faculty of Law, in addition to participating in several master's and doctoral programs. He has been a guest professor at multiple international universities, including Chile, Costa Rica, Ecuador, Bolivia, Germany, Finland, Portugal, and Canada. He has authored over 50 articles in international journals like Energies and Sustainability. Notable among his publications in Spanish journals are the Spanish Journal of Agricultural Research and the ITEA Journal.

Corporate Social Responsibility

Disclosure Practices for Tackling Climate Change in Large Spanish Listed Companies

María Mar Miralles-Quirós, José Luis Miralles-Quirós, and Lorena Leal-Espinosa

1 Introduction

Human-induced global warming has increased dramatically in recent decades, endangering life on Earth. Negative effects such as cold and heat waves, torrential rains, and melting glaciers and ice caps have become a reality. That is why combating climate change is crucial nowadays (Miralles-Quirós and Miralles-Quirós, 2022).

There is unanimity in stating that the Paris Agreement, adopted at the Climate Conference organized by the United Nations in Paris in 2015, was the turning point as it was the first universal and legally binding agreement on climate change. Since its entry into force, more and more public institutions are setting carbon neutrality targets. One example is the European Green Deal, an action plan to make the European Union climate-neutral by 2050, as well as laws in each member country such as the Climate Change and Energy Transition Law (Law 7/2021) of Spain. Specifically, this law has important implications for companies, especially in the most polluting sectors. This is because achieving this process requires drastic actions from the private sector as well.

As Ding et al. (2023) point out, combating climate change is one of the most pressing ethical challenges facing businesses today. Large companies in particular are major polluters and therefore have moral obligations to reduce the impact of greenhouse gas (GHG) emissions in the transition to sustainable, low-carbon economies.

M. M. Miralles-Quirós (✉) · J. L. Miralles-Quirós · L. Leal-Espinosa
Faculty of Economic and Business Sciences, University of Extremadura, Av. Elvas S/N, 06006 Badajoz, Spain
e-mail: marmiralles@unex.es

J. L. Miralles-Quirós
e-mail: miralles@unex.es

L. Leal-Espinosa
e-mail: llealesp@alumnos.unex.es

M. d. C. Valls Martínez and J. M. Santos-Jaén (eds.), *Environmentally Sustainable Production*, https://doi.org/10.1007/978-3-031-52656-5_13

In turn, stakeholders are increasingly demanding consistent and transparent information on how these companies are addressing climate change. This is why large listed companies have started to disclose climate-related information, but their methods remain largely unexplored. There are doubts as to whether this disclosure activity is a consequence of real corporate commitment or in response to greenwashing practices (Silva, 2021).

Studies conducted worldwide on the level of corporate commitment to environmental, social and corporate governance (ESG) issues coincide in analyzing the sustainability reports prepared and disclosed by the companies themselves (Ferrero-Ferrero et al., 2023; Heras-Saizarbitoria et al., 2022). However, as indicated by López and Monfort (2017) and López (2020), large companies usually follow a communication strategy with their stakeholders based on the use of various information channels or media to obtain greater legitimacy, as well as to be in more direct contact with all of them.

We understand, therefore, that a company committed to tackling climate change must provide information on its contribution to this overall objective in several ways. This should involve, at the very least, a dedicated sustainability tab on the home page of their official websites that leads directly to this type of information. In addition, they are expected to use their sustainability reports to provide information on their contribution to the fight against climate change. Finally, we understand that a company committed to this objective will use alternative communication channels to communicate its achievements to other stakeholders, such as the social network Twitter. Given that the use of this social network by a large company is voluntary, informing through this channel of its contribution to the fight against climate change can be understood as a greater commitment to these issues and an interest in reaching out to more stakeholders.

In this context, this chapter aims to provide information on how companies that are members of the Ibex-35, the selective index of the Spanish stock market, communicate their commitment to climate change. Previous studies document that these companies have high scores on environmental, social and governance issues and are pioneers when it comes to preparing and disclosing sustainability reports (Fernández-Feijoo-Souto et al., 2012; KPMG, 2022; Miralles-Quirós et al., 2021). Therefore, it is also to be expected that they will have a greater sensitivity to issues such as the 2030 Agenda or the Paris Agreements. Specifically, previous studies by López (2020) and Curtó-Pagés et al. (2021) have documented that Spanish listed companies have gradually increased their communication on the Sustainable Development Goals (SDGs) since 2016 and specifically on goal 13 called Climate Action. However, they also agree that it is necessary to continue shedding light on the characteristics of such communication to identify whether it is a real commitment or the result of external pressure from their stakeholders.

To address this research, a review of the literature is carried out in Sect. 2. Section 3 describes the database and the methodology. Section 4 presents the results of the content analysis carried out. Finally, the last section reflects the conclusions derived from the research as a whole.

2 Literature Review

As pointed out by Monfort et al. (2019) and López (2020), companies that create a communication strategy in various channels to communicate their commitments to stakeholders can be perceived as more transparent, responsible and credible. In this sense, and based on the rapidly growing number of users on social media platforms, many companies have started to share their sustainability activities and results through these (ElAlfy et al., 2020; Manetti & Bellucci, 2016). Social networks such as Facebook, Instagram, YouTube and Twitter are highly interactive communication and information platforms that allow companies to communicate or establish a direct dialogue with their stakeholders.

There are previous studies on the use of social networks for corporate social responsibility communication (Cho et al., 2017; Gómez-Carrasco et al., 2021). In relation to the SDGs, we have identified the work conducted by ElAlfy et al. (2020) and López (2020). Specifically, ElAlfy et al. (2020), unlike previous studies, only focus on the social network Twitter. These authors analyze the SDG-related tweets written up to mid-2019 by the 500 companies belonging to the Standard and Poor Index. The results show that companies tweet about the SDGs that are related to their core business, so the sector of activity influences the company's Twitter communication strategy. Finally, López (2020) conducted a case study of Spanish multinationals belonging to the Ibex-35 index and listed in the Dow Jones Sustainability Index in 2017 with data collected through corporate websites, reports, media and social networks. In total, they analyzed 14 companies. The results show that the sampled companies have integrated the SDGs into their communication, but they also note that it is difficult to quantify their contribution to the SDGs. They consider that further research is needed, using specific indicators to show whether companies are promoting the SDGs. However, we have not observed that similar studies exist to address climate change disclosure.

We believe that it is crucial to analyze the communication on climate change of large Spanish listed companies along the lines of what was previously done by López (2020), although incorporating additional questions that allow us to improve our knowledge of the communication strategy of these companies.

However, as indicated in the introductory section, most studies that address corporate commitment to ESG issues use sustainability reports prepared and disclosed by the companies themselves as a key tool. Within this field of research are studies by Hummel and Szekely (2020), Heras-Saizarbitoria et al. (2022), Aguado-Correa et al. (2023), Ferrero-Ferrero et al. (2023) and Lodhia et al. (2023), among others. Specifically, these studies address the dissemination of information on the achievement of the 17 global goals set by the United Nations 2030 Agenda, including Goal 13 on Climate Action. These studies highlight the scarce and heterogeneous information disseminated. However, they also note that the quality of the information increased significantly over time, but on many occasions, it was not quantitative and prospective information. Furthermore, these studies also coincide in highlighting objective 13 as one of the most pursued by the companies analyzed. This is why we consider it convenient to analyze exclusively climate action in the present research. Among these studies, it is necessary to highlight the one carried out by Ferrero-Ferrero et al. (2023). These authors reflect the importance of highlighting those companies that are leaders in SDG communication. In this way, it is possible to detect those sectors and/or companies that are most committed to sustainability, as well as those that need to be influenced to become more sustainable. Following this premise, in the final part of the empirical study, we will reflect a ranking of commitment to climate issues of the Ibex-35 companies based on the content analysis carried out. This will allow us to deduce which types of companies are the most committed and which should be the main target of public awareness-raising policies.

3 Database and Methodology

As a sample of large Spanish listed companies, we have selected the companies in the selective Ibex-35 index. As indicated by Bolsas y Mercados Españoles (2023), the Ibex-35 is an index that includes the 35 most liquid and largest capitalization stocks among all those listed on the Spanish Stock Exchange Interconnection System. This index is a national and international benchmark for many investors and analysts. It is also used as the underlying in the contracting of derivative products. Its composition changes every six months. Since the empirical work was carried out between November 2021 and January 2022, the companies that made up the Spanish selective index on that date were analyzed. The sample, therefore, is composed of companies belonging to the following sectors: consumer goods, consumer services, financials, industry and construction, energy and technology.

As for the methodology employed, we followed the previously documented literature and conducted an exploratory study based on a content analysis of the sustainability information provided by the companies in the sample through various communication channels: their official websites, their sustainability reports and the social network Twitter.

Companies communicate directly with stakeholders primarily through their websites, where they publish all kinds of information. We believe that a company that is truly committed to sustainable development will provide visible access to this information on its website and include activities to reduce the negative consequences of climate change. It is also expected that there will be direct access to their sustainability reports on the websites. Therefore, this has been the first aspect to be analyzed in our exploratory study.

These reports have also been analyzed in depth due to the comprehensive information they provide on the various areas of sustainability in which the company is involved: environmental, social and corporate governance. In addition, in the case of Ibex-35 companies, they are required to publish a sustainability report following the approval of Law 11/2018 on Non-Financial Information and Diversity, but it is not mandatory to include information on how they deal with climate change. However, these companies are expected to voluntarily report on this key aspect.

Finally, another of the platforms most used in recent years by companies to communicate with their stakeholders is social media (Cho et al., 2017; ElAlfy et al., 2020). This medium is a point of easy access, where less technical vocabulary, images and videos are used. All this facilitates communication and makes it much easier, as there is direct interaction between the company and its stakeholders. In addition, a wider and more plural audience can be reached in this way. Specifically, following ElAlfy et al. (2020), in this research we will only use the social network Twitter because, of all the social networks, it is the only one that is considered a means of communication. Specifically, through this social network, it is possible to reach the young society that mostly uses this medium as the main alternative for information. Therefore, the use of Twitter by large companies is very different from their use of sustainability reports. Firstly, these companies are not obliged to use this social network. Therefore, the use of Twitter to disclose their involvement in the Paris Agreement and the United Nations Climate Action Goal is purely voluntary, which can be understood as a greater commitment.

In short, in the first phase, we analyzed the use made by the Ibex-35 companies of these three communication channels in relation to their fight against climate change. The objective is not only to analyze whether they report on climate change but also whether the information they provide in these three communication channels is consistent.

Second, we focus on the information provided by these companies in their sustainability reports to analyze the quality of the information and not a descriptive study on what or simply how much is disclosed. As Silva (2021) points out, research that evaluates only the quantity of information has been criticized in the disclosure literature (Beattie et al., 2004; Kannenberg & Schreck, 2019). This is why subsequent analyses of sustainability disclosure have included additional criteria. Specifically, we take into account whether the disclosed information is both qualitative and quantitative in nature, i.e., whether it presents precise measures of its commitment to tackling climate change that are directly comparable with other companies in the market. Finally, we also consider the time orientation of the information disclosed, i.e. whether it reports retrospectively and even prospectively. This information is important to better understand both the past performance and the future direction of the company. According to Silva (2021), this is an indication of the level of commitment and implementation of information related to any aspect related to sustainability.

Finally, we will make a comparison of the results obtained in relation to whether climate change disclosure practices are consistent and complete and present a total score, following the studies of Pizzi et al. (2021) and Ferrero-Ferrero et al. (2023), among others. It will allow us to establish a mapping of the commitment to address climate change of the main companies in the Spanish stock market and will allow us to draw conclusions for the future.

4 Results

As indicated in the methodology section, the first phase of the empirical study consisted of analyzing whether large Spanish listed companies report on their practices to address climate change in various media platforms. Specifically, we have analyzed the content of their webpages, their Twitter accounts and the sustainability reports prepared and disseminated by them.

Firstly, we found that all the companies analyzed have an easily accessible sustainability section on their websites. However, 9 (26%) of these companies do not report on their practices to address climate change in this section (Table 1). Specifically, there are three companies in the pharmaceutical industry (Almirall, Grifols and Rovi), one in transportation (IAG), one in banking (Banco Sabadell), one in real estate (Merlin Properties), and one in telecommunications (Cellnex). In general, these companies limit themselves to talking about their contributions to the environment and do not go into issues related to climate change. The remaining 26 companies (74%) specifically report on their practices to address climate change. However, this information is mainly qualitative. They limit themselves to stating that their climate action plan or policy is based on reducing greenhouse gas (GHG) emissions, using renewable energy sources and sustainable fuels, etc.

Regarding the use of the social network Twitter to communicate on climate change, we have found that 24 of the 35 companies analyzed (69%) have tweeted at some point in the last few years on this subject, compared to the remaining 11 companies (31%). Most of these companies tweet news or newspaper articles highlighting their contribution to addressing climate change. This form of communication can elicit important signals of trust among the various stakeholders given the fact that it is an exponential and voluntary communication channel (Zarzycka et al., 2023). Undoubtedly, its use by the company enhances the idea that the company has a real commitment to this issue. However, it is also important to note that the companies that tweet the most about climate change are also those belonging to environmentally sensitive industries. That is, the companies that most use this medium to communicate with their stakeholders are companies that, by their nature, have to face serious demands for non-toxic packaging, less polluting processes, eco-friendly manufacturing practices, etc. The cases of Acciona, Enagás, Ferrovial and Iberdrola, among others, stand out.

Thirdly, we address the use made by these companies of sustainability reports to report on climate change. It should be noted that, although these companies are required by law to issue these reports, it is not mandatory to mention climate change in them. As shown in Table 1, all the companies prepare and disclose sustainability information annually and all mention climate change at least once. In the present research, the reports analyzed were those corresponding to the year 2020, since they were the latest published at the time of the empirical study.

The first aspect to highlight is the lack of homogeneity in the drafting of the information, with annual reports, integrated reporting, non-financial statements and sustainability reports coexisting. This implies that the length of the reports was also

Table 1 Disclosure practices about climate change in media platforms

Sector	Industry	Company	Webpage	Twitter	Sustainability report		
					Type	Pages	CC
Consumer goods	Pharmaceutical and biotechnology	Almirall			AR	55	17
		Grifols			IR	278	75
		PharmaMar	✔		NF	80	10
		Rovi			AR	98	6
	Textile	Inditex	✔		NF	210	29
Consumer services	Transportation	AENA	✔	✔	NF	268	81
		IAG			AR	230	39
	Tourism	Meliá	✔	✔	AR	413	74
Financial	Banks	B.Sabadell		✔	AR	282	32
		B.Santander	✔	✔	AR	872	125
		Bankinter	✔	✔	IR	134	29
		BBVA	✔	✔	AR	619	93
		CaixaBank	✔	✔	AR	401	60
	Insurance	Mapfre	✔	✔	IR	220	35
	Real estate	Colonial	✔	✔	AR	101	1
		Merlin			AR	96	1
Oil and energy	Electricity, gas, oil and renewable energies	Enagas	✔	✔	AR	306	63
		Endesa	✔	✔	NF	316	108
		Iberdrola	✔	✔	NF	556	187
		Naturgy	✔	✔	AR	706	121
		REC	✔	✔	AR	232	26
		Repsol	✔	✔	IR	200	100
		Solaria	✔	✔	SR	70	44
Raw materials, industry and construction	Construction	ACS group	✔		IR	287	64
		Acciona	✔	✔	IR	76	30
		Ferrovial	✔	✔	IR	290	40
	Engineering	Fluidra	✔		IR	136	7
	Manufacture	Siemens G	✔	✔	NF	191	59
	Metals and minerals	Acerinox		✔	IR	206	24
		ArcelorMittal	✔	✔	NF	51	11
		Cie Autom			AR	222	21
Technology	Electronic and software	Amadeus	✔	✔	AR	180	51
		Indra	✔	✔	AR	336	58
	Telecommunications	Cellnex			IR	592	50
		Telefónica	✔	✔	AR	474	105

very varied and, with it, the number of times that climate change was mentioned. Specifically, four companies from the energy sector (Endesa, Iberdrola, Naturgy and Repsol), one from the financial sector (Banco Santander) and one from the technology sector (Telefónica) mention climate change more than 100 times in their reports. In contrast, the companies that cite climate change the least are companies in the real estate industry (Inmobiliaria Colonial and Merlin Properties) and companies in the pharmaceutical industry (mainly Almirall, PharmaMar and Rovi). This does not surprise us since the vagueness of these companies in the use of this channel to communicate on climate change coincides with the use for the same purpose of the two channels considered above. These comparative results presented in Table 1 provide an initial idea of which companies are truly committed to climate change.

AR, annual report; IR, integrated reporting; NF, non-financial statement; SR, sustainability reporting. *Source* Own elaboration.

However, as indicated in the methodological section, it is not enough to analyze whether the company uses various communication channels or to count the number of times the company mentions climate change in its sustainability reports. This is not enough to be able to rigorously identify whether or not the company has a real commitment to climate change or whether its communication responds to greenwashing practices. It is advisable to complete the investigation by analyzing the content of the information provided in these reports. Undoubtedly, of the three channels analyzed, sustainability reports are the ones that provide the most information to stakeholders. Specifically, following Silva (2021), we have analyzed whether qualitative and quantitative information is disclosed in the reports, as well as prospective and retrospective information.

Regarding the quantitative information that companies can issue directly related to climate change, we have taken into account whether the company measures and disseminates its contribution to the carbon footprint. The basis for a correct measurement of the carbon footprint has its origin in Kyoto, specifically in the GHG Protocol. This protocol established the polluting gases ranging from carbon dioxide to methane and established a homogeneous and comparable measurement system based on three scopes. Scope 1 refers to direct emissions. That is, emissions caused by a company from the operation of the things it owns or controls. They may result from operating the machinery used to manufacture products, driving vehicles or simply heating buildings and powering computers. Scope 2, on the other hand, refers to indirect emissions, created by the production of the energy an organization purchases. Finally, Scope 3 also refers to indirect emissions. However, in this case, they are indirect emissions produced by customers who use the company's products or those produced by suppliers who manufacture the products used by the company (Fig. 1).

Usually, companies can easily measure their Scope 1 and 2 emissions and can control them by taking measures such as using renewable energy or electric vehicles. But Scope 3 emissions are under the control of suppliers or customers, so they are affected by decisions made outside the company. This means that measuring Scope 3 emissions involves tracking the activities of the entire business model or value chain, from suppliers to end users. Measuring and disclosing Scope 3 emissions therefore represents a greater commitment by the company to address climate change.

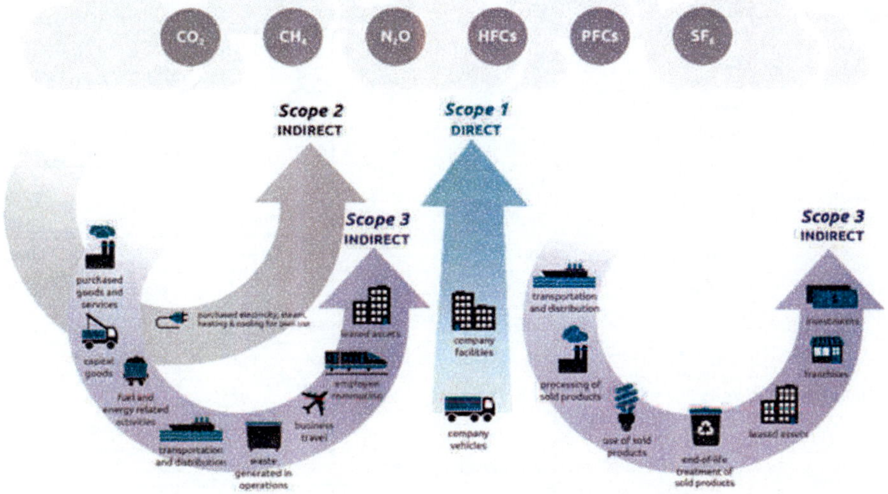

Fig. 1 GHG Protocol scopes and emissions across the value chain. *Source* www.wbcsd.org

Table 2 shows all the companies that measure and disclose information on scopes 1, 2 and 3. As might be expected, 31 companies report information on scopes 1 and 2 (89%). On the other hand, four companies (Almirall, PharmaMar, Enagás and ArcelorMittal) do not provide information directly or do not provide a breakdown or quantitative information. Only 23 companies provide information on Scope 3 emissions (66%). Specifically, all companies in the consumer services, financial and telecommunications sectors report on Scope 3 emissions. However, certain companies in the consumer goods (Almirall, PharmaMar, Rovi and Inditex), energy (Enagás and Red Eléctrica Española) and construction, manufacture, metals and minerals (ACS, Ferrovial, Siemens Gamesa, Acerinox, ArcelorMittal and Cie Automotive) sectors do not provide this information in their reports. These data are negative because many of these companies belong to key sectors in the fight against climate change. These data also reflect that policymakers should create strategies to help these companies from the public sector to measure and disclose their contribution to the carbon footprint throughout their value chain.

On the other hand, as shown in Table 2, we have also taken into account qualitative information that we consider key to assessing whether the company is truly committed to climate change. Specifically, we have considered two aspects: whether the company has a specific climate change plan and whether it has conducted an analysis of future risks and opportunities related to climate change. The first key information would reflect that the company has introduced into its corporate strategy concrete actions to address climate change. In contrast, the second key information would reflect that the company is considering future actions to address climate change.

Table 2 Quantitative and qualitative information provided by sustainability reports

Sector	Industry	Company	Quantitative			Qualitative	
			Scope 1	Scope 2	Scope 3	CC Policy	Risks analysis
Consumer goods	Pharmaceutical and biotechnology	Almirall				✔	
		Grifols	✔	✔	✔	✔	✔
		PharmaMar					
		Rovi	✔	✔		✔	
	Textile	Inditex	✔	✔		✔	
Consumer services	Transportation	AENA	✔	✔	✔	✔	✔
		IAG	✔	✔	✔	✔	✔
	Tourism	Meliá	✔	✔	✔	✔	✔
Financial	Banks	B.Sabadell	✔	✔	✔		✔
		B.Santander	✔	✔	✔		✔
		Bankinter	✔	✔	✔	✔	✔
		BBVA	✔	✔	✔	✔	✔
		CaixaBank	✔	✔	✔		✔
	Insurance	Mapfre	✔	✔	✔	✔	✔
	Real estate	Colonial	✔	✔	✔	✔	✔
		Merlin	✔	✔	✔		
Oil and energy	Electricity, gas, oil and renewable energies	Enagas				✔	
		Endesa	✔	✔	✔	✔	
		Iberdrola	✔	✔	✔	✔	
		Naturgy	✔	✔	✔	✔	
		REC	✔	✔		✔	✔
		Repsol	✔	✔	✔	✔	✔
		Solaria	✔	✔	✔		✔
Raw materials, industry and construction	Construction	ACS Group	✔	✔		✔	
		Acciona	✔	✔	✔	✔	✔
		Ferrovial	✔	✔		✔	✔
	Engineering	Fluidra	✔	✔	✔	✔	✔
	Manufacture	Siemens G	✔	✔		✔	✔
	Metals and minerals	Acerinox	✔	✔		✔	
		ArcelorMittal					
		Cie Autom	✔	✔		✔	✔
Technology	Electronic and software	Amadeus	✔	✔	✔	✔	✔
		Indra	✔	✔	✔	✔	✔
	Telecommunications	Cellnex	✔	✔	✔		✔
		Telefónica	✔	✔	✔	✔	✔

Source Own elaboration

As Table 2 shows, 27 companies (77%) document in their sustainability reports the climate change plan or policy they have implemented. In contrast, only 8 companies (23%) do not present a specific climate change policy. In these cases, these companies document the existence of an environmental action plan in their companies and, therefore, their commitment to environmental issues. However, they do not document a specific climate change plan.

Finally, 23 companies (66%) document in their reports that in 2020, they analyzed risks and opportunities related to climate change to adapt in the best possible way to this issue in the following years. This reflects that a high percentage of Ibex-35 companies have a real concern for climate change. On the other hand, 12 companies have not carried out a risk and opportunity analysis. Specifically, Almirall, PharmaMar and Rovi from the pharmaceutical industry, Inditex from the textile industry, Merlin Properties from the real estate industry, Enagás, Endesa, Iberdrola and Naturgy from the energy industry, Accerinox from the manufacturing industry and ArcelorMittal from the metals and minerals industry. Many of them are companies that also did not meet many of the aspects mentioned above. Thus, it is becoming clearer which companies have a real commitment to climate change and which do not.

Finally, the third objective of our research consisted of a climate change disclosure score of the Ibex-35 companies. The results are presented in Fig. 2 and Table 3. Specifically, a score of consistent disclosure practices, a score of full disclosure practices and a total score were obtained. The first tries to measure whether the company has used the three media and whether it has done so in a consistent manner, reflecting the same message to its stakeholders. The second tries to measure whether the information provided in the sustainability reports has been complete, i.e. whether it presents quantitative and qualitative, present and future information. Both scores range from 0 to 5 points. Finally, the total score is the sum of the first two scores and, therefore, has a range from 0 to 10 points. The purpose of this score is to reflect the conclusion about the real commitment of these companies in an intuitive way. As can be seen, three companies do not pass the test. These are Almirall and PharmaMar (score 3) and ArcelorMittal (score 4). On the other hand, 10 companies (29%) obtained the maximum score. Therefore, we can indicate that these are the companies that have a real commitment to addressing climate change. These companies are AENA (transportation), Meliá (Tourism), Bankinter and BBVA (Banks), Mapfre (Insurance), Repsol (Energy), Acciona (Construction), Amadeus and Indra (Electronic and Software), Telefónica (Telecommunications). These companies belong to a wide variety of industries. This suggests that the commitment to addressing climate change can be achieved by all companies and does not have to be specific to environmentally sensitive industries. All large companies, regardless of their business activity, can help to ameliorate this global problem.

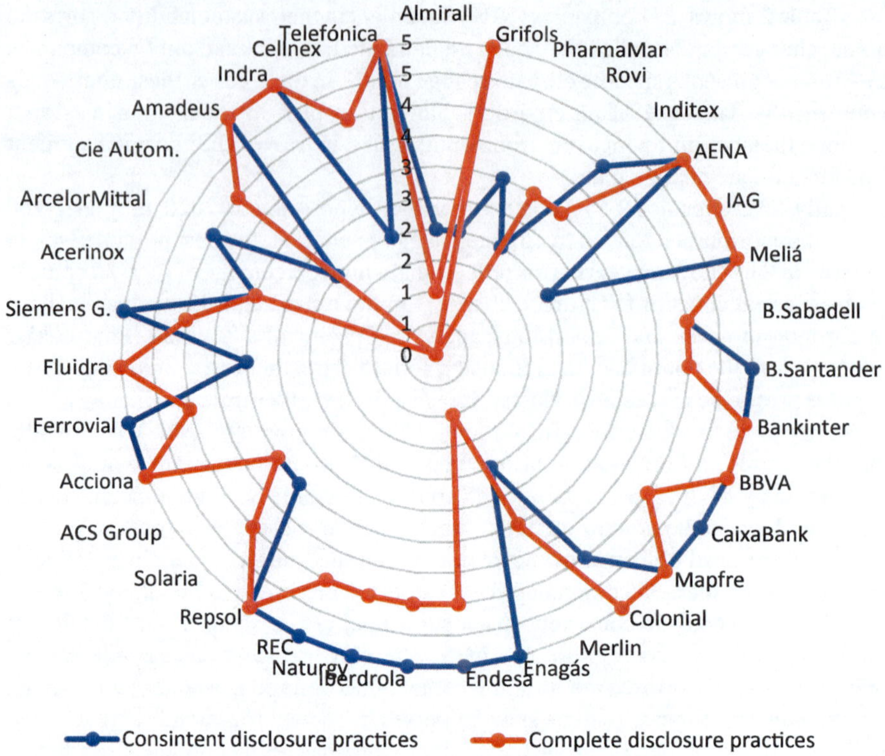

Fig. 2 Complete and consistent information comparison. *Source* Own elaboration

5 Conclusion

This research aimed to analyze how large Spanish companies that were part of the Ibex-35 index communicate their commitment to climate change. To this end, we have considered three communication channels: the official of each company, the latest sustainability report elaborated and disclosed through these websites during the preparation of the empirical study and, finally, the information provided through the social network Twitter. Finally, focusing on the information provided by the sustainability reports, we have prepared a score to detect the most committed companies.

Our overall results reveal that the information these companies provide regarding their climate change initiatives is very varied. Some companies show a real commitment to climate change, integrate climate change policies into their corporate strategy and inform stakeholders about their current and future actions in various ways. On the other hand, we also observe that a small percentage of companies belonging to the selective index of the Spanish stock market are not committed to climate change

Table 3 Disclosure practices scores

Sector	Industry	Company	Consistent disclosure practices	Complete disclosure practices	Total score
Consumer goods	Pharmaceutical and biotechnology	Almirall	2	1	3
		Grifols	2	5	7
		PharmaMar	3	0	3
		Rovi	2	3	5
	Textile	Inditex	4	3	7
Consumer services	Transportation	AENA	5	5	10
		IAG	2	5	7
	Tourism	Meliá	5	5	10
Financial	Banks	B.Sabadell	4	4	8
		B.Santander	5	4	9
		Bankinter	5	5	10
		BBVA	5	5	10
		CaixaBank	5	4	9
	Insurance	Mapfre	5	5	10
	Real estate	Colonial	4	5	9
		Merlin	2	3	5
Oil and energy	Electricity, gas, oil and renewable energies	Enagas	5	1	6
		Endesa	5	4	9
		Iberdrola	5	4	9
		Naturgy	5	4	9
		REC	5	4	9
		Repsol	5	5	10
		Solaria	3	4	7
Raw materials, industry and construction	Construction	ACS Group	3	3	6
		Acciona	5	5	10
		Ferrovial	5	4	9
	Engineering	Fluidra	3	5	8
	Manufacture	Siemens G	5	4	9
	Metals and minerals	Acerinox	3	3	6
		ArcelorMittal	4	0	4
		Cie Autom	2	4	6
Technology	Electronic and software	Amadeus	5	5	10
		Indra	5	5	10
	Telecommunications	Cellnex	2	4	6
		Telefónica	5	5	10

Source Own elaboration

since the disclosure of information to their stakeholders on this issue is excessively limited.

Finally, we believe that there is still much to be done and communicated to achieve the objectives established globally in the Paris Agreements and to reduce global warming. The business community must continue to carry out activities that contribute to meeting these challenges. To that end, we consider that policymakers should provide mechanisms to help those companies that are less committed. Moreover, the role of stakeholders in this field is essential. They should continue demanding homogeneous and truthful information from large companies on their contribution to tackling climate change.

References

Aguado-Correa, F., de la Vega-Jiménez, J. J., López-Jiménez, J. M., Padilla-Garrido, N., & Rabadán-Martín, I. (2023). Evaluation of non-financial information and its contribution to advancing the sustainable development goals within the Spanish banking sector. *European Research on Management and Business Economics, 29*(1), 100211. https://doi.org/10.1016/j.iedeen.2022.100211

Beattie, V., McInnes, B., & Fearnley, S. (2004). A methodology for analysing and evaluating narratives in annual reports: A comprehensive descriptive profile and metrics for disclosure quality attributes. *Accounting Forum, 28*(3), 205–236. https://doi.org/10.1016/j.accfor.2004.07.001

Bolsas y Mercados Españoles (2023) IBEX 35®: principal referencia para la economía, los mercados y las empresas españolas. Accesible en: https://www.bolsamadrid.es/esp/indices/ibex/PreguntasClaveIbex35.aspx

Cho, M., Furey, L. D., & Mohr, T. (2017). Communicating corporate social responsibility on social media: Strategies, stakeholders, and public engagement on corporate Facebook. *Business and Professional Communication Quarterly, 80*(1), 52–69. https://doi.org/10.1177/2329490616663708

Curtó-Pagès, F., Ortega-Rivera, E., Castellón-Durán, M., & Jané-Llopis, E. (2021). Coming in from the cold: A longitudinal analysis of SDG reporting practices by Spanish listed companies since the approval of the 2030 agenda. *Sustainability, 13*, 1178. https://doi.org/10.3390/su13031178

Ding, D., Liu, B., & Chang, M. (2023). Carbon emissions and TCFD aligned climate-related information disclosures. *Journal of Business Ethics, 182*, 967–1001. https://doi.org/10.1007/s10551-022-05292-x

ElAlfy, A., Darwish, K. M., & Weber, O. (2020). Corporations and sustainable development goals communication on social media: Corporate social responsibility or just another buzzword? *Sustainable Development, 28*(5), 1418–1430. https://doi.org/10.1002/sd.2095

Fernández-Feijóo-Souto, B., Romero, S., & Ruiz-Blanco, S. (2012). Measuring quality of sustainability reports and assurance statements: Characteristics of the high-quality reporting companies. *International Journal of Society Systems Science, 4*(1), 5–27. https://doi.org/10.1504/IJSSS.2012.045371

Ferrero-Ferrero, I., Muñoz-Torres, M. J., Rivera-Lirio, J. M., Escrig-Olmedo, E., & Fernández-Izquierdo, M. A. (2023). SDG reporting: An analysis of corporate sustainability leaders. *Marketing Intelligence and Planning, 41*(4), 457–472. https://doi.org/10.1108/MIP-07-2022-0332

Gómez-Carrasco, P., Guillamón-Saorín, E., García Osma, B. (2021). Stakeholders versus firm communication in social media: The case of Twitter and corporate social responsibility information. European Accounting Review, 30(1), 31–62. https://doi.org/10.1080/09638180.2019.1708428

Heras-Saizarbitoria, I., Urbieta, L., & Boiral, O. (2022). Organizations' engagement with sustainable development goals: From cherry-picking to SDG-washing? *Corporate Social Responsibility and Environmental Management, 29*(2), 316–328. https://doi.org/10.1002/csr.2202

Hummel, K., & Szekely, M. (2020). Disclosure on the sustainable development goals–evidence from Europe. https://ssrn.com/abstract=3411017

Kannenberg, L., & Schreck, P. (2019). Integrated reporting: Boon or bane? A review of empirical research on its determinants and implications. *Journal of Business Economics, 89*, 515–567. https://doi.org/10.1007/s11573-018-0922-8

KPMG (2022). The KPMG survey of corporate responsibility reporting 2022. KPMG. https://kpmg.com/sg/en/home/insights/2022/10/global-survey-of-sustainability-reporting-2022.html

Law 7/2021, of 20th May, of Climate Change and Energy Transition. https://www.boe.es/diario_boe/txt.php?id=BOE-A-2021-8447

Law 11/2018, of 28th December, about Non-Financial Information and Diversity. https://www.boe.es/eli/es/l/2018/12/28/11

Lodhia, S., Kaur, A., & Kuruppu, S. C. (2023). The disclosure of sustainable development goals (SDGs) by the top 50 Australian companies: Substantive or symbolic legitimation? *Meditari Accountancy Research, 31*, 1578–1605. https://doi.org/10.1108/MEDAR-05-2021-1297

López, B. (2020). Connecting business and SDGs in Spain. *Marketing Intelligence and Planning, 38*(5), 573–585. https://doi.org/10.1108/MIP-08-2018-0367

López, B., & Monfort, A. (2017). Creating shared value in the context of sustainability: The communication strategy of MNCs, corporate governance and strategic decision making. In: Lawrence Emeagwali, O. (Ed), InTech. https://www.intechopen.com/books/corporate-governance-and-strategic-decision-making/creating-shared-value-in-thecontext-of-sustainability-the-communication-strategy-of-mncs

Manetti, G., & Bellucci, M. (2016). The use of social media for engaging stakeholders in sustainability reporting. *Accounting, Auditing and Accountability Journal, 29*, 985–1011. https://doi.org/10.1108/AAAJ-08-2014-1797

Miralles-Quirós, M. M., & Miralles-Quirós, J. L. (2022). Decarbonization and the benefits of tackling climate change. *International Journal of Environmental Research and Public Health, 19*(13), 7776. https://doi.org/10.3390/ijerph19137776

Miralles-Quirós, M. M., Miralles-Quirós, J. L., & Daza-Izquierdo, J. (2021). The assurance of sustainability reports and their impact on stock market prices. *Management Letters/cuadernos De Gestión, 21*(1), 47–60.

Monfort, A., Villagra, N., & López-Vázquez, B. (2019). Exploring stakeholders' dialogue and corporate social responsibility (CSR) on Twitter. *El Profesional De La Información, 28*(5), e280513. https://doi.org/10.3145/epi.2019.sep.13

Pizzi, S., Rosati, F., & Venturelli, A. (2021). The determinants of business contribution to the 2030 agenda: Introducing the SDG reporting score. *Business Strategy and the Environment, 30*, 404–421. https://doi.org/10.1002/bse.2628

Silva, S. (2021). Corporate contributions to the sustainable development goals: An empirical analysis informed by legitimacy theory. *Journal of Cleaner Production, 292*, 125962. https://doi.org/10.1016/j.jclepro.2021.125962

Zarzycka, E., Krasodomska, J., Dobija, D., Grabowski, W., Jemielniak, D. (2023). Communication aimed at engendering trustworthiness: An analysis of CSR messages on Twitter. Business Ethics, the Environment and Responsibility

María Mar Miralles-Quirós She is a Full Professor of Financial Economics and Accounting and the Director of the Research Group on Markets and Financial Assets at the University of Extremadura in Spain. Her research has focused on providing shareholders' investment strategies that would beat the market using sophisticated mathematical approaches, as well as the analysis of companies' sustainable activities and their connection to shareholders' value creation. Her research works have been published in prestigious journals such as Business Strategy and the Environment, Corporate Social Responsibility and Environmental Management, Energy Economics, International Review of Economics and Finance, Journal of Business Ethics, Journal of International Financial Markets, Institutions & Money, Mathematics, and North American Journal of Economics and Finance, among others.

José Luis Miralles-Quirós He is an Associate Professor of Financial Economics and Accounting at the University of Extremadura in Spain. His research has primarily focused on providing shareholders' investment strategies that would beat the market in adverse situations using sophisticated mathematical approaches. Meanwhile, his research also includes the analysis of companies' environmental, social, and corporate governance activities and their connection to shareholders' value creation. These works have been published in prestigious journals such as Business Strategy and the Environment, Corporate Social Responsibility and Environmental Management, Energy Economics, International Review of Economics and Finance, Journal of Business Ethics, Journal of International Financial Markets, Institutions & Money, Mathematics, and North American Journal of Economics and Finance, among others.

Lorena Leal-Espinosa She has a degree in business administration and management and a degree in labor relations and human resources from the University of Extremadura in Spain. One of her final projects was carried out under the supervision of Professor María Mar Miralles-Quiros. The aim of this study was to analyze the climate change disclosure practices of companies included in the Ibex-35 index.

The Effect of Audit Committee Characteristics on Corporate Social Responsibility Practices. Evidence from Spain

José Manuel Santos-Jaén, Mercedes Palacios-Manzano, Ester Gras-Gil, and Ana León-Gómez

1 Introduction

In recent decades there has been a great increase in the role played by strategies focused on Corporate Social Responsibility (CSR) in the European Union and in all the countries and regions that make it up.

With the aim of consolidating the knowledge economy and turning its regions into much more dynamic societies, with greater social cohesion and more sustainable economies, the European Union has promoted a series of initiatives in the field of CSR. For this reason, Article 18 of the European Commission's Green Paper on CSR highlights the desirability of stimulating CSR quality and practices. This will be made possible through the development of a set of specific principles, approaches and instruments, as well as through the promotion of innovative ideas and good practices.

At the national level, the Spanish economy has been actively reviewing and improving its regulatory frameworks, in particular, corporate governance, transparency and CSR. In 2011, the Sustainable Economy Law was passed, following which the "Spanish CSR Strategy 2014–2020" was approved. These policies are

J. M. Santos-Jaén (✉) · M. Palacios-Manzano · E. Gras-Gil
Department of Accounting and Finance, University of Murcia, Murcia, Spain
e-mail: jmsj1@um.es

M. Palacios-Manzano
e-mail: palacios@um.es

E. Gras-Gil
e-mail: estergg@um.es

A. León-Gómez
Department of Accounting and Management, University of Malaga, Malaga, Spain
e-mail: ana.leon@uma.es

M. d. C. Valls Martínez and J. M. Santos-Jaén (eds.), *Environmentally Sustainable Production*, https://doi.org/10.1007/978-3-031-52656-5_14

designed for enterprises, and also for public administration and the rest of the organizations, with the aim of moving towards a more competitive, productive, sustainable and integrated society and economy, as well as creating a common framework to co-ordinate private and public CSR policies. The identified CSR gap will be reduced with these coordination policies.

Following stakeholders' demands, companies have enhanced their CSR involvement by considering their overall impact on society and applying more transparency in their CSR practices (Arif et al., 2020; Lee, 2011; Young & Marais, 2012).

As a result of the financial crises of the 1980s and 1990s, regulators have become increasingly interested in strengthening laws relating to the transparency and independence of audit committees (Samaha et al., 2012). According Pincus et al. (1989), one of the main functions of the audit committee is to assist boards of directors in overseeing the policies of the company in terms of the information they provide to their stakeholders. For this reason, the audit committee plays an essential role in providing investors with relevant, complete and clear information (Arcay & Vazquez, 2005). On the other hand, by monitoring the senior management bodies of companies, this committee also carries out a task that ensures a greater amount of voluntary information is disclosed, which allows for greater precision in the decisions and behaviour of senior management (Allegrini & Greco, 2013), while aligning the interests of these management bodies with those of the shareholders (Laksmana, 2008). This committee, acting as the main decision-making board, should minimize the information asymmetry between the company's management and stakeholders by monitoring the information (financial and non-financial) disclosed by the company (Karamanou & Vafeas, 2005; Sultana, 2015). This will not only improve the quality of the information provided by the company, but will also lead to more efficient systems for monitoring this information (Bédard et al., 2008; Carcello & Neal, 2003; Fama & Jensen, 1983; Li et al., 2012; Mangena & Pike, 2005; Song, 2022; Spira, 2003; Sultana, 2015).

Although most attention is focused on the role of the audit committee in improving information oversight processes and in the implementation of CSR practices, the literature analysing the relationship between various audit committee characteristics, such as independence or size for example, and CSR practices is scarce. There are studies that analyse the relationship between CSR practices and corporate governance mechanisms, focusing mainly on the ownership structure (e.g. Ghazali, 2007; Hossain et al., 1994; Raffournier, 1995). Other research studies the independence of the audit committee by analysing the existence of independent members and their proportion of the total number of independent directors on the audit committee (e.g. Forker, 1992; Malone et al., 1993). There are also some studies that focus on the analysis of the chairman of the audit committee, to see what differences can be found based on whether he is an executive director or not (Forker, 1992; Khemakhem & Fontaine, 2019). Other studies examine the effect of audit committee presence/absence on financial reporting (Beasley, 1996; Forker, 1992; Peasnell et al., 2001; Pucheta-Martínez et al., 2023; Uyar et al., 2023) and earnings management (Klein, 2002; Lin et al., 2006; Peasnell et al., 2005; Song, 2022). There are even studies that analyze the effect of the written charter of the audit committee in this committee (Böhm et al., 2016) or the effect of audit risk (Stewart & Munro, 2007). Despite the above-mentioned literature, it is currently unclear how much influence the audit committee has on the quality of CSR practices.

The conclusions drawn from these studies may not be directly relevant to the Spanish economy, given its distinct regulatory and cultural landscape. As Mahadeo et al. (2011) noted, various social, political, economic, and cultural factors shape the unique 'arrangements' in different countries. These arrangements encompass corporate ownership patterns, business laws, regulations, and attitudes toward philanthropy and social responsibility. Consequently, disparities in CSR practices across studies can often be attributed to specific institutional effects within individual countries (Wanderley et al., 2008).

In the case of Spain, significant changes have occurred within audit committees since 2002. Listed companies in Spain were mandated to establish audit committees for the first time that year. Since then, these committees have continued evolving, aligning with global trend, and emphasizing enhancing their supervisory functions, independence, and expertise. The legal framework, mainly Law 31/2014, dated December 3, amending the Capital Companies Law, now requires listed companies to form audit committees exclusively comprising non-executive directors with at least two independent directors. Moreover, the Unified Code of Good Governance for listed companies, established in June 2013, recommends that audit committees consist solely of external directors, with a minimum of three members. These developments underscore the ongoing efforts to strengthen corporate governance practices in Spain, distinctively shaping the country's approach to CSR.

The primary goal of this study is to investigate how the characteristics of audit committees impact the level of CSR practices among Spanish companies. CSR practices are evaluated using the MERCO rating, which identifies the top 100 Spanish companies with outstanding CSR initiatives. Given that the audit committee plays a pivotal role in a company's governance structure, ensuring transparency and credibility in all activities, its establishment and attributes are likely to influence managerial disclosure choices and, consequently, the extent of CSR practices (Appuhami & Tashakor, 2017). This research is motivated by the need to empirically examine and gain new insights into the specific traits of audit committees. These insights can enhance corporate accountability and reduce the risk of future corporate failures.

The remaining part of this document is organized as follows: Sect. 2 discusses the study's specific hypotheses. Section 3 outlines the methodology and sample characteristics. Section 4 presents the results of the hypothesis test. Section 5 analyzes and discusses the implications of the results. Finally, in Sect. 6, the main conclusions of this research are presented.

2 Background and Hypothesis Development

The audit committee plays a pivotal role in ensuring the quality of financial accounting and control systems (Collier, 1993). Positioned under the main board of directors, this committee is responsible for corporate reporting and, consequently, CSR practices (Bédard & Gendron, 2010). The connection between audit committee attributes and CSR practices stems from the fundamental premise that these committees serve as a corporate governance mechanism. They are established to monitor managerial behaviour on behalf of shareholders and oversee organizational reporting processes to enhance transparency. Through this enhanced transparency, audit committees play a vital role in reducing information asymmetry, which often gives rise to agency problems between management and shareholders (Akhtaruddin & Haron, 2010; Bamahros et al., 2022; Fama & Jensen, 1983; Koprowski et al., 2021). Consequently, by fulfilling this mission, audit committees effectively mitigate agency costs (Bédard & Gendron, 2010).

Many researchers have employed agency theory to investigate the impact of corporate governance on CSR. These scholars include Lambert (2001), Lim et al. (2007), Huafang and Jianguo (2007), Li et al. (2012) and Ho and Taylor (2013). In general, empirical investigations provide evidence of a positive correlation between audit committee attributes and CSR initiatives (Al-Shammari & Al-Sultan, 2010; Barako et al., 2006; Ika et al., 2017).

Audit committee effectiveness can be defined in various ways (DeZoort et al., 2002). Previous research focused on audit committee composition to measure quality (Carcello & Neal, 2003; Klein, 2002; Krishnan, 2005). All these studies are based on US companies. In 1999, major US stock exchanges introduced new requirements for audit committee composition, including size, independence, and financial expertise.

Over the years, various codes have been established, setting forth guidelines for the composition of audit committees. The quality of an audit committee is contingent upon factors such as its size, independence, and the frequency of its meetings (Madi et al., 2014). Major U.S. stock exchanges mandated that audit committees consist of at least three independent directors possessing financial literacy. Moreover, corporate governance codes, including Spain's Unified Report CNMV (2006), have recommended the inclusion of independent members in audit committees. There exists a plethora of codes and recommendations specifying the minimum number of members and the frequency of annual meetings for audit committees. For example, the Blue Ribbon Committee Report (1999) advocates for a minimum of three members and four meetings annually to ensure the audit committee's effectiveness. Notably, in Spain, there is no mandatory regulation regarding the composition of audit committees. Instead, companies themselves delineate these regulations within their statutes.

As far as the size of the audit committee is concerned, Allegrini and Greco (2013), based on resource dependency theory, larger audit committees invest more resources and authority to carry out responsibilities effectively. Bedard and Gendron (2010) highlight that a more significant number of directors on an audit committee increases the likelihood of incorporating diverse skills, perspectives, experiences, and expertise, thereby enhancing practical monitoring efforts. Therefore, Li et al. (2012) posit that the corporate reporting process's potential issues can be more readily resolved with increased audit committee members. Additionally, Othman et al. (2014) say that increasing the size of the audit committee enhances the company's monitoring and control, leading to more significant voluntary disclosures. Talpur et al. (2018) draw conclusions from their research conducted in the Malaysian setting, suggesting that the size of the audit committee is a determinant affecting the extent of voluntary corporate governance disclosure. It highlights the essential role played by the size of audit committees in ensuring effective oversight of companies' disclosure practices (Persons, 2009). In support of this perspective, it appears that having many directors on an audit committee enhances the level of voluntary practices, as Persons (2009), Musallam (2018), Appuhami and Tashakor (2017), and Aladwey et al. (2022) have found empirical evidence for. Following the results of previous research, this study puts forward its initial hypothesis, adhering to the established literature:

H_1: There is a positive relationship between the Audit Committee size and corporate social responsibility practices.

In relation to meeting frequency, there is no consensus about the best number of annual meetings to be held by the audit committee. Two of the biggest audit firms, PriceWaterhouseCoopers (1993) and (1999) consider a minimum of three.

Audit committee meetings offer an opportunity for each member to share their views and expertise. This contributes to an effective decision-making process within the committee. Consistent attendance at these meetings, as measured by the number of meetings, is considered an indication of the members' diligence and commitment (Talpur et al., 2018). In this regard, Karamanou and Vafeas (2005) investigate the correlation between the frequency of meetings and the effectiveness of monitoring. In the same line, according to Greco (2011), increasing the frequency of audit committee meetings enables members to voice their opinions regarding the company's selection of accounting principles, disclosures, and estimates.

Allegrini and Greco (2013) found a significant correlation between the frequency of audit committee meetings, set at a minimum of four times annually, and the extent of voluntary practices. Nevertheless, Madi et al. (2014) contend that the frequency of audit committee meetings does not show a significant correlation with corporate voluntary practices. In light of this debate, the study proposes its second hypothesis:

H_2: There is a positive relationship between frequency of Audit Committee meetings and corporate social responsibility practices.

As per García-Sánchez et al. (2012), Spanish firms encounter issues concerning the independence of their audit committees. The degree of autonomy in these audit committees is associated with including independent external directors (Appuhami & Tashakor, 2017). In this regard, Bedard and Gendron (2010) contend that independent directors on an Audit Committee are poised to prevent managerial interference due to their lack of personal or economic ties with management. This detachment enables committee members to perform their duties independently and impartially. According to agency theory, the presence of independent directors significantly enhances the oversight of management behaviour, fostering effective governance

(Fama & Jensen, 1983). Allegrini and Greco (2013) state that having independent directors can decrease the likelihood of management concealing information for personal gain. Patelli and Prencipe (2007), Akhtaruddin and Haron (2010), and Madi et al. (2014) all observed a strong and positive correlation between the audit committee's independence and the quality of CSR practices. The significance of the independence of audit committees is demonstrated through this relationship. Nevertheless, despite the research conducted by Li et al. (2012) and Othman et al. (2014), which revealed no substantial correlation between CSR and audit committee independence, this study proposes the formulation of the third hypothesis:

H_3: There is a positive association between the Audit Committee independence and corporate social responsibility practices.

Having taken into account the information provided earlier, it becomes apparent that the quality of audit committees is likely to influence corporate social responsibility (CSR) practices. It is reasonable to infer that a top-tier audit committee would be linked to favourable CSR practices. Consequently, the fourth hypothesis can be articulated as follows:

H_4: The quality of the audit committee is positively associated with corporate social responsibility practices.

3 Methodology

3.1 Sample and Measurements

The dataset consists of esteemed Spanish businesses from 2011 to 2022, meticulously selected from the MERCO index (Spanish Monitor of Corporate Reputation). Financial entities were deliberately omitted due to their regulatory oversight and unique financial reporting. Additionally, companies needing audit committee details and annual reports were excluded from the study. Audit committee data was gathered from the Corporate Governance Reports webpage of the CNMV (Spanish National Commission of Stock Exchange), while financial data was sourced from the SABI database. In the end, the unbalanced panel included 55 firms and 424 observations.

Dependent Variable

The variable to measure CSR has been constructed from the Spanish Monitor of Corporate Reputation (MERCO) index. This rate has been previously applied by research such as Delgado-García et al. (2013), Gras-Gil et al. (2016) and Palacios-Manzano et al. (2021).

This index is derived from a survey of the top 100 Spanish companies known for their exemplary practices. Managers are invited to assess these companies based on various criteria, including social and environmental responsibility. Subsequently, the businesses in the initial ranking undergo direct evaluation by diverse groups, including financial analysts, representatives from non-governmental organizations

(NGOs), consumer association members, economic journalists, and corporate social responsibility (CSR) experts. The evaluated companies are then assigned a score of 0 to 10,000 points.

Independent Variables

Audit Committee Quality

As discussed in the previous section, we use three measures to proxy for audit committee quality: size (ACSIZE), independence (ACINDEP), and meetings (ACMEET). ACSIZE is the total number of audit committee members. ACINDEP is the proportion of members that are independent directors. ACMEET is the number of audit committee meetings per year.

As an analysis extension, to study the effect of audit committee quality on CSR practices, we create a new dichotomy variable. The sample has been divided into two groups according to the quality level of the audit committee. We define a group of firms with higher-quality audit committee and the rest firms. To choose the first group of companies we take those that have the three variables (ACSIZE, ACINDEP and ACMEET) above the 75th percentile. In the second group will be the remaining companies. So, ACQUALITY is a dummy variable coded 1 if the firm is in the first group (higher-quality audit committee), and 0 otherwise.

Control Variables

Additionally we have included in the models a set of control variables that can influence CSR practices, used in the study of Brammer and Pavelin (2006), Delgado-García et al. (2010), Appuhami and Tashakor (2017), Patrisia and Dastgir (2017) and Khan et al. (2021). These variables are:

Total Assets: Numerous research papers have proven that the scale of companies influences CSR. Watts and Zimmerman (1978) assert that larger firms garner greater public attention and are more politically attuned, thus demonstrating a greater inclination to disclose extensive CSR details compared to smaller counterparts. This widely accepted perspective posits that larger companies, being more conspicuous, are inherently predisposed to exhibit enhanced social responsibility. To address this relationship, we incorporate the logarithm of total assets as a control variable (SIZE) in our analysis, ensuring the adjustment for firm size.

Profitability: Various research studies have explored the connection between CSR and a company's financial performance. However, these studies have yielded diverse results, indicating a complex relationship between CSR initiatives and corporate financial outcomes. The study of the correlation between CSR and corporate financial performance has produced conflicting results (Marom, 2006). Several studies have discovered a direct relationship between CSR and the financial performance of a business. (Cheung et al., 2010; Choi et al., 2013; McGuire et al., 1990; Waddock, 2000). Meanwhile, Aupperle et al. (1985), O'Neil and Saunders (1988), and Alexander and Bucholtz (1978) have found no relationship. The chosen profitability indicator is a return on assets (ROA) due to its widespread acceptance and lower likelihood of generating misleading results than other metrics (Aupperle et al., 1985).

Table 1 Dependent, independent and control variables

Variable		Description
Dependent variable		
CSR practices	CSR	Logarithm of Merco index
Independent variables		
Size of audit committee	ACSIZE	Total number of audit committee members
Independent members in the audit committee	ACINDEP	Proportion of independent members in the audit committee
Frequency of audit committee meetings	ACMEET	Total number of audit committee- meetings per year
Audit committee quality	ACQUALITY	1 if the firm is in the group of higher-quality audit committee, 0 otherwise
Control variables		
Firm size	SIZE	Logarithm of total assets
Firm profitability	ROA	Return on assets ratio
Financial leverage	LEV	Ratio of total debt to total assets
Company listed	LIST	1 if company listed, 0 otherwise
Industry	INDUSTRY	Dummies of industry type

In our analysis, we included ROA as one of the control variables. ROA was calculated by dividing the income before extraordinary items by the firms' total assets.

Leverage: previous research has found that in order to increase the likelihood of obtaining better terms from lenders, companies have increased their CSR practices (Appuhami & Tashakor, 2017). We measured financial leverage as the ratio of total debt to total assets.

List: a dummy variable has been included in order to differentiate between listed and unlisted companies. It is expected that CSR practices are improved when the company is listed.

Finally, we also controlled for industry and year by introducing temporal and industry dummies. Previous research has observed that CSR practices are influenced by the kind of the industry (Appuhami & Tashakor, 2017; Gray et al., 2001). Table 1 presents the description of the variables.

3.2 *Empirical Method*

The hypotheses formulated in the preceding section are examined through a comprehensive analysis employing multiple regression. The study's multiple regression models are calculated as follows:

$$\begin{aligned}
Model\ 1: \ CSR_{it} &= \beta_0 + \beta_1 ACSIZE_{it} + \beta_2 \mathrm{SIZE}_{it} \\
&\quad + \beta_3 ROA_{it} + \beta_4 LIST_{it} + \beta_5 LEV_{it} + \varepsilon_{it} \\
Model\ 2: \ CSR_{it} &= \beta_0 + \beta_1 ACINDEP_{it} + \beta_2 \mathrm{SIZE}_{it} \\
&\quad + \beta_3 ROA_{it} + \beta_4 LIST_{it} + \beta_5 LEV_{it} + \varepsilon_{it} \\
Model\ 3: \ CSR_{it} &= \beta_0 + \beta_1 ACMEET_{it} + \beta_2 \mathrm{SIZE}_{it} \\
&\quad + \beta_3 ROA_{it} + \beta_4 LIST_{it} + \beta_5 LEV_{it} + \varepsilon_{it} \\
Model\ 4: \ CSR_{it} &= \beta_0 + \beta_1 ACQUALITY_{it} + \beta_2 \mathrm{SIZE}_{it} \\
&\quad + \beta_3 ROA_{it} + \beta_4 LIST_{it} + \beta_5 LEV_{it} + \varepsilon_{it}
\end{aligned}$$

where:

CSR	is the natural logarithm of MERCO index.
ACSIZE	is the total number of audit committee members.
ACINDEP	is the proportion of independent members on the audit committee.
ACMEET	is the number of audit committee meetings per year.
ACQUALITY	is dummy, equal to 1 if the company has the three variables (ACSIZE, ACINDEP and ACMEET) above the 75th percentile, and 0 otherwise.
SIZE	is the logarithm of total assets.
ROA	is the return on assets ratio.
LEV	is the ratio of total debt to total assets.
LIST	is dummy variable, equal to 1 if the company listed, and 0 otherwise.

Following Petersen (2009), we employ t-statistics calculated from standard errors clustered at firm and year levels, ensuring robustness against heteroskedasticity and within-firm serial correlation.

4 Results

4.1 Descriptive Statistics

Table 2 provides the descriptive statistics for all the variables previously defined for the study period. The mean CSR practices is 8.677. The mean of audit committee size is 4.33 members. The mean of the proportion of independent members in the audit committee is 74.6%, and the mean of the committee meetings per year is 8.30. We observed that the companies of the sample fully comply with the recommendations of codes or laws.

Table 3 presents the correlation matrix of the variables. The results indicate that none of the variables exhibit a significant level of correlation. Hence, there is no issue

Table 2 Descriptive statistics

Variable	Mean	Median	STD	Min	Max
CSR	8.677	8.674	0.262	5.968	9.210
ACSIZE	4.332	4	1.113	2	7
ACINDEP	0.746	0.75	0.264	0	1
ACMEET	8.081	7	3.521	4	15
SIZE	14.686	14.793	2.681	7.673	27.456
ROA	6.668	5.855	13.085	− 7.320	13.889
LEV	0.628	0.663	0.236	0.024	1.367
	1 (%)	0 (%)			
LIST	71.6	28.4			
ACQUALITY	16.2	83.8			

Where CSR is the natural logarithm of Merco index in year t; ACSIZE is the total number of audit committee members; ACINDEP is the proportion of independent members on the audit committee; ACMEET is the number of audit committee meetings per year; ACQUALITY is a dummy variable (1 = company has higher audit committee quality, otherwise = 0); LEV: as end-of-year total liabilities divided by end-of-year total assets; SIZE: natural logarithm of total assets in year t; ROA: as net income divided by end-of-year total assets; LIST: Dummy variable (company listed = 1, else = 0)

of multicollinearity in these models. This conclusion is drawn by considering the variance inflation factor (VIF), as Marquardt (1970) recommended, which ensures correlations among the explanatory variables do not influence our results. Importantly, all VIF values are well within acceptable limits (Gujarati & Porter, 2003).

In Table 4, we carry out a test of means comparison to investigate whether CSR practices are different between firms with high/low level of ACSIZE, ACINDEP, ACMEET and ACQUALITY. We define a group of firms with ACSIZE over the ACSIZE median value and the group of firms with ACSIZE below the ACSIZE median value. We reply the same to ACINDEP, ACMEET and ACQUALITY.

We observe how the mean CSR is higher in firms with high levels of ACSIZE, ACINDEP, ACMEET and ACQUALITY than in firms with low levels. In all cases, except for ACSIZE, the differences are significantly different from zero. Therefore, in this first approximation, we would find state that audit committee quality does have an effect on CSR practices.

4.2 Regression Results

The outcomes of our models are displayed in Table 5, revealing a remarkable alignment with the descriptive analysis. We applied t-statistics utilizing standard errors clustered at the firm and year levels (Petersen, 2009), ensuring robustness against both heteroscedasticity and within-firm serial correlation. Substantial explanatory capability at the CSR level is evident in all the multiple regression models. All the

Table 3 Correlation matrix

	CSR	ACSIZE	ACINDEP	ACMEET	ACQUALITY	SIZE	ROA	LEV	LIST
CSR	1								
ACSIZE	0.054	1							
ACINDEP	0.2073**	0.171**	1						
ACMEET	0.243**	0.056	0.058	1					
ACQUALITY	0.123**	0.171*	0.252**	0.217**	1				
SIZE	0.014*	0.166**	– 0.066	– 0.162*	0.016	1			
ROA	0.207**	– 0.108*	– 0.042	0.112*	0.086	0.091*	1		
LEV	– 0.255**	0.355**	– 0.201	– 0.005	0.141	0.001	– 0.145*	1	
LIST	0.025	0.148*	0.323**	0.158*	0.131	0.060	– 0.001	– 0.043	1

**, *Significantly different from zero at the 0.01 and 0.05 levels, respectively

Where CSR is the natural logarithm of Merco index in year t; ACSIZE is the total number of audit committee members; ACINDEP is the proportion of independent members on the audit committee; ACMEET is the number of audit committee meetings per year; ACQUALITY is a dummy variable (1 = company has higher audit committee quality, otherwise = 0); LEV: as end-of-year total liabilities divided by end-of-year total assets; SIZE: natural logarithm of total assets in year t; ROA: as net income divided by end-of-year total assets; LIST: Dummy variable (company listed = 1, else = 0)

Table 4 Contrast between audit committee quality and CSR

Variable	Mean CSR		
	High	Low	*P*-value
ACSIZE	8.705	8.643	
ACINDEP	8.786	8.664	*
ACMEET	8.724	8.585	**
ACQUALITY	8.789	8.667	**

* $P < 0.1$, *** $P < 0.01$

Where CSR is the natural logarithm of Merco index in year t; ACSIZE is the total number of audit committee members; ACINDEP is the proportion of independent members on the audit committee; ACMEET is the number of audit committee meetings per year; ACQUALITY is a dummy variable (1 = company has higher audit committee quality, otherwise = 0)

multiple regression models indicate significant explanatory power at the CSR level. The adjusted R^2 of the four models range from 21 to 24%. Thus, ours models are valid and the independent variables are associated with CSR practices.

In Model 1 we test if audit committee size affects CSR practices. The findings show that the influence of ACSIZE on CSR practices is not significant, and therefore, hypothesis 1 is rejected.

Model 2 measures the effect of audit committee independence on CSR practices. The findings suggest that the influence of the rate of independent members on audit committee (ACINDEP) CSR practices is positive and significant. Therefore, hypothesis 2 is accepted, agreeing with previous studies (Appuhami & Tashakor, 2017; Mangena & Tauringana, 2007).

Model 3 measures the effects of audit committee meetings on CSR practices. The frequency of audit committee meetings (ACMEET) is significantly and positively associated with CSR. Thus, the finding supports hypothesis 3 and is consistent with the studies of Pucheta-Martínez and De Fuentes (2007), Appuhami and Tashator (2017) and Talpur et al. (2018).

Finally, with Model 4 we test if audit committee quality affects CSR practices. The coefficient of ACQUALITY is positive and significant. This result is in accordance with the results of the three previous models and suggests that audit committee quality affects and improves CSR practices.

As regards control variables included in the models, the results in Table 5 show that firm profitability (ROA) has a positive effect on CSR, while leverage (LEV) has a negative influence. These findings are consistent with previous studies (Appuhami & Tashakor, 2017; García-Meca & Palacio, 2018).

Table 5 Results of regression models

Variables	M1	M2	M3	M4
C	5.629 (0.999)***	5.229 (0.891)***	4.206 (1.117)***	6.117 (0.739)***
ACSIZE	0.010 (0.023)			
ACINDEP		0.359 (0.136)***		
ACMEET			0.111 (0.041)***	
ACQUALITY				1.276 (0.556)**
SIZE	0.007 (0.006)	0.046 (0.057)	0.072 (0.066)	0.042 (0.056)
ROA	0.381 (0.167)**	1.428 (0.967)*	2.228 (1.213)*	1.095 (0.980)
LEV	− 0.200 (0.079)***	− 1.656 (0.539)***	− 1.195 (0.510)**	− 1.771 (0.541)***
LIST	0.025 (0.034)	0.274 (0.338)	0.436 (0.296)	0.145 (0.366)
Industry Dummies	Yes	Yes	Yes	Yes
Year Dummies	Yes	Yes	Yes	Yes
N	424	424	424	424
F	5.25***	8.23***	9.74***	9.90***
R^2 (adjusted)	0.21	0.21	0.24	0.22

*, **, *** Significantly different from zero at the 0.10, 0.05 and 0.01 levels, respectively, (two-tailed) Where CSR is the natural logarithm of Merco index in year t; ACSIZE is the total number of audit committee members; ACINDEP is the proportion of independent members on the audit committee; ACMEET is the number of audit committee meetings per year; ACQUALITY is a dummy variable (1 = company has higher audit committee quality, otherwise = 0); LEV: as end-of-year total liabilities divided by end-of-year total assets; SIZE: natural logarithm of total assets in year t; ROA: as net income divided by end-of-year total assets; LIST: Dummy variable (company listed = 1, else = 0) Models include industry and year dummies. Regressions are run using two-way cluster standard errors (Petersen, 2009) at the time and firm level which are robust to both heteroscedasticity and within-firm serial correlation

4.3 Robustness Analysis

In order to verify the consistency of our results we repeated our analysis with another estimation method, OLS, and the results are totally similar to those presented in the previous section. Finally, we have removed the time effect and industry effect variables and we get the same results in all cases.

5 Discussions

The board typically forms the audit committee as a sub-committee entrusted with overseeing the financial reporting process. This committee's responsibilities encompass financial reporting and non-financial information disclosure.

Over the recent years, regulatory standards have promoted superior governance and transparency in corporate operations. This study explores the influence of audit committee attributes on CSR practices spanning eleven years (2011–2022). The examined audit committee characteristics include the ratio of independent members, the frequency of committee meetings, and the committee's size. CSR practices of Spanish companies are gauged using the Spanish MERCO ranking as a proxy. Utilizing the MERCO CSR ranking helps mitigate potential subjective biases in evaluating a company's CSR involvement.

An analysis of 100 leading Spanish companies reveals a strong correlation between the frequency of meetings and committee independence and the extent of CSR practices. The study also demonstrates a positive link between the quality of audit committees and CSR practices, indicating that higher-quality audit committees positively impact and enhance CSR initiatives. However, no conclusive evidence supports the notion that the size of audit committees affects CSR practices in Spanish firms.

According to the research results, it is affirmed that the presence of an audit committee positively influences the extent of CSR practices within companies. This enhancement in CSR activities occurs irrespective of legal mandates for the committee's existence. As Appuhami and Tashakor (2017) highlight, the study also indicates that audit committees with suitable traits can serve as a market indicator, signalling the quality of a company's internal oversight procedures and CSR initiatives.

6 Conclusions

The results are in line with Jamali et al. (2009), who stated that corporate governance, including audit committee, is crucial for a genuine and sustainable CSR strategy. In line with the global efforts to enhance the effectiveness of audit committees in promoting good corporate governance practices, the results have main empirical guidance for policymakers, regulators, investors and analysts. Last, our findings spread academic research and seek to improve our understanding of the role of audit committees in various aspects of CSR practices.

The results are subject to several limitations that need to be considered. Firstly, as mentioned earlier, MERCO index has been used as a proxy for CSR practices. Thus, our sample includes only the 100 companies included in MERCO, the sample choice was unavoidable. This data shortcoming may affect the generalizability of the results; future studies can solve this limitation by applying other procedures to assess

CSR, such as the Reputation Quotient (Fombrun et al., 2000). Secondly, the study focuses on a certain set of factors that affect CSR practices, but a number of other factors such as the engagement between the audit committee and external auditors have not been tested and may be an important factor that affects CSR. Finally, we acknowledge that that the relationship between CSR and the audit committee may reflect different outcomes in other countries as a result of the existence of a different legal framework, accounting system, institutions and CSR culture. These limitations provide an avenue for futures researches.

Acknowledgements Not applicable.

Funding This research received no external funding.

Data Availability Statement All data used in this research can be obtained from the SABI database and the MERCO index.

Conflicts of Interest The authors declare no conflict of interest.

References

Akhtaruddin, M., & Haron, H. (2010). Board ownership, audit committees' effectiveness and corporate voluntary disclosures. *Asian Review of Accounting*

Aladwey, L., Elgharbawy, A., & Ganna, M. A. (2022). Attributes of corporate boards and assurance of corporate social responsibility reporting: Evidence from the UK. *Corporate Governance: THe International Journal of Business in Society, 22*(4), 748–780. https://doi.org/10.1108/CG-02-2021-0066

Alexander, G. J., & Buchholz, R. A. (1978). Corporate social responsibility and stock market performance. *Academy of Management Journal, 21*(3), 479–486.

Allegrini, M., & Greco, G. (2013). Corporate boards, audit committees and voluntary disclosure: Evidence from Italian listed companies. *Journal of Management and Governance, 17*(1), 187–216.

Al-Shammari, B., & Al-Sultan, W. (2010). Corporate governance and voluntary disclosure in Kuwait. *International Journal of Disclosure and Governance, 7*(3), 262–280.

Appuhami, R., & Tashakor, S. (2017). The impact of audit committee characteristics on CSR disclosure: An analysis of Australian firms. *Australian Accounting Review, 27*(4), 400–420.

Arcay, M. R. B., & Vazquez, M. F. M. (2005). Corporate characteristics, governance rules and the extent of voluntary disclosure in Spain. *Advances in Accounting, 21*, 299–331.

Arif, M., Sajjad, A., Farooq, S., Abrar, M., & Joyo, A. S. (2020). The impact of audit committee attributes on the quality and quantity of environmental, social and governance (ESG) disclosures. *Corporate Governance: The International Journal of Business in Society*

Aupperle, K. E., Carroll, A. B., & Hatfield, J. D. (1985). An empirical examination of the relationship between corporate social responsibility and profitability. *Academy of Management Journal, 28*(2), 446–463.

Bamahros, H. M., Alquhaif, A., Qasem, A., Wan-Hussin, W. N., Thomran, M., Al-Duais, S. D., Shukeri, S. N., & Khojally, H. M. A. (2022). Corporate governance mechanisms and ESG reporting: Evidence from the Saudi stock market. *Sustainability, 14*(10), 6202. https://doi.org/10.3390/su14106202

Barako, D. G., Hancock, P., & Izan, H. Y. (2006). Factors influencing voluntary corporate disclosure by Kenyan companies. *Corporate Governance: An International Review, 14*(2), 107–125.

Beasley, M. S. (1996). An empirical analysis of the relation between the board of director composition and financial statement fraud. *Accounting Review, 14*, 443–465.

Bédard, J., Coulombe, D., & Courteau, L. (2008). Audit committee, underpricing of IPOs, and accuracy of management earnings forecasts. *Corporate Governance: An International Review, 16*(6), 519–535.

Bédard, J., & Gendron, Y. (2010). Strengthening the financial reporting system: Can audit committees deliver? *International Journal of Auditing, 14*(2), 174–210.

Blue Ribbon Committee. (1999). Report and recommendations of the blue ribbon committee on improving the effectiveness of corporate audit committees on JSTOR. https://www.jstor.org/stable/40687877?typeAccessWorkflow=login

Böhm, F., Bollen, L. H., & Hassink, H. F. (2016). Audit committee charter scope: Determinants and effects on audit committee effort. *International Journal of Auditing, 20*(2), 119–132.

Brammer, S. J., & Pavelin, S. (2006). Corporate reputation and social performance: The importance of fit. *Journal of Management Studies, 43*(3), 435–455.

Carcello, J. V., & Neal, T. L. (2003). Audit committee characteristics and auditor dismissals following "new" going-concern reports. *The Accounting Review, 78*(1), 95–117.

Cheung, Y.-L., Jiang, P., & Tan, W. (2010). A transparency disclosure index measuring disclosures: Chinese listed companies. *Journal of Accounting and Public Policy, 29*(3), 259–280.

Choi, B. B., Lee, D., & Park, Y. (2013). Corporate social responsibility, corporate governance and earnings quality: Evidence from Korea. *Corporate Governance: An International Review, 21*(5), 447–467. https://doi.org/10.1111/corg.12033

CNMV. (2006). Spain's unified report

Collier, P. (1993). Factors affecting the formation of audit committees in major UK listed companies. *Accounting and Business Research, 23*(sup1), 421–430.

Delgado-García, J. B., De Quevedo-Puente, E., & De La Fuente-Sabaté, J. M. (2010). The impact of ownership structure on corporate reputation: Evidence from Spain. *Corporate Governance: An International Review, 18*(6), 540–556.

Delgado-García, J. B., de Quevedo-Puente, E., & Díez-Esteban, J. M. (2013). The impact of corporate reputation on firm risk: A panel data analysis of Spanish quoted firms. *British Journal of Management, 24*(1), 1–20.

DeZoort, F. T., Hermanson, D. R., Archambeault, D. S., & Reed, S. A. (2002). Audit committee effectiveness: A synthesis of the empirical audit committee literature. *Audit Committee Effectiveness: A Synthesis of the Empirical Audit Committee Literature, 21*, 38.

Fama, E. F., & Jensen, M. C. (1983). Separation of ownership and control. *The Journal of Law and Economics, 26*(2), 301–325.

Fombrun, C. J., Gardberg, N. A., & Sever, J. M. (2000). The reputation quotient SM: A multi-stakeholder measure of corporate reputation. *Journal of Brand Management, 7*(4), 241–255.

Forker, J. J. (1992). Corporate governance and disclosure quality. *Accounting and Business Research, 22*(86), 111–124.

García-Meca, E., & Palacio, C. J. (2018). Board composition and firm reputation: The role of business experts, support specialists and community influentials. *BRQ Business Research Quarterly, 21*(2), 111–123.

García-Sánchez, I., Frias-Aceituno, J. V., & Garcia-Rubio, R. (2012). Determining factors of audit committee attributes: Evidence from Spain. *International Journal of Auditing, 16*(2), 184–213.

Ghazali, N. A. M. (2007). Ownership structure and corporate social responsibility disclosure: Some Malaysian evidence. *Corporate Governance: The International Journal of Business in Society*

Gras-Gil, E., Manzano, M. P., & Fernández, J. H. (2016). Investigating the relationship between corporate social responsibility and earnings management: Evidence from Spain. *BRQ Business Research Quarterly, 19*(4), 289–299.

Gray, R., Javad, M., Power, D. M., & Sinclair, C. D. (2001). Social and environmental disclosure and corporate characteristics: A research note and extension. *Journal of Business Finance and Accounting, 28*(3–4), 327–356.

Greco, G. (2011). Determinants of board and audit committee meeting frequency: Evidence from Italian companies. *Managerial Auditing Journal*

Gujarati, D., & Porter, D. (2003). Multicollinearity: What happens if the regressors are correlated. *Basic Econometrics, 143*, 363.

Ho, P., & Taylor, G. (2013). Corporate governance and different types of voluntary disclosure: Evidence from Malaysian listed firms. *Pacific Accounting Review*

Hossain, M., Tan, L. M., & Adams, M. (1994). Voluntary disclosure in an emerging capital market: Some empirical evidence from companies listed on the Kuala Lumpur Stock Exchange

Huafang, X., & Jianguo, Y. (2007). Ownership structure, board composition and corporate voluntary disclosure: Evidence from listed companies in China. *Managerial Auditing Journal*

Ika, S. R., Dwiwinarno, T., & Widagdo, A. K. (2017). Corporate social responsibility and corporate governance in Indonesian public listed companies. *SHS Web of Conferences, 34*, 1–11.

Jamali, D., Zanhour, M., & Keshishian, T. (2009). Peculiar strengths and relational attributes of SMEs in the context of CSR. *Journal of Business Ethics, 87*(3), 355–377.

Karamanou, I., & Vafeas, N. (2005). The association between corporate boards, audit committees, and management earnings forecasts: An empirical analysis. *Journal of Accounting Research, 43*(3), 453–486.

Khan, F. U., Zhang, J., Dong, N., Usman, M., Ullah, S., & Ali, S. (2021). Does privatization matter for corporate social responsibility? Evidence from China. *Eurasian Business Review, 11*(3), 497–515.

Khemakhem, H., & Fontaine, R. (2019). The audit committee chair's abilities: Beyond financial expertise. *International Journal of Auditing, 23*(3), 457–471.

Klein, A. (2002). Audit committee, board of director characteristics, and earnings management. *Journal of Accounting and Economics, 33*(3), 375–400.

Koprowski, S., Krein, V., Mazzioni, S., & Magro, C. B. D. (2021). Governança corporativa e conexões políticas nas práticas anticorrupção. *Revista De Administração De Empresas, 61*(2), 202. https://doi.org/10.1590/s0034-759020210202

Krishnan, J. (2005). Audit committee quality and internal control: An empirical analysis. *The Accounting Review, 80*(2), 649–675.

Laksmana, I. (2008). Corporate board governance and voluntary disclosure of executive compensation practices. *Contemporary Accounting Research, 25*(4), 1147–1182.

Lambert, R. A. (2001). Contracting theory and accounting. *Journal of Accounting and Economics, 32*(1–3), 3–87.

Lee, M.-D.P. (2011). Configuration of external influences: The combined effects of institutions and stakeholders on corporate social responsibility strategies. *Journal of Business Ethics, 102*(2), 281–298.

Li, J., Mangena, M., & Pike, R. (2012). The effect of audit committee characteristics on intellectual capital disclosure. *The British Accounting Review, 44*(2), 98–110.

Lim, S., Matolcsy, Z., & Chow, D. (2007). The association between board composition and different types of voluntary disclosure. *European Accounting Review, 16*(3), 555–583.

Lin, J. W., Li, J. F., & Yang, J. S. (2006). The effect of audit committee performance on earnings quality. *Managerial Auditing Journal*

Madi, H. K., Fries, C., & Jones, T. (2014). The impact of audit committee characteristics on corporate voluntary disclosure. *Procedia-Social and Behavioral Sciences, 164*, 486–492.

Mahadeo, J. D., Oogarah-Hanuman, V., & Soobaroyen, T. (2011). A longitudinal study of corporate social disclosures in a developing economy. *Journal of Business Ethics, 104*(4), 545–558.

Malone, D., Fries, C., & Jones, T. (1993). An empirical investigation of the extent of corporate financial disclosure in the oil and gas industry. *Journal of Accounting, Auditing and Finance, 8*(3), 249–273.

Mangena, M., & Pike, R. (2005). The effect of audit committee shareholding, financial expertise and size on interim financial disclosures. *Accounting and Business Research, 35*(4), 327–349.

Mangena, M., & Tauringana, V. (2007). Disclosure, corporate governance and foreign share ownership on the Zimbabwe stock exchange. *Journal of International Financial Management and Accounting, 18*(2), 53–85.

Marom, I. Y. (2006). Toward a unified theory of the CSP–CFP link. *Journal of Business Ethics, 67*(2), 191–200.

Marquardt, D. W. (1970). Generalized inverses, ridge regression and biased linear estimation. technometrics

McGuire, J. B., Schneeweis, T., & Branch, B. (1990). Perceptions of firm quality: A cause or result of firm performance. *Journal of Management, 16*(1), 167–180.

Musallam, S. R. M. (2018). The direct and indirect effect of the existence of risk management on the relationship between audit committee and corporate social responsibility disclosure. *Benchmarking: An International Journal*

O'Neill, H., & Saunders, C. (1988). Board member responsiveness and corporate profitability. *Unpublished Manuscript, Univ. of Connecticut, Storrs*

Othman, R., Ishak, I. F., Arif, S. M. M., & Aris, N. A. (2014). Influence of audit committee characteristics on voluntary ethics disclosure. *Procedia-Social and Behavioral Sciences, 145*, 330–342.

Palacios-Manzano, M., Gras-Gil, E., & Santos-Jaen, J. M. (2021). Corporate social responsibility and its effect on earnings management: An empirical research on Spanish firms. *Total Quality Management and Business Excellence, 32*(7–8), 921–937.

Patelli, L., & Prencipe, A. (2007). The relationship between voluntary disclosure and independent directors in the presence of a dominant shareholder. *European Accounting Review, 16*(1), 5–33.

Patrisia, D., & Dastgir, S. (2017). Diversification and corporate social performance in manufacturing companies. *Eurasian Business Review, 7*(1), 121–139.

Peasnell, K. V., Pope, P. F., & Young, S. (2001). The characteristics of firms subject to adverse rulings by the financial reporting review panel. *Accounting and Business Research, 31*(4), 291–311.

Peasnell, K. V., Pope, P. F., & Young, S. (2005). Board monitoring and earnings management: Do outside directors influence abnormal accruals? *Journal of Business Finance and Accounting, 32*(7–8), 1311–1346.

Persons, O. S. (2009). Audit committee characteristics and earlier voluntary ethics disclosure among fraud and no-fraud firms. *International Journal of Disclosure and Governance, 6*(4), 284–297.

Petersen, M. A. (2009). Estimating standard errors in finance panel data sets: Comparing approaches. *The Review of Financial Studies, 22*(1), 435–480.

Pincus, K., Rusbarsky, M., & Wong, J. (1989). Voluntary formation of corporate audit committees among NASDAQ firms. *Journal of Accounting and Public Policy, 8*(4), 239–265.

Pucheta-Martínez, M. C., Bel-Oms, I., & Gallego-Álvarez, I. (2023). Corporate social responsibility commitment of women directors through audit committees: Evidence from international firms. *Academia Revista Latinoamericana De Administración, 36*, 98–118. https://doi.org/10.1108/ARLA-07-2022-0139

Pucheta-Martínez, M. C., & De Fuentes, C. (2007). The impact of audit committee characteristics on the enhancement of the quality of financial reporting: An empirical study in the Spanish context. *Corporate Governance: An International Review, 15*(6), 1394–1412.

Raffournier, B. (1995). The determinants of voluntary financial disclosure by Swiss listed companies. *European Accounting Review, 4*(2), 261–280.

Samaha, K., Dahawy, K., Hussainey, K., & Stapleton, P. (2012). The extent of corporate governance disclosure and its determinants in a developing market: The case of Egypt. *Advances in Accounting, 28*(1), 168–178.

Song, B. (2022). The influence of audit-committee characteristics on the association between corporate social responsibility and earnings quality. *Sustainability, 14*(17), 10496. https://doi.org/10.3390/su141710496

Spira, L. F. (2003). Audit committees: Begging the question? *Corporate Governance: An International Review, 11*(3), 180–188.

Stewart, J., & Munro, L. (2007). The impact of audit committee existence and audit committee meeting frequency on the external audit: Perceptions of Australian auditors. *International Journal of Auditing, 11*(1), 51–69.

Sultana, N. (2015). Audit committee characteristics and accounting conservatism. *International Journal of Auditing, 19*(2), 88–102.

Talpur, S., Lizam, M., & Zabri, S. M. (2018). Do audit committee structure increases influence the level of voluntary corporate governance disclosures? *Property Management*

Uyar, A., Elbardan, H., Kuzey, C., & Karaman, A. S. (2023). Audit and CSR committees: Are they complements or substitutes in CSR reporting, assurance and GRI framework adoption? *International Journal of Accounting and Information Management, 31*(1), 1–36. https://doi.org/10.1108/IJAIM-04-2022-0086

Waddock, S. (2000). The multiple bottom lines of corporate citizenship: Social investing, reputation, and responsibility audits. *Business and Society Review, 105*(3), 323–345.

Wanderley, L. S. O., Lucian, R., Farache, F., & de Sousa Filho, J. M. (2008). CSR information disclosure on the web: A context-based approach analysing the influence of country of origin and industry sector. *Journal of Business Ethics, 82*(2), 369–378.

Watts, R. L., & Zimmerman, J. L. (1978). Towards a positive theory of the determination of accounting standards. *Accounting Review, 124*, 112–134.

Young, S., & Marais, M. (2012). A multi-level perspective of CSR reporting: The implications of national institutions and industry risk characteristics. *Corporate Governance: An International Review, 20*(5), 432–450.

José Manuel Santos-Jaén is PhD Associate Professor of Accounting and Finance at the University of Murcia, Spain. He is a graduate in Business Administration and Management and a graduate in Law. He has completed three masters (MBA, Financial Management and Auditing). He finished his PhD in 2011. He is currently teaching financial statement analysis, financial accounting and corporate accounting. His research is focused on several aspects of Business and Management, such as Firm Performance, Innovation, Sustainability or Corporate Social Responsibility, among others. He is strongly oriented to a quantitative, survey-based methodology, mainly Structural Equation Modeling (PLS-SEM). He has published in international journals, including *Journal of Cleaner Production, Corporate Social Responsibility and Environment, IEEE Transaction on Engineering Management, Sustainable production and consumption, The International Journal of Management Education,* and *Total Quality Management and Business Excellence.* He has also presented the progress of his research at national and international congress.

Mercedes Palacios-Manzano is Professor of Accounting and Finance at the University of Murcia, Spain. She finished her European PhD in 2004, and since then her research is focused on the fields of international accounting, non-financial reporting, corporate social responsibility and earnings management, collaborating in different research projects. She has published in international journals, including *Advances in International Accounting, Spanish Journal of Finance and Accounting, Business Research Quarterly, Journal of Cleaner Production, Total Quality Management and Business Excellence.*

Ester Gras-Gil is Professor in the Department of Financial Economics and Accounting at the University of Murcia. She has a PhD from the University of Murcia. She has a degree in Economics and Business Administration from the University of Murcia and a degree in Accounting and Finance from the University of Manchester. Master in Banking Economics, from the Business School of the University of Murcia. Master in Account Auditing, from the University of Murcia. She has extensive professional experience in financial institutions, which has allowed her to have many research contracts with companies in the financial sector. Main lines of research focused on banking, auditing and internal control, corporate social responsibility, and gender. She is the author of various articles in prestigious journals with a high impact index, book chapters, and speaker at national and international congress.

Ana León-Gómez is an assistant professor in the Department of Economics and Business Administration at the University of Málaga. She holds a PhD Cum Laude in Tourism, specialising in DSGE models. She received the award of Honorary Doctor from the Royal Academy of Doctors of Spain. She has a total of 28 scientific publications, where 80% of these publications are indexed in journals in the first quartiles of the Scopus and Web of Science databases. He has participated in 22 international conferences related to university education, economics and business management. In addition, she has received awards for the best international speaker at the Global Conference on Data Science, Big Data, Data Analytics and Linguistics held in Dubai and the best young researcher of the Year 2022 by the International Society for Research.

Unveiling Differences in ESG Adoption: A Comparative Analysis of the Big Four Auditors

María del Carmen Valls Martínez, José Manuel Santos-Jaén, and Gema Martín de Almagro Vázquez

1 Introduction

In recent years, there has been a growing concern for achieving sustainable development (Geng et al., 2022). The widely adopted concept of sustainability is defined as the capacity to maintain a balance between human needs and the preservation of natural resources (Ortiz-Martínez et al., 2023). This concept encompasses social, economic, and environmental dimensions and has become ubiquitous in society (Purvis et al., 2019). Sustainability, as per this definition, implies that all economic actors share some degree of responsibility in achieving this delicate yet imperative equilibrium. Auditors, in this context, are no strangers to this challenge.

Moreover, auditors can significantly influence a company's sustainability practices by evaluating and verifying information, ensuring regulatory compliance, providing guidance and recommendations, and advocating best practices (Del Giudice & Rigamonti, 2020). Consequently, they play a pivotal role in advancing corporate sustainability and accountability. Therefore, audit quality becomes a crucial determinant of auditors' contribution to sustainable development.

Concurrently, preserving auditor independence is essential for upholding audit quality and safeguarding stakeholders' interests (Jamal & Sunder, 2011). Consequently, there has been increasing scholarly interest in examining the primary challenges to auditor independence (Tepalagul & Lin, 2015). Puxty (1993) suggests that

M. C. Valls Martínez (✉)
Mediterranean Research Center on Economics and Sustainable Development, Economics and Business Department, University of Almería, Almería, Spain
e-mail: mcvalls@ual.es

J. M. Santos-Jaén
Department of Accounting and Finance, University of Murcia, Murcia, Spain

G. Martín de Almagro Vázquez
Sector 3 SAP, Murcia, Spain

M. d. C. Valls Martínez and J. M. Santos-Jaén (eds.), *Environmentally Sustainable Production*, https://doi.org/10.1007/978-3-031-52656-5_15

threats to auditor independence can stem from external factors such as rotations and fees, as well as internal factors related to the unique culture within each audit firm.

Meanwhile, fees have remained unregulated, and European Union (EU) policy-makers have expressed concerns about long-term relationships between audit firms and their clients, which they believe pose a significant risk to audit quality. This perspective, outlined in the European Commission's Green Paper, resulted in the implementation of mandatory rotation requirements under the 2014 EU Regulation (Garcia-Blandon et al., 2020a, 2020b). In contrast, the United States currently lacks legislation mandating auditor rotation. However, the Public Company Accounting Oversight Board (PCAOB) has previously explored this possibility and proposed regulations that would require periodic auditor rotation.

Previous literature contains numerous studies that primarily examine external threats to independence, particularly the impact of rotations and audit fees (Gandía & Huguet, 2018; Garcia-Blandon et al., 2020a, 2020b; Salehi et al., 2022). However, these studies have generated inconclusive findings. Some researchers, such as Zgarni et al. (2016) and Choi et al. (2022), have specifically focused on analyzing these threats within the context of the Big Four audit firms. However, few studies have taken into account that auditor independence is also influenced by internal factors, including corporate culture. Consequently, audit firms of similar size operating in the same market, such as the Big Four in the United States, may yield different outcomes concerning external threats to auditor independence and, as a result, the influence of auditors on the ESG practices implemented by the companies they audit.

This analysis is the primary gap addressed by this research. To this end, we have formulated the following research questions: (1) Do auditor rotation and audit fees influence the ESG practices conducted by audited companies? (2) Do the impacts of these threats vary for each audit firm operating within the same economic and legal environment? This study specifically concentrates on the United States, which has held a prominent position in auditing since the aftermath of the Wall Street Crash in 1929.

To address our research inquiries, we conducted an empirical examination using data from companies listed in the S&P 500 index from 2012 to 2021. For this purpose, we utilized panel data analysis with fixed effects regressions to account for endogeneity and mitigate potential issues related to omitted variables.

This study adds to the existing body of scientific literature by examining the impact of auditor rotation and audit fees on the implementation of ESG practices by audited companies. Furthermore, a significant contribution of this research is the analysis of these effects specifically within each Big Four audit company.

The results demonstrate that higher audit fees are associated with greater implementation of ESG practices by the audited companies. The same applies to the tenure of the auditor in the audited company. Regarding the differences due to the corporate culture of each audited entity, distinctions have been found in the social and governance aspects of sustainability but not in environmental aspects.

The remaining sections of this article are structured as follows: the Sect. 2 comprises a review of the relevant literature and the formulation of the hypotheses; the Sect. 3 outlines the data analysis and methodology employed; the Sect. 4 presents the obtained results; and the Sect. 5 includes a discussion of the results, along with the conclusions drawn from the study.

2 Literature Review

2.1 *Audit Fees and ESG Practices*

The scientific literature has extensively addressed the relationship between audit fees and audit quality, primarily through the lens of the impact of fees on auditor independence (Alareeni, 2017). Audit market segmentation, a concept proposed by Gandía and Huguet (2018), has highlighted that different audit firms employ distinct approaches, leading to varying levels of audit quality. Eshleman and Guo (2014) suggest a positive association between higher audit fees and improved audit quality, supporting this theory. Abnormal audit fees, those exceeding estimated fees based on company attributes and the audit itself, have also been studied. Asthana and Boone (2012) and Alhadab (2018) argue that increased abnormal audit fees are a significant contributing factor to enhanced audit quality. This perspective finds further support in the research of Hossain and Wang (2022), who demonstrated that low abnormal audit fees have a negative influence on audit quality. Examining Spanish SMEs, Gandía and Huguet (2021) investigated the impact of mandatory audit firm rotations on both voluntary and mandatory audits. Their findings revealed that for voluntary audits, higher audit quality is associated with lower fees paid for auditing work. However, in the context of mandatory audits, quality improves as fees increase. Given the multitude of factors influencing audit fees, including client-related aspects (e.g., size, audit complexity, litigation risk, and corporate governance) and auditor-related factors (e.g., tenure, expertise, and quality), Chang et al. (2021) conducted an analysis that considered all these factors and confirmed a positive relationship between audit fees and audit quality, aligning their results with those of many other researchers (Alareeni, 2017).

External financial audits encompass a comprehensive evaluation of risks associated with various aspects of a company that could impact the accuracy of financial information. These aspects include ESG practices, corporate reputation risk, allocation of corporate resources, and profitability (Zahid et al., 2023). According to agency theory, auditing is a fundamental control mechanism designed to mitigate information asymmetry, prevent exploitative practices, and enhance performance and transparency in ESG practices (Jensen & Meckling, 1976). Therefore, financial auditors are implicitly equipped with knowledge about ESG practices and strategies, as well as ethical business conduct. This equips them to assist companies in improving their ESG practices and reducing reputational risk (Asante-Appiah & Lambert, 2022). Additionally, maintaining high audit quality entails having ample resources available for conducting audits and making substantial investments in human capital and technology. This investment, in turn, enhances the reliability of the provided information, including non-financial data. Consequently, higher audit quality leads to reduced information asymmetry and improved transparency. Consequently, the success of ESG practices and the transparency of business activities are closely linked to higher audit quality, which enhances the legitimacy and reputation of the company (Hammami & Hendijani Zadeh, 2019).

In summary, this discussion highlights the interplay between audit quality, audit fees, and ESG practices. On the one hand, audit quality is influenced by audit fees, and on the other hand, audit quality significantly impacts the ESG practices implemented by audited companies. Based on these relationships, we propose the following hypothesis:

$\mathbf{H_1}$: *There exists a positive relationship between audit fees and the ESG practices implemented by audited companies.*

2.2 Auditor Tenure and ESG Practices

The accounting community has engaged in a prolonged and intense debate about the mandatory implementation of audit firm rotation (Chi et al., 2011) to ensure auditor independence and, consequently, audit quality. However, the outcomes of this debate have been mixed (Lin & Yen, 2022). While extended auditor tenure may enhance auditors' familiarity with their client's operations, potentially improving audit quality, it can also lead to closer and more amicable relationships with management, raising concerns about auditor independence and, by extension, audit quality (Chi et al., 2011). Therefore, there are two opposing perspectives on auditor rotation, with both proponents and critics of the practice (Salehi et al., 2022).

Proponents of longer auditor tenure often rely on the theory of influence or bias. This theory suggests that a strong and long-standing relationship between an auditor and their client can inadvertently lead to favoritism towards the client during the auditing process (Alhadab, 2018). Conversely, new auditors who are unfamiliar with their clients are more prone to overlooking financial misreporting. Carcello and Nagy (2004) propose that mandatory audit firm rotation could have adverse effects on audit quality, as their research indicates a higher likelihood of fraudulent financial reporting during the initial three years of the auditor–client relationship. Another argument against auditor rotation is based on the belief that it brings new perspectives and ideas that can enhance the quality of financial reporting. However, these fresh viewpoints are often assumed rather than clearly evident (Lin & Yen, 2022). In our literature review, we found studies opposing auditor rotation as a means to improve audit quality. For instance, Lin and Yen (2022) argue that auditor rotation is not significantly linked to the quality of accruals when there is no change in key audit matters following the rotation. In this line, Kuang et al. (2020) found no evidence to support the idea that mandatory auditor rotation enhances audit quality. Instead, their findings suggest a higher likelihood of significant misstatements occurring after a compulsory rotation of the audit partner, particularly when the audit firm's tenure is short. Similarly, Garcia-Blandon et al. (2020a, 2020b) demonstrate that companies with an auditor tenure exceeding ten years do not exhibit lower auditing quality compared to other firms. In fact, their research even suggests the potential for superior auditing quality in such companies.

Therefore, the above suggests that auditor tenure positively influences audit quality. Moreover, based on the earlier section where we established that audit quality impacts the ESG practices conducted by audited companies, we propose the following hypothesis:

H_2: *There is a positive relationship between audit tenure and the ESG practices conducted by audited companies.*

2.3 Differences Between Audit Companies

Puxty (1993) suggests that laws and regulations are insufficient in ensuring the independence of audit firms and, therefore, the audit quality. Auditor independence can be jeopardized by external factors such as auditor tenure and fees, as well as internal factors specific to the audit firm itself. Therefore, researchers have been interested in identifying the mechanisms that maintain auditor independence despite these threats (Bauer, 2015). According to Puxty (1993), cultural and socio-economic factors have a significant impact on the concept of auditor independence. In the same vein, Hudaib and Haniffa (2009) discovered that auditors perceive independence based on their social interactions across three levels: personal, organizational, and societal. In this study, our focus will be on the organizational level, with a particular emphasis on examining the ethical culture within the societal context of the United States.

Ethical culture, a subset of organizational culture, plays a significant role in various aspects of an audit company. While organizational culture affects areas such as group innovation, job satisfaction, and work ethics, ethical culture specifically becomes crucial in assessing discretionary matters involving ethical implications (Treviño et al., 2001). According to Svanberg and Öhman (2016), ethical culture comprises a complex blend of formal and informal systems that can promote either ethical or unethical behavior. Additionally, Kung and Li Huang (2013) found evidence indicating that auditors, due to the practical nature of their auditing profession, tend to give precedence to relativism over idealism when it comes to ethical considerations. Consequently, relativist auditors may be less inclined to condemn their clients' unethical actions, highlighting the potential mismatch between the theoretical ideal of auditor independence and its practical implementation. Correspondingly, Bauer (2015) suggests that exposure to a robust ethical culture may contribute to enhanced auditor impartiality.

Similarly, Svanberg and Öhman (2016) suggest that auditors employed in organizations that prioritize ethical practices are more effective in preserving their objectivity when compared to auditors in organizations that do not prioritize ethics. In addition, several authors such as Albaqali and Kukreja (2017), Barrainkua and

Espinosa-Pike (2018), and Kaptein (2008) have supported the notion that cultural ethics promote auditor objectivity. This implies that audit firms should strive to foster a strong ethical culture to reduce the risk of auditors being influenced in their judgments. Finally, Zhang and Wei (2022) demonstrated a strong negative correlation between the level of ethical culture within an audit firm and the extent of ethical misconduct.

Furthermore, other factors specific to the audit firm, such as processes and formal structure, the audit as a business, working papers, and image management, influence auditor independence and quality (Reiter & Williams, 2004). For this reason, we believe that the influence of audit fees and auditor tenure on audit quality differs for each audit firm. Therefore, the effect on ESG practices is different for each audit firm.

Based on the above, we establish the following hypotheses:

H_3: *The influence of audit fees on ESG practices performed by audited companies differs among audit firms.*

H_4: *The influence of audit tenure on ESG practices performed by audited companies differs among audit firms.*

3 Research Methods

The study population consists of large companies in the US market included in the S&P 500 index. The study period covers a total of ten years, from 2012 to 2021, to ensure the robustness of the results. The data was obtained from the Eikon database, a source used by both professional investors and scientists, thereby validating its reliability (Valls Martínez et al., 2022c). The final sample, comprising only observations for which data were available for all study variables, consisted of 3ss008 observations. Table 1 shows the sample description categorized by auditing company. Notably, there are only six audit firms for the S&P 500 companies, with only four being relevant. Specifically, these firms, in order of importance, are Ernest & Young, Price Waterhouse Coopers, Deloitte, and KPMG, responsible for auditing 34.14%, 30.72%, 20.61%, and 13.76% of the listed companies, respectively.

Table 1 Sample description

Auditor	Frequency	Percentage
Ernest & Young	1027	34.14
Price waterhouse coopers	924	30.72
Deloitte	620	20.61
KPMG	414	13.76
Grant Thornton	20	0.66
BDO international	3	0.10

To test the research hypotheses, the study employed a multiple linear regression methodology. To mitigate potential bias stemming from relevant omitted variables, panel data analysis was utilized. Specifically, the study applied panel data with fixed effects, which were selected after conducting the Hausman test (Hausman, 1978).

The dependent variable chosen for analysis was the overall ESG score assigned by Eikon (ESG) to each company. The independent variables considered included the audit fees (FEE), with a specific focus on the logarithm of the fees, transforming the scale to align with the ESG score values. Additionally, in recognition of the relationship established in the literature between board characteristics and ESG practices (Gallego-Álvarez et al., 2010; Giannarakis, 2014; Sial et al., 2018; Velte, 2017), the following variables were included: the percentage of women on the board of directors (BGD) as a proxy for gender diversity (Valls Martínez et al., 2022a); the percentage of non-executive board members (NEM); the number of years after which directors must be re-elected (TBM); the size of the board, measured by the number of members (BSZ); the average overall attendance percentage of board committee meetings (BMA); a binary variable indicating whether the chairman of the board of directors is also the CEO of the company (DUA); and, lastly, the compensation of board members, measured as a logarithm (BCO). Furthermore, financial variables were included as control variables (Valls Martínez et al., 2022b): the logarithm of the company's sales (LSA) as an indicator of company size; the debt ratio (IND); and the market-to-book ratio (MTB) as a measure of financial performance valued according to the market. Table 2 shows the definitions of each research variable.

Next, to examine the influence of the independent variables on each of the three ESG pillars, and analyze their similarities or differences, the previous model was replicated, substituting the overall ESG score for each of its components: environmental score (ESC), social score (SSC), and governance score (GSC).

Finally, to test the stability or different behavior among the Big Four audit firms, individualized models for ESC, SSC, and GSC were replicated for each audit firm.

4 Results

Table 3 presents the main descriptive statistics for the variables included in the study. Among the three pillars of ESG practices, the environmental pillar exhibits the lowest average value and the highest standard deviation, suggesting a need for greater emphasis and effort in this area in the future. Auditors, on average, have a tenure of just over 16 years auditing the same company, while audit fees tend to concentrate around their mean value, indicating minimal fee disparities. Regarding the characteristics of the board of directors, it is noteworthy that gender parity in top management positions remains a distant goal for U.S. companies; only a small proportion of board members hold executive positions (averaging less than 15%); the re-election of board members typically occurs within one to four years; the average board size is eleven members; board meeting attendance exceeds 80%; most CEOs

Table 2 Definition of variables

Abbreviation	Variable	Definition
ESG	ESG score	ESG score assigned by Eikon, ranging from 0 to 100
ESC	Environmental score	Environmental score assigned by Eikon, ranging from 0 to 100
SSC	Social score	Social score assigned by Eikon, ranging from 0 to 100
GSC	Governance score	Environmental score assigned by Eikon, ranging from 0 to 100
FEE	Audit fees	Logarithm of audit fees
TIM	Auditor time	Average number of years during which the auditor has provided continuous service to the company
BGD	Board gender diversity	Percentage of women on board of directors
NEM	Non-executive board members	Percentage of non-executive board members
TBM	Tenure of board members	Term, expressed in years, for the board members' reelection
BSZ	Boar size	Number of board members
BMA	Board meeting attendance	The average overall attendance percentage of board committee meetings
DUA	Duality	Dummy variable, 1 if the CEO is a board member, and 0 otherwise
BCO	Board members' compensation	Logarithm of total compensation of the board members
LSA	Ln sales	Logarithm of total company sales
IND	Indebtedness	Percentage of total debt to total equity
MTB	Market to book	Company's market value divided by book value

Note Monetary amounts are expressed in thousands of dollars

also serve as chairpersons of the board of directors; and board member remuneration is fairly uniform, showing low dispersion.

Table 4 displays the Pearson correlation matrix between the regressors and the dependent variable ESG score, as well as with each of its three components, also used as explained variables. It is found that all the regressors show a significant correlation with the dependent variables, which is indicative of their explanatory power.

Table 5 also shows the Pearson correlation matrix between the regressors, indicating that there are no high correlations that could lead to issues of collinearity in the regression model.

Table 6 presents the estimation of the main ESG model in terms of the regressors considered, as well as the decomposition in relation to each of its components.

Table 3 Descriptive statistics

Variable	Mean	Median	SD	Minimum	Maximum
ESG	59.11687	61.70500	17.81636	6.07000	93.84000
ESC	51.76519	57.20500	26.83345	0.00000	98.55000
SSC	61.26267	62.94500	20.45628	5.85000	99.56000
GSC	61.40438	64.02000	19.76995	3.80000	99.62000
FEE	8.98014	8.95893	0.90696	6.03068	11.86990
TIM	16.39096	18.00000	6.88507	1.00000	33.00000
BGD	22.87479	22.22000	10.05990	0.00000	69.23000
NEM	85.87793	88.89000	6.94302	50.00000	100.00000
TBM	1.37599	1.00000	0.78601	1.00000	4.00000
BSZ	10.88065	11.00000	3.57001	4.00000	138.00000
BMA	80.76563	75.00000	9.80532	6.00000	100.00000
DUA	0.98637	1.00000	0.11597	0.00000	1.00000
BCO	14.81389	14.85221	0.55035	10.50780	21.38070
LSA	16.11829	16.07177	1.25053	11.00558	20.03080
IND	0.31801	0.30258	0.22897	0.00000	3.91589
MTB	−11.02691	2.32107	766.41630	−42001.150000	727.58170

Number of observations: 3008

Regarding the independent variables under study, it is verified that there exists a positive relationship between audit fees and the ESG practices of the audited companies, thereby confirming H1. Nevertheless, when considering each of the ESG dimensions separately, a positive and statistically significant relationship is observed for the environmental and social dimensions. However, for the governance domain, a negative and statistically significant relationship is found. Consequently, while H1 is validated individually for the environmental and social dimensions, it does not hold for the governance aspect. Similar results are observed for auditor tenure. Thus, H2 is confirmed for the overall ESG dimension, as well as for the individual environmental and social domains, but not for the governance component.

Regarding the variables related to the board of directors, the results indicate that gender diversity on the board, a greater presence of non-executive board members, and higher compensation for board members are all positively and significantly associated with increased ESG practices, both in the overall context and in each of its three components. However, the longer the term of re-election for directors, the fewer ESG practices the company tends to engage in, both in the global context and when considering the individual components. Additionally, higher attendance at board meetings and CEO-Chairman duality show positive relationships for the overall ESG dimension and the individual environmental and governance components, but the relationship is not significant for the social aspect.

Table 4 Pearson's correlation matrix between dependent variables and regressors

Variable	ESG	ESC	SSC	GSC
FEE	0.4503*** (0.0000)	0.4766*** (0.0000)	0.4147*** (0.0000)	0.1814*** (0.0000)
TIM	0.2224*** (0.0000)	0.2202*** (0.0000)	0.2259*** (0.0000)	0.0670*** (0.0000)
BGD	0.4131*** (0.0000)	0.3395*** (0.0000)	0.3405*** (0.0000)	0.2984*** (0.0000)
NEM	0.3293*** (0.0000)	0.2365*** (0.0000)	0.2330*** (0.0000)	0.3490*** (0.0000)
TBM	−0.2410*** (0.0000)	−0.1933*** (0.0000)	−0.1337*** (0.0000)	−0.2855*** (0.0000)
BSZ	0.1842*** (0.0000)	0.2284*** (0.0000)	0.1817*** (0.0000)	0.0252 (0.1664)
BMA	0.1635*** (0.000)	0.1152*** (0.0000)	0.0930*** (0.0000)	0.2104*** (0.0000)
DUA	0.0713*** (0.0001)	0.0614*** (0.0008)	0.0792*** (0.0000)	0.0232 (0.2031)
BCO	0.3205*** (0.0000)	0.3096*** (0.0000)	0.3027*** (0.0000)	0.1508*** (0.0000)
LSA	0.4501*** (0.0000)	0.4850*** (0.0000)	0.3704*** (0.0000)	0.2581*** (0.0000)
IND	0.0569*** (0.0018)	0.0823*** (0.0000)	0.0407** (0.0257)	0.0289 (0.1134)
MTB	0.0474*** (0.0093)	0.0355* (0.1040)	0.0479*** (0.0087)	0.0297 (0.1040)

p-value in parentheses

***, ** and * indicate a significance of less than 1%, less than 5% and less than 10%, respectively

Number of observations: 3008

Finally, concerning the financial variables, it is observed that larger companies tend to have more extensive ESG practices, both globally and in each of its dimensions: environmental, social, and governance. Additionally, indebtedness shows a significant positive relationship only with the governance component.

The means of each of the dependent variables (global ESG and its individual dimensions) and independent target variables (FEE and TIM) exhibit significant differences based on the audit firm (see Table 7). Therefore, to test H3 and H4, regressions for the three ESG dimensions were conducted individually for each of the Big Four audit firms.

In the environmental pillar, audit fees and auditor tenure exhibit a positive and significant relationship in all companies. The same holds true for the variables related to gender diversity on the board of directors and company size (see Table 8). Therefore, H3 and H4 are not confirmed in the environmental dimension.

Table 5 Pearson's correlation matrix between regressors

Variable	FEE	TIM	BGD	NEM	TBM	BSZ	BMA	DUA	BCO	LSA	IND
TIM	0.1414*** (0.0000)										
BGD	0.2529*** (0.0000)	0.1514*** (0.0000)									
NEM	0.2439*** (0.0000)	0.0760*** (0.0000)	0.2331*** (0.0000)								
TBM	−0.2346*** (0.0000)	−0.1698*** (0.0000)	−0.0991*** (0.0000)	−0.0705*** (0.0001)							
BSZ	0.2618*** (0.0000)	0.0500*** (0.0061)	0.0798*** (0.0000)	0.1409*** (0.0000)	−0.1121*** (0.0000)						
BMA	0.1146*** (0.0000)	0.0001 (0.9971)	0.0860*** (0.0000)	0.0712*** (0.0001)	−0.0642*** (0.0004)	0.0588*** (0.0012)					
DUA	0.0364** (0.0459)	0.0554*** (0.0024)	0.0278 (0.1276)	0.0265 (0.1462)	−0.0021 (0.9070)	0.0483*** (0.0081)	−0.0478*** (0.0087)				
BCO	0.3053*** (0.0000)	0.1399*** (0.0000)	0.1789*** (0.0000)	0.1685*** (0.0000)	−0.1294*** (0.0000)	0.2446*** (0.0000)	0.0064 (0.7265)	0.0722*** (0.0001)			
LSA	0.6960*** (0.0000)	0.1912*** (0.0000)	0.02242*** (0.0000)	0.2073*** (0.0000)	−0.2881*** (0.0000)	0.2645** (0.0000)	0.1190*** (0.0000)	−0.0144 (0.4312)	0.2953*** (0.0000)		
IND	0.0293 (0.1076)	−0.0063 (0.7297)	0.0375** (0.0398)	0.1439*** (0.0000)	−0.0286 (0.1169)	0.0416** (0.0227)	0.0291 (0.1104)	−0.0008 (0.9661)	−0.0039 (0.8314)	−0.0078 (0.6702)	
MTB	0.0093 (0.6110)	0.0159 (0.3829)	0.0418** (0.0218)	0.0145 (0.4281)	0.0097 (0.5938)	0.0043 (0.8150)	0.0110 (0.5462)	−0.0020 (0.9119)	0.0454** (0.0127)	0.0188 (0.3030)	−0.0557*** (0.0023)

p-value in parentheses
***, ** and * indicate a significance of less than 1%, less than 5% and less than 10%, respectively
Number of observations: 3008

Table 6 Panel data regressions (fixed effects)

Variable	ESG	ESC	SSC	GSC
Intercept	−168.439500*** (0.000)	−198.6991*** (0.000)	−152.893600*** (0.000)	−158.296000*** (0.000)
FEE	3.176334*** (0.000)	6.433800*** (0.000)	4.651900*** (0.000)	−1.970885** (0.041)
TIM	0.474262*** (0.000)	0.760680*** (0.000)	0.728542*** (0.000)	−0.163563** (0.028)
BGD	0.367102*** (0.000)	0.363524*** (0.000)	0.365389*** (0.000)	0.388617*** (0.000)
NEM	0.344634*** (0.000)	0.117812** (0.014)	0.136007*** (0.001)	0.834518*** (0.000)
TBM	−1.837574*** (0.000)	−1.815756*** (0.005)	−1.118655** (0.044)	−2.699358*** (0.000)
BSZ	0.000772 (0.989)	0.077390 (0.364)	0.050087 (0.490)	−0.139381* (0.086)
BMA	0.134747*** (0.000)	0.068360** (0.036)	0.035490 (0.202)	0.323999*** (0.000)
DUA	3.813223** (0.019)	6.373711*** (0.008)	1.574235 (0.443)	5.993930*** (0.009)
BCO	2.222228*** (0.000)	2.524575*** (0.000)	2.622819*** (0.000)	1.288012** (0.021)
LSA	6.618947*** (0.000)	7.059507*** (0.000)	6.011048*** (0.000)	6.877343*** (0.000)
IND	4.827424 (0.003)	1.408685 (0.561)	3.884970* (0.060)	8.628345*** (0.000)
MTB	0.000020 (0.918)	−0.000097 (0.741)	0.000041 (0.868)	0.000077 (0.783)
Adjusted R^2	0.8203	0.8257	0.7829	0.7091
F-statistic	152.84*** (0.0000)	87.63*** (0.0000)	100.13*** (0.0000)	87.15*** (0.0000)
Observations	3008	3008	3008	3008
Hausman test	54.32*** (0.0000)	22.31** (0.0221)	56.05*** (0.0000)	48.24*** (0.0000)
Breush Pagan test	18.70*** (0.0000)	21.164*** (0.000)	18.11*** (0.0000)	12.133*** (0.0000)

***, ** and * indicate a significance of less than 1%, less than 5% and less than 10%, respectively

In the social pillar (see Table 9), audit fees demonstrate a positive relationship with corporate responsibility practices, but they are not significant for Ernest & Young and have low significance for Deloitte. On the other hand, auditor tenure also exhibits a positive relationship with the social practices of the audited companies, but this relationship is not significant for Deloitte. Therefore, H3 and H4 are confirmed in the social dimension of corporate responsibility. Furthermore, the percentage of women

Table 7 Mean of dependent and independent variables by auditor

Auditor	Audit fees	Auditor time	ESG score	Environmental score	Social score	Governance score
Ernest & Young	8.93	17.16	58.59	51.22	60.76	60.73
Price Waterh	9.09	15.29	61.55	54.66	64.72	61.81
Deloitte	9.03	17.18	57.15	51.20	58.33	60.60
KPMG	8.86	16.12	59.33	49.57	60.25	64.54
Grant Thornton	7.06	9.80	29.28	7.20	38.16	36.72
BDO Int	9.08	7.67	66.05	64.08	68.91	63.19
ANOVA test	24.26^{***} (0.0000)	13.97^{***} (0.0000)	16.89^{***} (0.0000)	14.31^{***} (0.0000)	13.61^{***} (0.0000)	8.96^{***} (0.0000)

Table 8 Panel data regressions by auditor (fixed effects) in environmental pillar

Variable	Ernest & Young	Price waterhouse coopers	Deloitte	KPMG
Intercept	−132.090600*** (0.000)	−203.082900*** (0.000)	−237.542400*** (0.000)	−281.613700*** (0.000)
FEE	6.181929*** (0.001)	5.452105*** (0.009)	5.278928*** (0.008)	8.585241*** (0.000)
TIM	1.319971*** (0.000)	0.658075*** (0.000)	0.335689** (0.011)	0.897226*** (0.000)
BGD	0.272429*** (0.000)	0.407983*** (0.000)	0.466526*** (0.000)	0.278849*** (0.000)
NEM	0.159081** (0.031)	0.053049 (0.602)	−0.049033 (0.662)	0.172349 (0.129)
TBM	−1.548757 (0.226)	0.898493 (0.452)	−5.205580*** (0.000)	−1.555576 (0.240)
BSZ	−0.133784 (0.709)	0.073878 (0.446)	−0.323729 (0.465)	0.792123* (0.095)
BMA	0.137198** (0.030)	0.069224 (0.246)	0.002454 (0.969)	0.044338 (0.578)
DUA	3.950034 (0.315)	7.985803 (0.104)	9.260673* (0.059)	1.035220 (0.863)
BCO	0.508292 (0.570)	3.873927*** (0.000)	3.954788** (0.017)	4.522987** (0.013)
LSA	4.076446*** (0.005)	6.863304*** (0.000)	10.664700*** (0.000)	8.791506 (0.000)
IND	1.958723 (0.671)	−7.405702 (0.164)	−2.165963 (0.779)	2.729973 (0.477)
MTB	−0.000046 (0.872)	0.847979*** (0.000)	0.118179 (0.721)	0.000036 (0.996)
Adjusted R^2	0.8459	0.7823	0.8311	0.8565
F-statistic	31.90*** (0.0000)	26.93*** (0.0000)	21.11*** (0.0000)	18.65** (0.0000)
Observations	1027	924	620	414
Breush Pagan test	24.317*** (0.0000)	13.171*** (0.0000)	20.936*** (0.0000)	31.069*** (0.0000)

***, ** and * indicate a significance of less than 1%, less than 5% and less than 10%, respectively

on the board of directors, the compensation of board members, and company size are positively and significantly related to the social dimension.

Finally, in the governance pillar, audit fees exhibit a negative and significant relationship with corporate responsibility practices for the audit firms Ernest & Young and KPMG, but not for Price Waterhouse Coopers and Deloitte. Additionally, auditor tenure shows a significant negative relationship only in the case of Deloitte, but not for the other audit firms. Thus, H3 and H4 are confirmed for the governance dimension

Table 9 Panel data regressions by auditor (fixed effects) in social pillar

Variable	Ernest & Young	Price waterhouse coopers	Deloitte	KPMG
Intercept	−129.980300*** (0.000)	−131.301500*** (0.000)	−165.564800*** (0.000)	−200.874700*** (0.000)
FEE	2.433533 (0.129)	5.248625*** (0.003)	2.915171* (0.063)	9.415193*** (0.000)
TIM	1.344441*** (0.000)	0.727450*** (0.000)	0.155798 (0.135)	1.130688*** (0.000)
BGD	0.306741*** (0.000)	0.374207*** (0.000)	0.501641*** (0.000)	0.211062*** (0.002)
NEM	0.140831** (0.028)	0.043707 (0.616)	0.200003** (0.025)	0.1485889 (0.133)
TBM	−0.933639 (0.403)	−0.660754 (0.518)	−2.568188** (0.019)	0.599001 (0.604)
BSZ	−0.066947 (0.830)	0.052823 (0.524)	−0.135864 (0.699)	0.139572 (0.735)
BMA	0.093108* (0.091)	0.071177 (0.163)	−0.064494 (0.191)	−0.071386 (0.305)
DUA	5.365040 (0.118)	−3.905987 (0.353)	4.112868 (0.291)	−1.252946 (0.810)
BCO	2.296291*** (0.003)	2.256887** (0.016)	3.899656*** (0.003)	4.193228*** (0.008)
LSA	4.933208*** (0.000)	5.479454*** (0.001)	6.957922*** (0.000)	5.318358** (0.014)
IND	6.708276* (0.096)	0.563808 (0.901)	8.720026 (0.155)	−0.990121 (0.767)
MTB	0.000037 (0.884)	0.376245*** (0.008)	−0.131456 (0.616)	0.000599 (0.917)
Adjusted R^2	0.7882	0.7595	0.7933	0.8112
F-statistic	41.49*** (0.0000)	27.42*** (0.0000)	22.01*** (0.0000)	18.96** (0.0000)
Observations	1027	924	620	414
Breush Pagan test	17.289*** (0.0000)	13.807*** (0.0000)	18.172*** (0.0000)	21.959*** (0.0000)

***, ** and * indicate a significance of less than 1%, less than 5% and less than 10%, respectively

(refer to Table 10). Furthermore, board gender diversity, the percentage of non-executive board members, attendance at board committee meetings, and company size are positively and significantly linked to corporate governance practices.

Table 11 provides a summary of the results obtained and described above.

Table 10 Panel data regressions by auditor (fixed effects) in governance pillar

Variable	Ernest & Young	Price waterhouse coopers	Deloitte	KPMG
Intercept	−139.189300*** (0.000)	−156.125700*** (0.000)	−253.097700*** (0.000)	−147.786400*** (0.001)
FEE	−3.643988** (0.042)	1.188415 (0.519)	−1.869353 (0.327)	−5.353330** (0.028)
TIM	0.124098 (0.446)	−0.081930 (0.532)	−0.421444*** (0.001)	−0.346991* (0.098)
BGD	0.416717*** (0.000)	0.288842*** (0.000)	0.544781*** (0.000)	0.244149*** (0.002)
NEM	0.805228*** (0.000)	0.941258*** (0.000)	0.944615*** (0.000)	0.752894*** (0.000)
TBM	−5.805290*** (0.000)	−2.773459*** (0.009)	1.025577 (0.443)	−3.885628*** (0.004)
BSZ	−1.662099*** (0.000)	0.031224 (0.716)	−1.201977*** (0.005)	−1.481000*** (0.002)
BMA	0.220894*** (0.000)	0.283639*** (0.000)	0.345143*** (0.000)	0.454088*** (0.000)
DUA	5.905875 (0.123)	5.062372 (0.245)	13.063430*** (0.006)	−1.148012 (0.850)
BCO	2.979975*** (0.001)	−0.576956 (0.551)	1.011359 (0.527)	6.070219*** (0.001)
LSA	6.674013*** (0.000)	6.262396*** (0.000)	12.163000*** (0.000)	5.515778** (0.028)
IND	6.782414 (0.131)	10.970310** (0.020)	12.669600* (0.090)	6.918123* (0.076)
MTB	0.000085 (0.763)	0.076561 (0.604)	−0.517267 (0.106)	0.001783 (0.789)
Adjusted R^2	0.7166	0.6795	0.7118	0.7710
F-statistic	34.45*** (0.0000)	25.57*** (0.0000)	24.79*** (0.0000)	13.51** (0.0000)
Observations	1027	924	620	414
Breush Pagan test	10.690*** (0.0000)	10.118*** (0.0000)	11.970*** (0.0000)	16.808*** (0.0000)

***, ** and * indicate a significance of less than 1%, less than 5% and less than 10%, respectively

5 Discussion and Conclusion

This research used a sample of companies from the S&P 500 index spanning the years 2012–2021. Its primary objective was to assess how auditor tenure and audit fees impact the implementation of ESG practices by the audited companies. Additionally, the study provided separate insights for each of the four major audit firms, collectively known as the Big Four.

Table 11 Summary of results by auditor (sign and significance)

Variable	Auditor	Environmental	Social	Governance
Auditor fees	Total sample	(+) ***	(+) ***	(−) **
	Ernest & Young	(+) ***	NS	(−) **
	Price waterhouse	(+) ***	(+) ***	NS
	Deloitte	(+) ***	(+) *	NS
	KPMG	(+) ***	(+) ***	(−) **
Auditor time	Total sample	(+) ***	(+) ***	(−) **
	Ernest & Young	(+) ***	(+) ***	NS
	Price waterhouse	(+) ***	(+) ***	NS
	Deloitte	(+) **	NS	(−) ***
	KPMG	(+) ***	(+) ***	(−) *

***, ** and * indicate a significance of less than 1%, less than 5% and less than 10%, respectively
NS denotes not significant

Overall, the results indicate that higher audit fees correspond to greater implementation of ESG practices by the audited companies. These findings align with previous research which has shown that increased fees are associated with higher quality work (Asthana & Boone, 2012; Eshleman & Guo, 2014; Gandía & Huguet, 2021). This relationship can be attributed to the ability to allocate more resources when audit fees are higher. In line with the perspective of Asante-Appiah and Lambert (2022), auditing firms can invest these additional resources in enhancing control over various aspects of audited companies, such as non-financial information.

Regarding auditor tenure, the results confirm that longer auditor tenure correlates with increased ESG practices by the audited companies. This relationship suggests that as auditor tenure lengthens, there is a tendency for audit quality to improve, which is consistent with previous studies (Garcia-Blandon et al., 2020a, 2020b; Kuang et al., 2020; Lin & Yen, 2022).

Moreover, in line with previous research (Barrainkua & Espinosa-Pike, 2018; Kaptein, 2008; Zhang & Wei, 2022), the findings indicate that the influence of audit fees and auditor tenure on ESG practices is mediated by internal factors within auditing firms (Albaqali & Kukreja, 2017; Barrainkua & Espinosa-Pike, 2018; Kaptein, 2008). This suggests that even under similar economic and regulatory conditions, such as those found within the Big Four firms in the US, different audit firms may report varying outcomes regarding the effect of fees and tenure on ESG practices, except for the environmental aspect of sustainability.

This study contributes to the existing literature by supporting the notion that imposing restrictions on audit fees or enforcing mandatory auditor rotation may not be advisable. Such requirements could potentially hinder the implementation of ESG practices by companies. Moreover, a significant contribution of this study is the analysis of how internal aspects, particularly the ethical culture of auditing companies, influence the adoption of ESG practices.

The findings of this study carry important implications for policymakers and legislators. They suggest that implementing restrictions on audit fees or enforcing mandatory auditor rotation, similar to the European approach, may not be a recommended course of action. The results also highlight the need for further research into the internal aspects of audit firms, particularly their ethical culture, in order to enhance the ESG practices of audited companies. This study demonstrates that the professionalism of auditors plays a pivotal role in determining the direction and strength of the impact of external variables (fees and rotation) on independence, thereby influencing the quality of financial reports and the level of engagement of audited companies in sustainable development.

While this study provides valuable insights, it also has certain limitations that could serve as avenues for future research. The sample used here consisted solely of US-listed companies, and as a result, cultural and legal variations in different regions might yield differing results. Future research could consider including companies from diverse global regions in their study samples to gain a more comprehensive perspective on this subject.

Acknowledgements This research was funded by PPIT-UAL, Junta de Andalucía-ERDF 2021-2027. Programme: 54.A to the Consolidated Research Group on Ethics, Gender, and Sustainability (SEJ-647).

References

Alareeni, B. (2017). The association between audit firm characteristics and audit quality: A meta-analysis. *SSRN Electronic Journal*. https://doi.org/10.2139/ssrn.2952836

Albaqali, Q., & Kukreja, G. (2017). The factors influencing auditor independence: The perceptions of auditors in Bahrain. *Corporate Ownership and Control, 14*(2), 369–382. https://doi.org/10.22495/cocv14i2c2p10

Alhadab, M. (2018). Abnormal audit fees and accrual and real earnings management: Evidence from UK. *Journal of Financial Reporting and Accounting, 16*(3), 395–416. https://doi.org/10.1108/JFRA-07-2017-0050

Asante-Appiah, B., & Lambert, T. A. (2022). The role of the external auditor in managing environmental, social, and governance (ESG) reputation risk. *Review of Accounting Studies*. https://doi.org/10.1007/s11142-022-09706-z

Asthana, S. C., & Boone, J. P. (2012). Abnormal audit fee and audit quality. *AUDITING: A Journal of Practice and Theory, 31*(3), 1–22. https://doi.org/10.2308/ajpt-10294

Barrainkua, I., & Espinosa-Pike, M. (2018). *The influence of auditors' commitment to independence enforcement and firms' ethical culture on auditors' professional values and behaviour* (pp. 17–52). https://doi.org/10.1108/S1574-076520180000021002

Bauer, T. D. (2015). The effects of client identity strength and professional identity salience on auditor judgments. *The Accounting Review, 90*(1), 95–114. https://doi.org/10.2308/accr-50863

Carcello, J. V., & Nagy, A. L. (2004). Audit firm tenure and fraudulent financial reporting. *AUDITING: A Journal of Practice and Theory, 23*(2), 55–69. https://doi.org/10.2308/aud.2004.23.2.55

Chang, H., Ho, L. C. J., Liu, Z., & Ouyang, B. (2021). Income smoothing and audit fees. *Advances in Accounting, 54*, 100547. https://doi.org/10.1016/j.adiac.2021.100547

Chi, W., Lisic, L. L., & Pevzner, M. (2011). Is enhanced audit quality associated with greater real earnings management? *Accounting Horizons, 25*(2), 315–335. https://doi.org/10.2308/acch-10025

Choi, S. U., Lee, K. C., & Na, H. J. (2022). Exploring the deep neural network model's potential to estimate abnormal audit fees. *Management Decision, 60*(12), 3304–3323. https://doi.org/10.1108/MD-07-2021-0954

Del Giudice, A., & Rigamonti, S. (2020). Does audit improve the quality of ESG scores? Evidence from corporate misconduct. *Sustainability, 12*(14), 5670. https://doi.org/10.3390/su12145670

Eshleman, J. D., & Guo, P. (2014). Abnormal audit fees and audit quality: The importance of considering managerial incentives in tests of earnings management. *AUDITING: A Journal of Practice and Theory, 33*(1), 117–138. https://doi.org/10.2308/ajpt-50560

Gallego-Álvarez, I., García-Sánchez, I. M., & Rodríguez-Dominguez, L. (2010). The influence of gender diversity on corporate performance. *Revista De Contabilidad-Spanish Accounting Review, 13*(1), 53–88. https://doi.org/10.1016/S1138-4891(10)70012-1

Gandía, J. L., & Huguet, D. (2018). Differences in audit pricing between voluntary and mandatory audits. *Academia Revista Latinoamericana De Administración, 31*(2), 336–359. https://doi.org/10.1108/ARLA-01-2016 0007

Gandía, J. L., & Huguet, D. (2021). Audit fees and earnings management: Differences based on the type of audit. *Economic Research-Ekonomska Istraživanja, 34*(1), 2628–2650. https://doi.org/10.1080/1331677X.2020.1836990

Garcia-Blandon, J., Argiles, J. M., & Ravenda, D. (2020a). On the relationship between audit tenure and fees paid to the audit firm and audit quality. *Accounting in Europe, 17*(1), 78–103. https://doi.org/10.1080/17449480.2019.1669808

Garcia-Blandon, J., Argilés-Bosch, J. M., & Ravenda, D. (2020b). Audit firm tenure and audit quality: A cross-European study. *Journal of International Financial Management and Accounting, 31*(1), 35–64. https://doi.org/10.1111/jifm.12098

Geng, Y., Zhu, R., & Maimaituerxun, M. (2022). Bibliometric review of carbon neutrality with CiteSpace: Evolution, trends, and framework. *Environmental Science and Pollution Research, 29*(51), 76668–76686. https://doi.org/10.1007/s11356-022-23283-3

Giannarakis, G. (2014). Corporate governance and financial characteristic effects on the extent of corporate social responsibility disclosure. *Social Responsibility Journal, 10*(4), 569–590. https://doi.org/10.1108/SRJ-02-2013-0008

Hammami, A., & Hendijani Zadeh, M. (2019). Audit quality, media coverage, environmental, social, and governance disclosure and firm investment efficiency. *International Journal of Accounting and Information Management, 28*(1), 45–72. https://doi.org/10.1108/IJAIM-03-2019-0041

Hausman, J. A. (1978). Specification tests in econometrics. *Econometrica, 46*(6), 1251–1271. https://doi.org/10.2307/1913827

Hossain, S., & Wang, J. J. (2022). Abnormal audit fees and audit quality: Australian evidence. *Australian Journal of Management*, 031289622210938. https://doi.org/10.1177/03128962221093831

Hudaib, M., & Haniffa, R. (2009). Exploring auditor independence: An interpretive approach. *Accounting, Auditing and Accountability Journal, 22*(2), 221–246. https://doi.org/10.1108/09513570910933951

Jamal, K., & Sunder, S. (2011). Is mandated independence necessary for audit quality? *Accounting, Organizations and Society, 36*(4–5), 284–292. https://doi.org/10.1016/j.aos.2011.03.001

Jensen, M. C., & Meckling, W. H. (1976). Theory of the firm: Managerial behavior, agency costs and ownership structure. *Journal of Financial Economics, 3*(4), 305–360. https://doi.org/10.1016/0304-405X(76)90026-X

Kaptein, M. (2008). Developing and testing a measure for the ethical culture of organizations: The corporate ethical virtues model. *Journal of Organizational Behavior, 29*(7), 923–947. https://doi.org/10.1002/job.520

Kuang, H., Li, H., Sherwood, M. G., & Whited, R. L. (2020). Mandatory audit partner rotations and audit quality in the United States. *AUDITING: A Journal of Practice and Theory*, *39*(3), 161–184. https://doi.org/10.2308/ajpt-18-152

Kung, F., & Li Huang, C. (2013). Auditors' moral philosophies and ethical beliefs. *Management Decision, 51*(3), 479–500. https://doi.org/10.1108/00251741311309616

Lin, H.-L., & Yen, A.-R. (2022). Auditor rotation, key audit matter disclosures, and financial reporting quality. *Advances in Accounting, 57*, 100594. https://doi.org/10.1016/j.adiac.2022.100594

Ortiz-Martínez, E., Marín-Hernández, S., & Santos-Jaén, J.-M. (2023). Sustainability, corporate social responsibility, non-financial reporting and company performance: Relationships and mediating effects in Spanish small and medium sized enterprises. *Sustainable Production and Consumption, 35*, 349–364. https://doi.org/10.1016/j.spc.2022.11.015

Purvis, B., Mao, Y., & Robinson, D. (2019). Three pillars of sustainability: In search of conceptual origins. *Sustainability Science, 14*(3), 681–695. https://doi.org/10.1007/s11625-018-0627-5

Puxty, A. G. (1993). *The social and organizational context of management accounting* (London Academic Press in association with The Chartered Institute of Management Accountants (ed.)). London Academic Press in association with The Chartered Institute of Management Accountants.

Reiter, S. A., & Williams, P. F. (2004). The philosophy and rhetoric of auditor independence concepts. *Business Ethics Quarterly, 14*(3), 355–376. https://doi.org/10.5840/beq200414329

Salehi, M., Zimon, G., Tarighi, H., & Gholamzadeh, J. (2022). The effect of mandatory audit firm rotation on earnings management and audit fees: Evidence from Iran. *Journal of Risk and Financial Management, 15*(3), 102. https://doi.org/10.3390/jrfm15030102

Sial, M. S., Zheng, C., Cherian, J., Gulzar, M. A., Thu, P. A., Khan, T., & Khuong, N. V. (2018). Does corporate social responsibility mediate the relation between boardroom gender diversity and firm performance of Chinese listed companies? *Sustainability, 10*(10), 1–18. https://doi.org/10.3390/su10103591

Svanberg, J., & Öhman, P. (2016). Does ethical culture in audit firms support auditor objectivity? *Accounting in Europe, 13*(1), 65–79. https://doi.org/10.1080/17449480.2016.1164324

Tepalagul, N., & Lin, L. (2015). Auditor independence and audit quality. *Journal of Accounting, Auditing and Finance, 30*(1), 101–121. https://doi.org/10.1177/0148558X14544505

Treviño, L. K., Butterfield, K. D., & McCabe, D. L. (2001). *The ethical context in organizations: Influences on employee attitudes and behaviors* (pp. 301–337). https://doi.org/10.1016/S1529-2096(01)03018-8

Valls Martínez, M. C., Martín Cervantes, P. A., & Miralles Quirós, M. M. (2022). Sustainable development and the limits of gender policies on corporate boards in Europe. A comparative analysis between developed and emerging markets. *European Research on Management and Business Economics, 28*(1), 1–15. https://doi.org/10.1016/j.iedeen.2021.100168

Valls Martínez, M. C., Santos Jaén, J. M., Soriano Román, R., & Martín Cervantes, P. A. (2022b). Are gender and cultural diversities on board related to corporate CO_2 emissions? *Journal of Cleaner Production, 363*, 1–17. https://doi.org/10.1016/j.jclepro.2022.132638

Valls Martínez, M. C., Soriano Román, R., & Martín Cervantes, P. A. (2022). Should risk-averse investors target the portfolios of socially responsible companies? *Oeconomia Copernicana, 13*(2), 1–36. https://doi.org/10.24136/oc.2022.014

Velte, P. (2017). Does board composition have an impact on CSR reporting? *Problems and Perspectives in Management*, *15*(2), 19–35. https://doi.org/10.21511/ppm.15(2).2017.02

Zahid, R. M. A., Taran, A., Khan, M. K., & Chersan, I.-C. (2023). ESG, dividend payout policy and the moderating role of audit quality: Empirical evidence from Western Europe. *Borsa Istanbul Review, 23*(2), 350–367. https://doi.org/10.1016/j.bir.2022.10.012

Zgarni, I., Hlioui, K., & Zehri, F. (2016). Effective audit committee, audit quality and earnings management. *Journal of Accounting in Emerging Economies, 6*(2), 138–155. https://doi.org/10.1108/JAEE-09-2013-0048

Zhang, Y., & Wei, L. (2022). Philanthropy, audit firms culture and auditor independence. *PLoS ONE, 17*(11), e0277567. https://doi.org/10.1371/journal.pone.0277567

María del Carmen Valls Martínez is a PhD in Economics and Business. In 1995 she joined the Economics and Business Department at the University of Almería (Spain), where she is an Associate Professor in Financial Economics and Accounting and leads the Ethics, Gender and Sustainability Research Group. Her current research interests include Financial Operations, Ethical Banking, Gender Economics Studies, Health Economics and Sustainability. She has published numerous books and book chapters in reputable publishers such as Springer, Dykinson, Pirámide, McGraw-Hill, etc. She is the author of a large number of articles in prestigious scientific journals, such as Journal of Cleaner Production, Corporate Social Responsibility and Environmental Management, Risk Management, Physica A: Statistical Mechanics and its Applications, European Research on Management and Business Economics, Mathematics, Plos One, Women's Studies International Forum, International Journal of Environmental Research and Public Health, Frontiers in Psychology, Occonomia Copernicana, etc. She is also an editor at Plos One, Frontiers in Public Health and Mathematics.

José Manuel Santos-Jaén is PhD Associate Professor of Accounting and Finance at the University of Murcia, Spain. He is a graduate in Business Administration and Management and a graduate in Law. He has completed three masters (MBA, Financial Management and Auditing). He finished his PhD in 2011. He is currently teaching financial statement analysis, financial accounting and corporate accounting. His research is focused on several aspects of Business and Management, such as Firm Performance, Innovation, Sustainability or Corporate Social Responsibility, among others. He is strongly oriented to a quantitative, survey-based methodology, mainly Structural Equation Modeling (PLS-SEM). He has published in international journals, including Journal of Cleaner Production, Corporate Social Responsibility and Environment, IEEE Transaction on Engineering Management, Sustainable production and consumption, The International Journal of Management Education, and Total Quality Management and Business Excellence. He has also presented the progress of his research at national and international congress.

Gema Martín de Almagro Vázquez is an Auditor at Sector 3, S.A.P. She holds a degree in Business Administration and Management from the University of Murcia, Spain. She has a Master's Degree in Account Auditing. She is Auditor of Accounts by the Instituto de Contabilidad y Auditoría de Cuentas (Accounting and Auditing Institute) in 2005 and is member nº 20.193 of the Official Registry of Auditors (ROAC). She completed her doctorate courses in 2000. She is a member of the Standards and Procedures Committee, of the Advisory Board of the Register of Economist Auditors (REA) and of the consultation resolution team of the Register of Expert Accountants (REC) of the General Council of Economists of Spain. Her professional activity includes audits of public and private sector accounts. She has made publications and lectures at national and international level, as in OEconomia, Revista Hispanoamericana de Historia de las Ideas Políticas y Sociales, being currently her research focused on the influence of the auditor's work in the value of companies.

Does Reducing Carbon Emissions Affect Business Profitability? An Analysis of Family and Non-family Businesses

José L. Gallizo-Larraz, Jordi Moreno-Gené, and Laura Sánchez-Pulido

1 Introduction

Companies have a leading role to play in achieving the emissions reduction target. Specifically, Spain has a series of companies committed to the United Nations in relation to climate change. *The United Nations CSR Report* analyses the advances in sustainability reported by the companies that have adhered to the initiative and indicates that 87% of the companies surveyed include in their strategic plan actions related to the climate change (CSR Commitment, 2023). In general, investment in emission efficiency is one of the solutions to the problem of GHG pollution adopted by companies, observing the existence of a relationship between the investments associated with eco-efficient commitments and the financial economic profitability of companies (Pérez-Calderón et al., 2021).

In particular, family businesses (FBs) pay great attention to the environment and demonstrate that they care about the situation, and FBs have unique characteristics that differentiate their governance, structures and behaviour from non-family businesses (Claub et al., 2022). FBs tend to be more principled and have long-term investment objectives because they are less constrained by shareholder accountability pressures than non-family businesses (Miller & Le Breton-Miller, 2006). In addition, FBs have the potential, resources and market position to lead sustainable change in the key global sectors they dominate. Indeed, it is noteworthy that more than a fifth

J. L. Gallizo-Larraz (✉) · J. Moreno-Gené · L. Sánchez-Pulido
Department of Economics and Business, University of Lleida, C/. Jaume II, 73, 25001 Lleida, Spain
e-mail: joseluis.gallizo@udl.cat

J. Moreno-Gené
e-mail: jordi.moreno@udl.cat

L. Sánchez-Pulido
e-mail: laura.sanchez@udl.cat

M. d. C. Valls Martínez and J. M. Santos-Jaén (eds.), *Environmentally Sustainable Production*, https://doi.org/10.1007/978-3-031-52656-5_16

of the global shipping industry is run by a handful of family businesses. In the automotive industry, just 36 families control companies that account for 55% of the total global market. Large FBs also dominate many other sectors such as clothing retailing, engineering and construction (Englisch, 2021). The widespread presence of FBs in a variety of activities represents a major opportunity for positive change in limiting GHG emissions.

This research is focused on the carbon emissions of firms and the influence on their economic profitability, for which it establishes a relationship between GHG emissions and the type of ownership of the firm. In previous work there is significant evidence suggesting that family ownership affects pollution levels (Borsuk et al., 2023). Many family firms prioritise passing the business onto the next generation over maximising profits. Because the climatic change change threatens both families and the long-term survival of firms, family-owned businesses might place a higher value on preserving future climatic conditions and ensuring a sustainable future for their descendants (Chrisman et al., 2012; Naldi et al., 2013). Upholding a good name and reputation encourages many family firms to protect environmental, social and governance (ESG) elements across industries and geographies (Sun et al., 2023).

The data used was obtained from the Carbon Footprint Registry *(Ministry for the Ecological Transition of Spain, MITECO)*. Here the *Carbon Footprint* is established as a measure of the number of tons of CO_2 emitted per year, which allows each company to evaluate greenhouse gas emissions, a concept adopted and certified in 2008 by the Carbon Trust (2008) of the United Kingdom. In our research, these data are supplemented with financial data obtained from the balance sheets and profit and loss accounts of the manufacturing companies from the SABI database.

The question is whether it pays to be an environmentally friendly company or whether it is more advantageous to ignore the recommendations and not become a green company (i.e. one with a green business model and a commitment to environmental responsibility). The literature is mixed. For example, there are authors who find that the most polluting companies are in a better financial position; the reason being that being green requires investments that are not certain to be recouped through a company's productive activity (Busch et al., 2022). Other authors have found evidence that companies with lower emissions have better financial performance (Gallego-Álvarez et al., 2015), Likewise, other authors find that corporate carbon emissions have a negative relation with the market value of equity (Desai & Raval, 2022; Saka & Oshika, 2014). Certainly, compliance costs and reputational risks increase the credit risk of carbon-intensive companies by creating contingent liabilities. In this respect, FBs show a commitment to continuity with subsequent generations, which forces them to avoid such risks (Sun et al., 2023). In our study, we introduce the moderating effect that operates in FBs compared to non-family businesses from the perspective of Socio-Emotional Wealth (SEW). Some authors argue that the protection of this SEW leads FBs to have better environmental performance than their non-family counterparts (Berrone et al., 2010). For this reason, we have adopted an approach that aims to deliver results for FBs compared to other businesses.

Our study contributes to the progress made on the influence of carbon emissions on return on assets (ROA). The distinguishing characteristic of this study is that it approaches the problem from an accounting perspective, distinguishing between the family and non-family nature of the companies. This means that from the accounting data of SABI database, the influence of CO_2 emissions on profitability ratios is observed together with financial characteristics that are introduced into the model as control variables, such as: sales figure, financial leverage, or variation in fixed assets among others. This approximation provides a more realistic picture of the real impact of carbon emission reductions on profitability and consequently on business continuity.

To this end, we developed a regression model that relates the GHG emissions released into the atmosphere to the type of ownership of the company and other variables that characterise the companies in the sample. Our results show that GHG emissions are negatively associated with the level of ROA, which means that reducing a company's emissions helps to improve its ROA. The results also show that family ownership of the company has a significant negative impact on total emissions. These results confirm the more environmentally responsible behaviour of FBs compared to non-family businesses, and a positive relationship between this behaviour and economic profitability, from which we conclude that there are economic benefits to be gained from environmental investment by FBs.

The rest of this paper is structured as follows: Sect. 2 discusses the relevant literature that allowed us to formulate the working hypotheses. Section 3 explains the data and methodology, while Sect. 4 interprets the results. Finally, Sect. 5 draws conclusions and offers recommendations to those who have to take decisions in the field of carbon emissions, such as policy makers.

2 Literature Review and Hypotheses Development

Little is known about the influence of family ownership and management on the social and environmental performance of businesses. Improving sustainability through the simultaneous pursuit of economic, environmental and social goals has become a key requirement for FBs across industries and countries (Claub et al., 2022). Research shows that FBs are more socially responsible than non-family businesses on several dimensions. This is likely due to family concerns about image and reputation and the desire to protect family wealth (Borsuk et al., 2023; Dyer & Whetten, 2006).

Some scholars have suggested that FBs are unlikely to act in a socially responsible manner (Craig & Dibrell, 2006), while others have pointed out that socially responsible behaviour by the FBs protects the family's wealth and their reputation (Berrone et al., 2010). This negative effect is more pronounced in primary studies that measure environmental performance in terms of the environmental operational practices adopted and in those that define the FBs in terms of family ownership and management (Miroshnychenko et al., 2022).

The mixed empirical evidence as to whether the effect of corporate environmental investment on financial performance is positive, negative or insignificant has been explained in terms of the different conditions and contexts that facilitate or hinder the ability to generate a profit situation. This explanation has gradually led the academic debate to consider the factors and conditions that moderate such a relationship (Gárces-Ayerbe et al., 2021).

Previous research has found that the positive effect of family ownership on environmental performance persists regardless of whether the CEO is a family member or serves as both CEO and chairman of the Board (Berrone et al., 2010). Sun et al. (2023) used stock market data to test shareholder reaction to the adoption of social, environmental and governance (ESG) criteria and found that both family ownership and control are positively related to ESG fundamentals, but that market competition negatively moderates family influence on the adoption of ESG criteria. From the previous studies, the relationship between the ownership of the company and its environmental behaviour is intuitive, so we introduce the component of the company being family or non-family in the belief that the characteristics of the family business moderate the value placed on the environment, and we establish the first hypothesis of this study, which takes into account the family ownership of the company.

H1. Family businesses are more environmentally friendly, which results in lower GHG emissions.

Meeting the expectations of society and the other stakeholders of organisations regarding the impact of business activities on climate change must go hand in hand with the necessary economic and financial viability (Pérez-Calderón et al., 2021). Given the importance of this issue to society, the relationship between carbon emissions and financial performance has been studied before, with mixed results. Some papers found that companies that are unconcerned about carbon emissions outperform companies with lower emissions (Busch et al., 2022; García-Sánchez et al., 2020; Thomas, 2001). The environmental impact of business activities is strongly related to the technology that companies use in their production processes (Kneller & Manderson, 2012; Przychodzen & Przychodzen, 2015; Teece, 2010), and it is unclear to what extent it is economically beneficial for companies to invest in new technologies that lead to lower GHG emissions. Delmas et al. (2015) showed that going green reduces a company's profitability in the short-term, but compensates for this in the long term. This reduction in short-term profitability may affect managers' decision making due to short-term performance goals. However, Lewandowski (2017) found evidence that companies with lower emissions have better financial performance in the long run. In a similar way, other papers found a positive relation between reduction of GHF and financial performance (Gallego-Álvarez et al., 2015; Misani & Pogutz, 2015; Iwata & Okada, 2011; Wagner, 2015). This positive relation suggests that a prevention approach entails competitive advantages that increase demand and improvements in productivity (Nishitani et al., 2011). This is in line with the Porter hypothesis (Porter, 1991) according to which companies must be able to generate capacities and abilities to reduce GHG emissions and at the same time maintain both economic and financial benefits. Environmental performance is an opportunity for firms because if they are able to adapt their operations to the requirements of reducing their environmental impact, they will become more competitive and improve their economic and financial performance (Porter & Van der Linde, 1995a, 1995b). Based on the above, we also propose the following hypothesis:

H2. The reduction of GHG emissions has a positive effect on corporate performance as measured by return on assets (ROA).

3 Data and Methodology

The data corresponding to the CO_2 emissions of the companies were obtained from the Carbon Footprint Register. This register, which is voluntary, depends on the Ministry for Ecological Transition of the Spanish government. It is a tool that aims to promote the fight against climate change, thus contributing to the reduction of GHG emissions. The carbon footprint identifies the amount of GHG emissions released into the atmosphere as a result of the development of an activity. With registration, the Ministry of Ecological Transition urges companies to follow such a sustainable path and take measures to try to reduce calculated emissions.

The companies included in the register have received a certificate of registration and the right to use a seal, which makes it possible to distinguish the level of participation of the company in the register and the achievements made in the attempt to reduce the carbon footprint. It is therefore assumed that the companies in the sample already have a certain level of environmental awareness, as demonstrated by their voluntary registration in the register.

The economic and financial data of the companies, as well as the information related to the type of ownership, were obtained from the Iberian Balance Analysis System (SABI) database, prepared by Bureau Van Dijk, which contains information on more than 2,600,000 Spanish companies, including financial profiles of the companies, information on their activities, annual accounts and financial ratios.

The final sample is made up of all Spanish manufacturing companies that have voluntarily registered in the Carbon Footprint Register and provided information on their emissions for at least one year during the period 2014–2021. In total, the sample consists of 593 companies providing a total of 1271 observations.

To explore the differences in carbon emissions between family and non-family businesses, the following regression model was estimated:

$$GHG\ emissions_{it} = \alpha + \beta_1 Family_{it} + \beta_{2-6} Control\ Variables_{it} + \varepsilon_{it} \quad (1)$$

In Eq. (1), the following variables were taken as the dependent variable:

- Total GHG emissions (measured in tonnes of CO_2): Total GHG emissions released into the atmosphere by an organisation as a result of its activities, both directly and indirectly. Due to its high variability, this variable has been taken in form of logarithm to minimise asymmetry.

Next, in order to check the robustness of the results, we also used as dependent variable two other proxies related to GHG emissions, taking into account whether they are directly or indirectly generated as a result of the activity of the companies in the sample. These proxies are the following:

- Scope 1 + Scope 2 carbon footprint (in tonnes of CO_2): GHG emissions from sources owned or controlled by the organisation, and indirect GHG emissions associated with the generation of electricity purchased and consumed by the organisation.
- Scope 3 carbon footprint (in tonnes of CO_2): Emissions resulting from the organisation's activities but caused by sources owned or controlled by another company, and indirect GHG emissions associated with the generation of electricity purchased and consumed by the first organisation.

To distinguish between family and non-family businesses we focus on the control and management of the company, which has already been used in the literature on the decision to adopt ESG criteria or on carbon emissions by authors such as Sun et al. (2023) and Garcés-Ayerbe et al. (2021). It is the family dummy, a dichotomous variable that takes the value 1 if the company is family owned and 0 if it is non-family owned. In classifying businesses as family-owned or non-family-owned, the definition proposed by the Family Business Institute (2015) was used. This definition is based on the percentage of capital controlled by the owner family, taking into account that it is inappropriate to apply the same percentage to all companies, since in companies with more dispersed ownership, it is not necessary to have such a high percentage of ownership to exercise control over the company (Sánchez-Pulido et al., 2022). Based on this consideration, the definition proposed by the Family Business Institute (2015) establishes that a company is considered a FB in the following cases:

- Dispersed ownership (no shareholder owns more than 50% of the capital). The company is classified as family-owned if any one person or family owns more than 5% individually or 20% collectively, and in addition the individual shareholder is a member of the board of directors, or there are shareholders with more than 20% of the capital who hold a directorship. Otherwise, the company is classified as non-family.
- Concentrated ownership structure (one shareholder owns more than 50% of the capital). The company is classified as family-owned if the family shareholder controls the ownership with a high percentage (50.01%) or if there are shareholders-directors with a shareholding of more than 50.01%. Otherwise, the company is classified as non-family.

We used the following control variables that are specific to the company:

Size (Ln Sales)

We define company size on the basis of turnover and assume that size influences the relationship between GHG emissions and profitability. In fact, large companies face higher pressures or social and political demands, as well as the monitoring and application of strict regulations; therefore, the larger companies are forced to take environmental responsibility, including the implementation of measures to limit carbon emissions (Deantari et al., 2019). Large companies face higher requirements or pressures to disclose carbon emissions because their operational activities generate large emissions, which makes large companies subject to public supervision (Deantari et al., 2019) and more likely to implement ESG practices (Sun et al., 2023). These observations led us to investigate the effect of firm size, and whether the benefits of environmental performance differ between large and small companies.

Variation in Fixed Assets

The change in fixed assets of the companies in the sample expresses the increase in investment in fixed assets and can lead to the reduction of GHG (Debbarma et al., 2022). This variable is related to the internal generation of resources and the existence of financial constraints at the time of obtaining these resources. Small companies and FBs resort to self-financing to a greater extent, probably because they have difficulties in carrying out capital increases and also because they are close to their credit limits; therefore, the execution of their investment projects depends to a greater extent on the generation of their own resources (Hernando & Villanueva, 2014). In terms of new investment projects, we expect the change in fixed assets to be significant. There are prospective studies where investments in climate action mean an increase in infrastructure that can open up economic and employment opportunities in the coming years (World Resources Institute, 2018). Therefore, we assume that the companies in our sample characterised by a high level of environmental concern will implement new eco-efficient investment projects.

Growth (Variation in Sales)

Sales growth can be influenced by actions taken by companies to reduce their carbon footprint. Integrating circular economy principles into the design of products and services, and engaging customers through marketing, are actions that convey an image of reputation and business innovation to the market, which can have a positive effect on sales (SBT, 2018).

Age (years Since Founding)

The age of the company has been shown to have a positive effect on carbon disclosure. That is, companies with more experience are more used to disclosing carbon emissions and participating in environmental protection is a way for the company to gain legitimacy (Solikhah et al., 2021). This greater knowledge of the company's reputation within the financial community leads the longest-standing companies to adopt policies that limit GHG emissions.

Leverage (Calculated as Debt/Equity Ratio)

Companies with high levels of leverage have higher financial obligations to pay debt and interest. A high level of leverage can lead to a higher risk of default, threatening the continuity of the firm. The risk arises because the company with high leverage has a capital structure with higher debt and a high dependence on that debt (Hernando & Villanueva, 2014). A firm's default risk could be negatively affected by a lack of environmental sustainability (Eichholtz et al., 2019). Firms with higher environmental sustainability have lower regulatory risks, as they are less likely to be fined for environmental misconduct and are better prepared to adopt any regulatory changes in environmental matters (Höck Klein et al., 2020).

Indeed, research has been conducted to determine whether the leverage of a company has a moderating effect on the relationship between environmental sustainability and credit risk premium. The results show that companies that are more sustainable have lower credit risk premiums along with higher creditworthiness (Dorfeitner et al., 2015). Investments to limit carbon emissions are likely to increase a company's leverage. However, these efforts do not prevent or control the risk of non-compliance with the company's obligations related to environmental protection regulations (Lewandowski, 2017). In a situation of capital scarcity, a company that decides to limit its carbon emissions will require a high level of debt to be able to make the necessary investments (Wang et al., 2022).

The relationship between profitability, GHG emissions, the type of ownership and the control variables was analysed using a panel data methodology. The Breusch-Pagan test was used to select the most appropriate regression technique. This has led us to reject the null hypothesis (X = 362.00 p-value 0.0000), indicating that in this case that the Random Effects model is more appropriate for estimation than Ordinary Least Squares for estimation. Then we run a Hausman test to compare Random Effects Model and the Fixed Effects Model. According to this latter test, the model had some significant exogeneity issue with the unobserved error. Thus, the random-effects model is preferred in this case. Therefore, the results presented in this study correspond to this best-fit model. The estimation of these models was carried out using the statistical software package Stata.

To analyse the influence that GHG emissions exert on the performance of these companies, the following model was estimated:

$$\begin{aligned} Economicperformance_{it} &= \alpha + \beta_1 TotalGHGEmissions_{it} \\ &\quad + \beta_{2-6} ControlVariables_{it} + \varepsilon_{it} \end{aligned} \tag{2}$$

Return on assets (ROA) was used in this Eq. (2) as a measure of business performance; it shows the return on the capital invested in the asset. ROA is calculated as earnings before interest and tax (EBIT) divided by total assets (Stickney et al., 2004). We use this variable as an approximation of the financial performance of the company, in line with other works in the field of environmental action research (Alvarez, 2012; Angelia & Suryaningsih, 2015; Dixon-Fowler et al., 2013; Garcés-Ayerbe et al., 2021).

4 Results

4.1 Descriptive Statistics

Table 1 summarises the descriptive statistics corresponding to the variables used in the model.

In terms of total GHG emissions, the results show that the manufacturing companies registered in the Carbon Footprint Register were responsible, directly or indirectly, for the annual emission of an average of 226,389 tonnes of CO_2 per company over the period 2014–2021. It can be seen that around 79% of the total GHG emissions generated are Scope 3 emissions, i.e., emissions resulting from the organisation's activities but occurring in sources owned or controlled by another organisation, with the exception of emissions associated with the generation of electricity purchased and consumed by the first organisation. It should be noted, however, that there are significant differences between the organisations included in the register, as evidenced by the high dispersion. This indicates that there are large differences between the amounts of GHG emitted by these companies.

Regarding the characteristics of the companies included in the register, we observe that, on average, they were quite profitable throughout the period (7.59%). This is an idea echoed in previous research, which suggests that corporate virtue in the form of social responsibility and, to a lesser extent, environmental responsibility, is likely to pay off (Dixon-Fowler et al., 2013; Orlitzky et al., 2003). However, the coexistence of companies with high economic returns and companies with negative returns is again observed. In terms of ownership, about 74% of these companies meet the criteria to be classified as FB, while the remaining 26% are classified as non-family businesses.

Table 1 Descriptive statistics

Variable	Mean	Std. dev.	Min.	Max.
Total GHG emissions (in tonnes of CO_2)	226,389	3,825,588	0	86,158,033
Scope 1 + Scope 2 carbon footprint (in tonnes of CO_2)	47,428	502,374	0	11,005,389
Scope 3 carbon footprint (in tonnes of CO_2)	178,959	3,526,522	0	80,516,949
Return on assets (ROA)	0.0759	0.1028	−0.3820	0.9423
Family businesses	0.7384	0.4397	0	1
Size (sales in thousands of euros)	125,441	1,099,546	33.76	21,546,136
Variation in fixed assets	0.2700	3.6287	−0.8657	119.45
Growth (variation in sales)	0.1322	0.6382	−0.8121	17.35
Age (years since its constitution)	30.68	17.45	1	119
Leverage	1.8728	3.2482	0.0358	66.01

There are also significant differences in the size of the businesses, with both large manufacturing businesses and micro-enterprises included in the register. Nevertheless, the results show that, on average, the enterprises in the sample experienced significant annual growth throughout the period, both in terms of fixed assets (27% per year) and turnover (13% per year). The average age of the companies analysed throughout the study was 30.68 years. As for the level of debt, it exceeds equity by about 87%.

A correlation matrix was used to test for multicollinearity of the model variables (see Table 2). As expected, there is a high correlation between the different Footprints, but this is not a problem as these variables do not coincide in any model. In general, the correlation between most variables is low and would not be significant as it does not exceed 0.4. At the same time, a higher correlation (0.56) is observed between sales and 1 + 2 footprint and also between sales and Total GHG emissions (0.55). To complement the analysis of multicollinearity, the variance inflation factor (VIF) is also shown in Table 2. We can see that the VIF is clearly below 2 in all cases, meaning the results are not biased due to mulicollinearity (Sheather, 2009).

4.2 Regression Results

Table 3 shows the influence of ownership type and other control variables on GHG emissions for total GHG emissions, those included in the Scope 1 + 2 footprint and those included in the Scope 3 footprint.

By incorporating the moderating effect of "family and non-family firm" into the general model, the results indicate that family ownership causes a negative and significant influence on total GHG emissions, with a significance level of 1%. In the following models, when we analyze its influence on the most direct emissions of the firm, measured on Scope 1 + Scope 2, the results obtained are confirmed and when we incorporate the effect of indirect emissions caused by the organization's activities (Scope 3) we also find the confirmation. Therefore, considering Hypothesis 1 which suggests that the Family businesses are more environmentally friendly, which results in lower GHG emissions, our results confirm that, in any of the carbon footprint measures, family firms are more environmentally responsible than non-family firms.

In a previous analysis we expected that the control variables would have a direct influence on GHG emissions. Once the calculations have been carried out, our results indicate the existence of a significant relationship between GHG emissions and the size of the company (at the 1% level of significance), which confirms that it is the largest companies that emit more GHG into the atmosphere as a result of their activities. A significant positive relationship is also observed between a company's debt level and its total GHG emissions (at the 1% level of significance). On the other hand, no significant relationships are observed for the rest of the control variables (growth, variation in fixed assets and age) and GHG emissions in any of their scope classifications 1, 2 and 3.

Table 2 Pairwise correlation matrix and variation inflation factor (VIF)

	(1)	(2)	(3)	(4)	(5)	(6)	(7)	(8)	(9)	(10)	VIF
(1) ROA	1										
(2) LnTotal GHG	0.0065	1									1.51
(3) LnScope1 + 2	0.0050	0.9604*	1								
(4) LnScope 3	0.0071	0.5699*	0.3884*	1							
(5) Family business	0.0064	−0.2170*	−0.2067*	−0.1317*	1						1.07
(6) Size	0.1507*	0.5573*	0.5631*	0.2553*	−0.1724*	1					1.57
(7) Var. Fixed assets	0.0389	−0.0420	−0.0431	−0.0148	0.0300	−0.0588*	1				1.04
(8) Growth	0.1557*	−0.0183	−0.0548	0.0671*	−0.0378	0.0140	0.1130*	1			1.03
(9) LnAge	−0.0196	0.2523*	0.2720*	0.0286	0.0298	0.3412*	−0.1598*	−0.1317*	1		1.27
(10) Indebtedness	−0.1336*	0.0333	0.0088	0.0692*	−0.0842*	−0.0981*	0.0387	0.0421	−0.2645*	1	1.10

* Significance at 1%

Table 3 Influence of family ownership on the emission of CO_2

	Total GHG emissions		Footprint Scope 1 + 2		Footprint Scope 3	
	Coef.	z	Coef.	z	Coef.	z
Independent variable						
Family company	−1.0944	−5.25***	−0.9892	−5.12***	−0.5400	−2.15**
Control variables						
Size	0.3151	8.49***	0.3131	9.03***	0.1372	2.49**
Var. fixed assets	0.0026	0.20	−0.0027	0.23	−0.0039	−0.21
Growth	−0.0441	−0.99	−0.0631	−1.51	0.0280	0.36
Age	0.1518	1.47	0.1155	1.20	0.0299	0.21
Indebtedness	0.0452	2.73***	0.0301	2.00**	0.0481	2.00**
Constant	3.1735	6.92***	3.0735	−7.20***	−1.5257	−0.40
Observations	1271		1271		1271	
Wald Chi^2	135.84***		141.13***		28.30***	
R^2	0.2802		0.2884		0.0844	

*** and ** indicate significance at 1 and 5% levels, respectively

Table 4 shows the relationship between GHG emissions and economic profitability for manufacturing companies registered in the Carbon Footprint Register.

Table 4 shows that GHG emissions have a weak and negative relationship (10%) with the level of economic profitability of companies. This suggests that reducing GHG emissions contributes to improving the profitability of companies.

Table 4 Influence of GHG emissions on economic profitability

	ROA	
	Coef.	z
Independent variable		
Ln total Footprint	−0.0028	−1.74*
Control variables		
Size	0.1326	5.46***
Var. fixed assets	0.0009	1.10
Growth	0.0215	5.32***
Ln age	−0.0141	−2.53**
Indebtedness	−0.0040	−4.11***
Constant	0.0167	0.71
Observations	1271	
Wald Chi^2	86.10***	
R-square	0.1152	

***, ** and * indicate significance at 1, 5 and 10% levels, respectively

Regarding the control variables, almost all of them show a statistically significant influence (with the only exception of changes in fixed assets) and with the expected signs. A positive influence on ROA is observed (at the 1% level of significance) from both the volume and the growth of sales.

On the contrary, a significant negative effect on ROA is observed for both age (at the 5% level of significance) and debt (at the 1% level of significance). It can be seen that the higher the level of debt, the lower the profitability giving the appearance of negative leverage. Reducing emissions requires an increase in debt, which is used to finance investments that limit carbon emissions. Therefore, a higher level of debt implies a limitation of carbon emissions. The companies in the sample, being environmentally aware, make the decision to invest with a long-term return, with the strategic priority of reducing GHG emissions. At the same time, older companies are found to be less profitable due to higher energy consumption as a result of being less active in energy efficiency measures.

5 Discussion and Conclusions

This paper develops information on the relationship between carbon emissions and firm performance. In a review of previous studies in the literature on the same topic, we observe a trend change in the profitability of environmentally committed companies.

Thus Delmas et al. (2015) suggest they obtain that higher carbon emissions are associated with higher ROA. Later, Busch et al. (2022) revisited this particular study and found inverse results for EU companies. The disparity is seen to have become especially pronounced since the 2015 Paris Agreement, suggesting that a shift in regulatory policy has had a particular impact on businesses. In this chapter, our results are in the same direction, where GHG emissions are negatively associated with the level of profitability, but we also find that for the companies included in our study, GHG emissions have an inverse relationship with family ownership. Keeping ownership and decision-making in the hands of the family reduces GHG emissions.

Busch et al. (2022) suggest that while reducing carbon emissions may initially hurt financial performance in the short run, it would pay off in the long run. It has been proven that the competitive advantages derived from environmental investments improve corporate image, stakeholder relations, product quality and the market share (Garcés-Ayerbe et al., (2021). In a similar vein, the results of our study show a negative relationship between GHG emissions and the level of ROA, which means that the reduction of these emissions in a company contributes to the improvement of its ROA. When we distinguish the family character of the companies, Craig and Dibrell (2006) have shown that FBs are better able than their non-family counterparts to facilitate environmentally friendly business policies, which are associated with better business innovation and financial performance. These results refer to a general concept of environmental protection and financial performance, which is confirmed in our results when we circumscribe it to the phenomenon of GHG emissions. Indeed, our results also show that family ownership of the organisation has a significant negative effect on total emissions.

In short, empirical evidence is obtained that confirms the more environmentally responsible behaviour of FBs compared to non-family businesses, and a positive relationship between this behaviour and economic profitability, from which we conclude that economic benefits can be obtained from the investment of family businesses in improving the environment. The policy implication of these findings is the importance of stricter environmental regulations to achieve the goals of the United Nations Climate Change Conference (COP21). Such regulations can encourage companies to develop and implement an effective carbon reduction strategy and promote the leverage effect of financial markets for a low-carbon economy.

References

Alvarez, I. G. (2012). Impact of CO_2 emission variation on firm performance. *Business Strategy and the Environment, 21*(7), 435–454. https://doi.org/10.1002/bse.1729

Angelia, D., & Suryaningsih, R. (2015). The effect of environmental performance and corporate social responsibility disclosure towards financial performance. *Procedia—Social and Behavioral Sciences, 211*, 348–355. https://doi.org/10.1016/j.sbspro.2015.11.045

Berrone, P., Cruz, C., & Gómez-Mejía, L. R. (2010). Socioemotional wealth and corporate responses to institutional pressures: Do family-controlled firms pollute less? *Administrative Science Quarterly, 55*(1), 82. https://www.jstor.org/stable/27856089

Borsuk, M., Eugster, N., Klein, P., & Kowalewski, O. (2023). Family ownership and carbon emissions. In *Working Papers 2023-ACF-01*. IESEG School of Management.

Busch, T., Bassen, A., Lewandowski, S., & Sump, F. (2022). Corporate carbon and financial performance revisited. *Organitzation and Environment*, *35*(1), 5–18. https://doi.org/10.1177/1086026620935638

Carbon Trust. (2008). Introduction to the carbon trust. https://www.carbontrust.com

Chrisman, J. J., Chua, J. H., Pearson, A. W., & Barnett, T. (2012). Family involvement, family influence, and family-centered non-economic goals in small firms. *Entrepreneurship Theory and Practice, 36*(2), 267–293. https://doi.org/10.1111/j.1540-6520.2010.00407.x

Claub, T., Kraus, S., & Jones, P. (2022). Sustainability in family business: Mechanisms, technologies and business models for achieving economic prosperity, environmental quality and social equity. *Technological Forecasting and Social Change, 176*, 121450. https://doi.org/10.1016/j.techfore.2021.121450

Craig, J., & Dibrell, C. (2006). The natural environment, innovation and firm performance: A comparative study. *Family Business Review, 19*(4), 275–288. https://doi.org/10.1111/j.1741-6248.2006.00075.x

CSR Commitment. (2023). Corporate, social responsibility commitment on progress report. ICTS UK & Ireland.

Deantari, S. A. O., Pinasti, M., & Herwiyanti, E. (2019). Faktor-Faktor yang Memengaruhi Pengungkapan Gas Rumah Kaca dari Perspektif Akuntansi Hijau. *Journal Equilibrium, 7*(2), 88–111. https://doi.org/10.21043/equilibrium.v7i1.5225

Debbarma, J., Choi, Y., Yang, F., & Lee, H. (2022). Exports as a new paradigm to connect business and information technology for sustainable development. *Journal of Innovation and Knowledge,* 7, 100233. https://doi.org/10.1016/j.jik.2022.100233

Delmas, M., Nairn-Birch, N., & Lim, J. (2015). Dynamics of environmental and financial performance: The case of greenhouse gas emissions. *Organitzation and Environment, 28*(4), 374–393. https://doi.org/10.1177/1086026615620238

Desai, R., & Raval, A. (2022). Examining the relation between market value and CO_2 emission: Study of Indian firms. *Copernican Journal of Finance and Accounting, 11*(3), 9–25. https://doi.org/10.12775/CJFA.2022.011

Dixon-Fowler, H. R., Slater, D. J., Johnson, J. L., Ellstrand, A. E., & Romi, A. M. (2013). Beyond "does it pay to be green?" A meta-analysis of moderators of the CEP–CFP relationship. *Journal of Business Ethics, 112*, 353–366. https://doi.org/10.1007/s10551-012-1268-8

Dorfeitner, G., Halbritter, G., & Nguyen, M. (2015). Measuring the level and risk of corporate responsibility: An empirical comparison of different ESG rating approaches. *Journal of Asset Management, 16*(7), 450–466. https://doi.org/10.1057/jam.2015.31

Dyer, W. G., & Whetten, D. A. (2006). Family firms and social responsibility: Preliminary evidence from the S&P 500. *Entrepreneurship Theory and Practice, 30*(6), 785–802. https://doi.org/10.1111/j.1540-6520.2006.00151.x

Eichholtz, P., Holtermans, R., Kok, N., & Yönder, E. (2019). Environmental performance and the cost of debt: Evidence from commercial mortgages and REIT bonds. *Journal of Banking and Finance,102*, 19–32. https://doi.org/10.1016/j.jbankfin.2019.02.015

Englisch, P. (2021). Family businesses have an opportunity to lead on ESG. In *Climate Change.* PWC

Family Business Institute. (2015). *La Empresa Familiar en España.* Edited and published by Instituto de la Empresa Familiar. Madrid. https://www.iefamiliar.com/publicaciones/la-empresa-familiar-en-espana-2015/

Gallego-Álvarez, I., Segura, L., & Martínez-Ferrero, J. (2015). Carbon emission reduction: The impact on the financial and operational performance of international companies. *Journal of Cleaner Production, 103*, 149–159. https://doi.org/10.1016/j.jclepro.2014.08.047

Garcés-Ayerbe, C., Rivera-Torres, P., Murillo-Luna, J., & Suarez-Galvez, C. (2021). Does it pay more to be green in family firms than in non-family firms? *Review of Managerial Science, 16*, 1365–1386. https://doi.org/10.1007/s11846-021-00475-8

García-Sánchez, I. M., Gallego-Álvarez, I., & Zafra-López, J. L. (2020). Do the ecoinnovation and ecodesing strategies generate value added in munificent environments? *Business Strategy and the Environment, 29*(3), 1021–1033. https://doi.org/10.1002/bse.2414

Hernando, I., & Villanueva, E. (2014). The recent slowdown in bank lending in Spain: Are supply-side factors relevant? *SERIE, 5*, 245–285. https://doi.org/10.1007/s13209-014-0117-7

Höck Klein, C., Landau, A., & Zwergel, B., (2020). The effect of environmental sustainability on credit risk. *Journal of Asset Management,* 22(5). https://doi.org/10.1057/s41260-020-00155-4

Iwata, H., & Okada, K. (2011). How does environmental performance affect financial performance? Evidence evidence from Japanese manufacturing firms. *Ecological Economics, 70*(9), 1691–1700. https://doi.org/10.1016/j.ecolecon.2011.05.010

Kneller, R., & Manderson, E. (2012). Environmental regulations and innovation activity in UK manufacturing industries. *Resource. Energy Economy, 34*(2), 211–235. https://doi.org/10.1016/j.reseneeco.2011.12.001

Lewandowski, S. (2017). Corporate carbon and financial performance: The role of emission reductions. *Business Strategy and the Environment, 26*, 1196–1211. https://doi.org/10.1002/bse.1978

Miller, D., & Le Breton-Miller, I. (2006). Family governance and firm performance: Agency, stewardship, and capabilities. *Family Business Review, 19*, 73–87. https://doi.org/10.1111/j.1741-6248.2006.00063.x

Miroshnychenko, I., De Massis, A., Barontini, R., & Testa, F. (2022). Family frms and environmental performance: A meta-analytic review. *Family Business Review, 35*(1), 68–90. https://doi.org/10.1177/08944865211064409

Misani, N., & Pogutz, S. (2015). Unraveling the effects of environmental outcomes and processes on financial performance: A non-linear approach. *Ecological Economics, 109*, 150–160. https://doi.org/10.1016/j.ecolecon.2014.11.010

Naldi, L., Cennamo, C., Corbetta, G., & Gomez-Mejia, L. (2013). Preserving socioemotional wealth in family firms: Asset or liability? The moderating role of business context. *Entrepreneurship Theory and Practice, 37*(6), 1341–1360. https://doi.org/10.1111/etap.12069

Nishitani, K., Kaneko, S., Fujii, H., & Komatsu, S. (2011). Effects of the reduction of pollution emissions on the economic performance of firms: An empirical analysis focusing on demand and productivity. *Journal of Cleaner Production, 19*(17–18), 1956–1964. https://doi.org/10.1016/j.jclepro.2011.06.021

Orlitzky, M., Schmidt, F., & Rynes, S. (2003). Corporate social and financial performance: A meta-analysis. *Organization Studies, 24*(3), 403–441. https://doi.org/10.1177/0170840603024003910

Pérez-Calderón, E., Pache-Durán, M., & Milanés-Montero, P. (2021). Eco-efficient investment: Effects on the economic and financial performance of Dow Jones sustainability world index companies. *Revista De Contabilidad—Spanish Accounting Review, 24*(2), 220–230. https://doi.org/10.6018/rcsar.403061

Porter, M. E. (1991). America's green strategy. *Scientific American, 264*(4), 168–179. https://doi.org/10.1038/scientificamerican0491-168

Porter, M. E., & Van der Linde, C. (1995a). Green and competitive: Ending the stalemate. *Harvard Business Review, 73*(5), 120–134.

Porter, M. E., & Van der Linde, C. (1995b). Toward a new conception of the environment competitiveness relationship. *Journal of Economic Perspectives, 9*(4), 97–118. https://doi.org/10.1257/jep.9.4.9

Przychodzen, J., & Przychodzen, W. (2015). Relationships between eco-innovation and financial performance–evidence from publicly traded companies in Poland and Hungary. *Journal of Cleaner Production, 90*, 253–263. https://doi.org/10.1016/j.jclepro.2014.11.034

Saka, C., & Oshika, T. (2014). Disclosure effects, carbon emissions and corporate value. *Sustainability Accounting, Management and Policy Journal, 5*(1), 22–45. https://doi.org/10.1108/SAMPJ-09-2012-0030

Sánchez-Pulido, L., Moreno-Gené, J., & Gallizo-Larraz, J. L. (2022). Internationalization of family firms: The effect of CEO attributes. *Journal of Management and Governance, 26*, 1123–1154. https://doi.org/10.1007/s10997-021-09597-3

SBT. (2018). *Value change in the value chain, best practices.* Science Based Targets. https://sciencebasedtargets.org/resources/files/SBT_Value_Chain_Report-1.pdf

Stickney, C., Brown, P., & Wahlen, J. (2004). *Financial reporting and statement analysis: A strategic perspective* (5th ed.). Dryden Press.

Sheather, S. (2009). *A modern approach to regression with R.* Springer.

Solikhah, B., Wahyuningrum, F. S., Yulianto, A., Sarwono, E., & Widiatami, A. K. (2021). Carbon emission report: A review based on environmental performance, company age and corporate governance. *IOP Conference Series: Earth and Environmental Science, 623*(2021), 012042. https://doi.org/10.1088/1755-1315/623/1/012042

Sun, J., Pellegrini, M. M., Dabić, M., Wang, K., & Wang, C. (2023). Family ownership and control as drivers for environmental, social, and governance in family firms. *Review of Managerial Science.* https://doi.org/10.1007/s11846-023-00631-2

Teece, D. J. (2010). Business models, business strategy and innovation. *Long Range Planning, 43*(2–3), 172–194. https://doi.org/10.1016/j.lrp.2009.07.003

Thomas, A. (2001). Corporate environmental policy and abnormal stock price returns: An empirical investigation. *Business Strategy and the Environment, 10*, 125–134. https://doi.org/10.1002/bse.281

Wagner, M. (2015). The link of environmental and economic performance: Drivers and limitations of sustainability integration. *Journal of Business Research, 68*(6), 1306–1317. https://doi.org/10.1016/j.jbusres.2014.11.051

Wang, Y., Wu, Z., & Zhang, G. (2022). Firms and climate change: A review of carbon risk in corporate finance. *Carbon Neutrality, 1*(6), 1–10. https://doi.org/10.1007/s43979-022-00005-9

World Resources Institute. (2018). *New climate economy. the global commission on the economy and climate.* Washington, DC 20002, USA. www.newclimateeconomy.report

Jose L. Gallizo-Larraz is a Professor of Financial Economics and Accounting at the University of Lleida (Spain). Previously he was an associate professor at the University of Zaragoza. He has been a visiting researcher in the School of Accounting, Banking and Economics at Bangor University (UK) and visiting lecturer at Paul Sabatier University, Toulouse (France). He is director of the Chair of Family Business at the UdL. He is co-author of European Financial Reporting Spain (Routledge) and he has published in Journal of Applied Econometrics, Emerging Markets Finance and Trade, Applied Stochastic Models in Business and Industry, the International Journal of Management Science-Omega, European Accounting Review, Journal of Business and Economics, Journal of Forecasting, Journal of the Royal Statistical Society and several other journals.

Jordi Moreno-Gené is a Serra Hunter Fellow (Senior Lecturer) at the Faculty of Law, Economics and Tourism at the University of Lleida (Spain). He holds a PhD in Business Administration, which was awarded the Extraordinary Prize for Doctoral Thesis in Economics by the University of Lleida. His main research interests include banking efficiency, family businesses and the economic sustainability of service companies. He has participated in several research projects and has published numerous papers in national and international journals, such as the Journal of Business Economics and Management, International Finance, Baltic Journal of Economics, Spanish Journal of Finance and Accounting, Spanish Journal of Agricultural Research, Applied Economics or the International Journal of Gastronomy and Food Science, among others.

Laura Sánchez-Pulido is a Lecturer at the Faculty of Law, Economics and Tourism at the University of Lleida (Spain). She holds a PhD in Business Administration, for which she was awarded the Extraordinary Prize for Doctoral Thesis in Economics by the University of Lleida. He has a Master's degree in Accounting, Auditing and Management Control, in which he obtained the Extraordinary Master's Award. His research is developed around the knowledge of the family business and the economic sustainability of snow tourism. He has participated in national and international conferences, presenting papers about different economic and financial aspects of the family business. He has published in national and international journals such as Applied Economics, Sustainability, Intangible Capital and Journal of Management and Governance, among others. She has more than ten years of professional experience in accounting and taxation.

Gender Diversity in Top Management for Greater Sustainability

Carbon Performance and Board Gender Diversity: The Moderating Effect of Patriarchal Attitudes

Sara Corral

1 Introduction

In recent years, society's interest in issues related to the environment and gender equality has significantly increased, positioning them as issues of great relevance.

The Paris Agreement, established in 2015, set the foundation for addressing climate change by making it the primary objective to limit global temperature rise to 1.5 °C and, at most, to stay below 2 °C. To achieve this goal, signatory countries were urged to "reduce GHG emissions by at least 40% by 2030 compared to 1990 levels" (European Commission, 2020). In this regard, the European Union (EU) is firmly committed to the fight against climate change by taking measures to reduce Greenhouse Gas (GHG) emissions. These emissions originating from human activities are considered to be the main causes of climate change and global warming. For this reason, European authorities are developing a regulatory framework that pushes companies to decrease their carbon emissions.

Regarding gender diversity, in the past decade, achieving gender equality has become a pressing global concern, strongly advocated for by both societal movements and regulatory bodies. This drive for gender equality has been particularly prominent in advanced economies and international organizations. As an example of this, the European Parliament released in (2017) a resolution on gender equality within the European Union for the years 2014–2015. This worldwide societal shift has opened doors for women to access positions of power within organizations on equal terms with their male counterparts, including opportunities for higher-ranking roles with economic influence. The inclusion of women in leadership positions serves as a potent catalyst for gender integration, as demonstrated by research such as that conducted

S. Corral (✉)
Facultad de Ciencias Económicas y Empresariales, Universidad de León, Campus de Vegazana S/N, 24071 León, Spain
e-mail: scorf@unileon.es

M. d. C. Valls Martínez and J. M. Santos-Jaén (eds.), *Environmentally Sustainable Production*, https://doi.org/10.1007/978-3-031-52656-5_17

by Huffman et al. (2010), highlighting the positive impact of women's participation on boards on a firm's overall performance.

The convergence of these two issues is particularly relevant as both gender equality and environmental protection are part of the 2030 Agenda, being the 5th and 13th goals. Companies with gender-diverse boards may be better positioned to address sustainability and carbon emission reduction-related issues. The inclusion of women in leadership roles can lead to increased attention to sustainable business actions and the adoption of policies that promote emission reduction (Atif et al., 2021; Elmagrhi et al., 2019; Nuber & Velte, 2021).

One variable that affects awareness of both the environment and gender diversity is the culture present in society. Stereotypes and cultural gender norms within a society establish the expected behavioural standards for each gender (Heilman, 2001). These norms can introduce biases into subjective evaluation processes. Consequently, patriarchal stereotypes create obstacles for women seeking to join boards of directors. Limited empirical evidence on this matter appears to align with this rationale, indicating that cultural factors within a country could potentially moderate gender disparities in board careers and influence the trajectory of gender diversity within boardrooms. Furthermore, cultural aspects have been found to influence the effect of the presence of women on the board of directors (Castro et al., 2023) by mitigating or reversing its positive effects, and this could also affect the environmental policies implemented by the firm.

Our results indicate a positive effect of board gender diversity on environmental performance measured through GHG emissions for an international sample of 665 listed firms from 17 European countries over the period 2005–2019. Similarly, we found a negative moderating effect of patriarchal attitudes on the baseline relationship. That is, patriarchal attitudes measured across four different variables reduce and even turn negative the relationship between gender diversity and carbon performance. Our results remain robust when using two alternative measures of gender culture and a proxy measure of environmental performance in terms of carbon emissions.

This study offers two main contributions to board gender diversity research. The first lies in the use of a concrete measure of environmental performance such as carbon performance, measured by taking into account all GHG emissions. As a second and more important contribution, for the first time, the moderating effect of culture on the above-mentioned base relationship is analysed. It should also be noted that in this work the effect of culture has been measured through patriarchal attitudes in the country. Using data from the World Values Survey, four cultural variables measuring different aspects of patriarchal attitudes were used. Culture, as measured by patriarchal attitudes, has a strong influence on achieving gender equality in society and in workplaces such as boards of directors. Finally, we provide strong evidence of the negative effect that patriarchal attitudes have on the positive influence of women's presence on boards on the carbon performance of firms.

The rest of the chapter consists of the following sections. The second section presents the theoretical framework that addresses the proposed hypotheses. The third section includes the sample, the research model and the methodology employed. The results obtained are presented in the fourth section. Finally, the fifth section presents the conclusions.

2 Literature Review and Hypotheses

Since achieving workplace gender equality is both a social and ethical imperative, it is crucial to recognize that women's representation in leadership roles plays a substantial role in fostering gender integration within an organization. Consequently, there has been a substantial focus on assessing the desirability of achieving gender equality at the highest levels of the corporate hierarchy (Dezsö & Ross, 2012).

Within the organisational chart of firms, most previous studies have focused on analysing the effect of board gender diversity on firm performance. This is because the board of directors plays a central role as the highest management and representative body of a firm, setting its goals, policies and strategies, and exercising oversight over top management (Johnson et al., 1996). The influence that boards of directors have within corporate governance is crucial. Its effect can be observed from three perspectives: corporate performance, social and environmental progress, and the promotion of socially responsible behaviour (Ben-Amar et al., 2017; Fernández-Gago et al., 2018; Rao & Tilt, 2016). Therefore, gender diversity can be considered as one of the main dimensions of corporate governance mechanisms (Zaid et al., 2020).

2.1 Gender Diversity and CSR

Previous literature has found a potential effect of gender diversity on corporate social responsibility (CSR) in firms, and specifically its environmental dimension, that can be explained by an array of different theories. Diversity within the board of directors, according to social identity theory, generates different perspectives that become unique and valuable tools for addressing complex situations (Hillman et al., 2007). Men and women hold different positions toward decisions regarding integrity or competitive issues (Yarram & Adapa, 2021). According to Liu (2018), organizations that include women in their board composition demonstrate a reduced propensity for engaging in fraudulent activities, tax evasion, or unethical conduct. Women typically exhibit a greater sense of responsibility when it comes to environmental concerns and are more inclined to promote the adoption of renewable energy sources within companies (Atif et al., 2021), and enhance a company's environmental performance by reducing total carbon emissions (Nuber & Velte, 2021).

Social role theory posits that the societal division of labor between genders also produces differences in social roles. By altering work roles to include more women in top corporate leadership positions, their social role is also modified, breaking traditional stereotypes and leading to the exchange and sharing of gender-specific capabilities (Eagly et al., 2000). Women tend to exhibit greater concern and interest towards issues raised by stakeholders, with environmental concerns being one of the main ones (Bernardi & Threadgill, 2010; Liao et al., 2019).

Providing resources to companies is one of the main tasks of the board of directors according to the resource dependency theory. Companies with higher gender diversity on their boards have more varied human resources, which may be key to generating linkages with the environment (relational capital) (Pfeffer, 1973). Having greater and more diverse resources may enable companies to access valuable and critical resources to help them meet their social and environmental obligations (Elmagrhi et al., 2019).

Stakeholder theory suggests that the presence of women in management teams puts greater pressure on firms to achieve stakeholder expectations (Elmagrhi et al., 2019). This is because women bring new perspectives and experiences to decision-making processes based on different gender-identified values (Jeong & Harrison, 2017; Loyd et al., 2013; Post & Byron, 2015). Women in managerial roles tend to exhibit higher self-transcendence, a trait related to benevolence and collectivism (Adams & Funk, 2012; Adams et al., 2011). Bruckmüller and Branscombe (2010) define self-transcendence as the manifestation of interpersonal qualities such as empathy and sensitivity to others' feelings. As a result, board diversity boosts relationships between firms and a broader range of external stakeholders. In this context, the presence of women on boards has been shown to drive firms' commitment to corporate social responsibility (Romano et al., 2020; Valls Martínez et al., 2019; Shaukat et al., 2016).

The Gender Integration and CSR Theory suggests that women in managerial roles may be more inclined to promote CSR policies and actions due to their heightened concern to social and gender-related issues (Post et al., 2011). Furthermore, companies with diverse boards may exhibit greater transparency in disclosing CSR-related information (Terjesen et al., 2015).

The agency theory (Carter et al., 2010) analyses the relationship between the parties involved in a contract between a principal and an agent, and how conflicts arising from their divergent interests and priorities affect that relationship. A gender-diverse board of directors can play a significant role in resolving agency conflicts as diversity enhances its oversight function, reducing information asymmetries and increasing transparency. This is also relevant when implementing environmental policies that involve incurring in costs that may result in lower profitability and, consequently, an agency problem.

2.2 *Gender Diversity and Environment*

In addition to the previously mentioned theories that link board diversity to environmental policies, such as carbon emissions reduction or improved carbon performance, legitimacy theory could be considered the main theory addressing this issue.

Legitimacy theory posits that organizations seek to maintain social legitimacy and public support through practices perceived as socially responsible and acceptable (Suchman, 1995). Stakeholders consider a company legitimate when its actions conform to the rules, principles and convictions established by society (Scherer & Palazzo, 2007). While firms are obligated to comply with regulations, they must also adhere to societal norms in order to gain widespread acceptance. The amount of emissions a company produces or its carbon performance can be a signal of the company's moral legitimacy (Zhang et al., 2013).

From the perspective of legitimacy theory, gender diversity in boards of directors can influence the amount of greenhouse gas emissions produced by a company. The inclusion of women in top management may be related with higher sensitivity to social and environmental issues, which can lead to the adoption of more sustainable business actions and a reduction in GHG emissions (Aguilera et al., 2007). Although implementing such measures may be costly for companies, there are numerous benefits stemming from social acceptance. Notable among these benefits is the positive impact of environmental policies on company profitability (measured through accounting variables such as ROA or market variables like Tobin's Q) (Busch & Lewandowski, 2018; Trumpp & Guenther, 2017). Additionally, there is a negative effect on financing costs and increased ease of access to credit. This is because companies with better carbon performance exhibit lower environmental risks and lower compliance costs (Herbohn et al., 2019; Kim et al., 2015).

In summary, gender diversity in boards of directors contributes to improving the legitimacy of companies. On one hand, this is due to the unique characteristics and leadership style of women. Women tend to be more responsible for the environment and inclined toward socially responsible actions, often opting for long-term and selfless leadership, thus enhancing a firm's CSR performance. This leadership style promotes more proactive and extensive CSR strategies, enabling companies to improve their social and environmental performance (Shaukat et al., 2016). On the other hand, the resources and relational capital they bring favour the implementation of policies to reduce emissions. In other words, women's concern for the environment does not translate only in the implementation of environmental strategies, but it also pushes for its maintenance over time, promoting environmentally responsible activities, eliminating those that are harmful, and disseminating environmental information about the companies. The positive effect of women on these three aspects improves the environmental performance of companies as a whole (Elmagrhi et al., 2019). Therefore, gender diversity would operate as a legitimacy catalyser, reinforcing the legitimacy of firms by fulfilling equality societal expectations and driving/pushing for environmental policies. According to the above, we establish the following hypothesis:

H1: There is a positive relationship between gender diversity and carbon performance.

2.3 Gender Diversity and Patriarchal Attitudes

Culture has an important influence on women's activities within a firm, playing a key role in the perception of gender roles at work. Different cultures have different expectations and social norms as to what is considered appropriate for women and men in the work environment (Eagly & Wood, 2012). In cultures where traditional gender values prevail, women may face greater resistance and challenges in advancing in their careers. Whereas in cultures where patriarchal values are less pronounced and women are allowed to compete on equal terms with men, they have been shown to be able to rise to any managerial position and demonstrate their competitiveness and preparedness. Even so, they are always more scrutinised and need to constantly prove that they can perform any kind of responsibility in the company. Cultural differences can therefore exert a considerable influence on how women are perceived, treated and advance in the workplace.

This influence of culture on the relationship between board gender diversity and profitability has been widely analysed in previous studies (Cabeza-García et al., 2019; Uribe-Bohorquez et al., 2019), using Hofstede (2011) dimensions to determine country cultural factors. However, this proxy comes with limitations, namely the lack of year-to-year variation of the data and its lack of updated information. In addition, gender-specific cultural traits are addressed only through one variable (masculinity). For this reason, in this chapter the effect of culture will be measured via patriarchal attitudes by country and year, consisting of four different dimensions. These variables have been obtained from the World Values Survey (WVS) which allows through its various waves to obtain more current and diverse values over the study period (Inglehart et al., 2020a, 2020b, 2020c).

Using this measure for patriarchal attitudes, Castro et al. (2023) find out that in environments with high patriarchal attitudes, the positive effect of board gender diversity on the economic performance of the firm is mitigated or even reversed. This may be caused by women's voice being overlooked or because these women on board adopt the roles of men (Diehl et al., 2020; Uribe-Bohorquez et al., 2019). Accordingly, this could also affect the effect of gender diversity on the implementation and effectiveness of environmental policies within the firm, leading to a worse carbon performance. According to the above, we establish the following hypothesis:

H2: A context with high patriarchal attitudes mitigates the positive association between gender diversity and carbon performance.

3 Empirical Research

3.1 Data

Our sample consists of 4361 firm-year observations of 665 listed firms from 17 European countries (Austria, Cyprus, Czech Republic, Denmark, Finland, France, Germany, Greece, Hungary, Italy, Netherlands, Norway, Poland, Portugal, Spain, Sweden and United Kingdom) over the period 2005–2019. Data for financial and corporate governance firms were collected from Thomson Reuters Eikon and the European Commission provided carbon emissions data via the European Union Transaction Log (EUTL) data viewer. As for the data used to analyse culture, patriarchal attitudes were calculated from data collected from the World Values Survey.

3.2 Model and Methodology

We employ a dynamic approach in our research framework, considering it the most appropriate methodology to analyse the effect of board gender diversity on a firm's carbon performance. To address unobserved heterogeneity, mitigate possible collinearity problems between the variables used in our study and take into account possible reverse causality between the dependent variable (carbon performance) and the explanatory variables of the model, we apply the Generalised Method of Moments (GMM) following the methodology proposed by Arellano in (2003). More specifically, we use the two-step system approach of the GMM that allows us to control for potential endogeneity (Arellano & Bover, 1995; Blundell & Bond, 1998).

In order to test H1, we use the baseline model, which includes the relationship between board gender diversity and carbon performance as well as several control variables [1]. The model would be as follows:

(1) $$CEP_{it} = \alpha_0 + \alpha_1 CEP_{it-1} + \alpha_2 WOM\_BOARD_{it} + \alpha_3 PROF_{it} + \alpha_4 TOBIN_{it} + \alpha_5 DEBT_{it} + \alpha_6 LIQ_{it} + \alpha_7 TANG_{it} + \alpha_8 RDA_{it} + \alpha_9 NDTS_{it} + \alpha_{10} AGE_{it} + \alpha_{11} BOARD\_SIZE_{it} + \alpha_{12} BOARD\_IND_{it} + \alpha_{13} CEO_{it} + \alpha_{14} GENDER\_QUOTA_{it} + \sum_{k=1}^{48} S_k + \sum_{k=1}^{17} C_k + \sum_{t=2005}^{2019} Y_t + \varepsilon_{it}$$

In the model, a firm's carbon performance (CEP) is the dependent variable. In order to avoid potential endogeneity issues and to fit the methodology, the carbon performance variable is lagged by one period. Gender diversity on the board of directors, measured through the proportion of women on a company's board of directors (WOM_BOARD) is the variable of interest (Abad et al., 2017; Li & Chen, 2018).

We include profitability (PROF), Tobin's Q (TOBIN), firm leverage (DEBT), liquidity (LIQ), tangibility of assets (TANG), research and development expenses (RDA), nondebt tax shield (NDTS) and firm age (AGE) as firm characteristic control variables (Fan et al., 2023; Valls Martínez et al., 2022). We also include board size (BOARD_SIZE), board independence (BOARD_IND) and the duality chief executive officer–chairperson (CEO) as governance control variables (Botta, 2020; Brahma et al., 2020; Nguyen et al., 2015). In addition, we include the gender quota law (GENDER_QUOTA) as a regulatory control variable.

Finally, we use dummies to control for effects related to sector, country and year. In addition, to avoid outliers, we winsorize all continuous variables at the 1st and 99th percentiles. The definitions of the variables are reported in Table 1.

To test H2, we modified the previous model by introducing four patriarchal attitude variables as additional moderators, that are interacted with the gender board diversity variable:

$$(2)\quad CEP_{it} = \alpha_0 + \alpha_1 CEP_{it-1} + \alpha_2 WOM\_BOARD_{it} + \alpha_3 PA_{it} + \alpha_4 WOM\_BOARD_{it} * PA_{it} + \alpha_5 PROF_{it} + \alpha_6 TOBIN_{it} + \alpha_7 DEBT_{it} + \alpha_8 LIQ_{it} + \alpha_9 TANG_{it} + \alpha_{10} RDA_{it} + \alpha_{11} NDTS_{it} + \alpha_{12} AGE_{it} + \alpha_{13} BOARD\_SIZE_{it} + \alpha_{14} BOARD\_IND_{it} + \alpha_{15} CEO_{it} + \alpha_{16} GENDER\_QUOTA_{it} + \sum_{k=1}^{48} S_k + \sum_{k=1}^{17} C_k + \sum_{t=2005}^{2019} Y_t + \varepsilon_{it}$$

These patriarchal attitudes are measured following Davis and Williamson (2019), establishing four levels of patriarchy across countries. In order to address each patriarchal attitude individually, we analyse the relationship between gender diversity and the four different measures representing patriarchal attitudes in four separate regressions. The four patriarchal attitude variables include the proportion of people who agree or strongly agree with the following statements: "University is more important for a boy than for a girl" (University_boy), "Job scarce: Men should have more right to a job than women" (Men_job_right), "Men make better business executives than women do" (Men_executives), and "Men make better political leaders than women do" (Men_politician), analysing patriarchal attitudes from the lowest to the highest rung of the power pyramid.

4 Empirical Results

4.1 Descriptive Results

The main statistics of the variables used in the study can be seen in Table 2. The dependant variable (CEP) had a mean value of −0.3758, with some firms achieving zero direct GHG emissions as shown by its maximum value. In terms of gender diversity on boards, we found that 13.04% of board members are women (WOM_

Table 1 Definition of variables

Variable	Definition	Source
CEP	The negative total verified direct GHG emissions produced by the firm to total sales. Calculated as CEP = −log(emissions)/log(sales)	EUTL
WOM_BOARD	Percentage of board members who are women	Eikon
Men_job_right	Perception that men have more right to a job than women	WVS
Men_executives	Perception that men are better executives than women	WVS
Men_politicians	Perception that men are better politicians than women	WVS
University_boy	Perception that university is more important for boys than girls	WVS
Control variables		
PROF	Operating income before depreciation to total assets. Operating income is measured as earnings before interests, taxes, depreciation, and amortization (EBITDA)	Eikon
TOBIN	Total assets minus the value of common equity plus the market value of the firm relative to total assets	Eikon
DEBT	The sum of short-term and long-term debt relative to total assets	Eikon
LIQ	Current assets to current liabilities	Eikon
TANG	Net property, plant, and equipment to total assets	Eikon
RDA	Research and development expenses relative to total assets	Eikon
NDTS	Depreciation and amortization to total assets	Eikon
AGE	Number of years since the firm appeared in the Eikon database	Eikon
Control variables		
BOARD_SIZE	Number of directors on a firm´s board	Eikon
BOARD_IND	Percentage of independent directors on the board	Eikon
CEO	Dummy variable, equal to 1 if the CEO is the board chairperson, and 0 otherwise	Eikon
GENDER_QUOTA	Dummy variable, equal to 1 if the following year the country mandatorily applies the gender quotas in firms' boardrooms, and 0 otherwise	Gender Statistics of the European Institute for Gender Equality

BOARD), in line with previous studies (Atif et al., 2019; Valls Martínez et al., 2019). Regarding patriarchal attitudes, we can observe that the mean value of each of them increases in relation to the degree of power and influence they hold over society. In other words, society presents greater barriers and obstacles to women as the power pyramid ascends. The lowest echelon of the pyramid, related to education, is represented by the patriarchal attitude "University is more important for boys than for girls," which has the lowest mean value of the four patriarchal attitudes (5.52%). The second and third echelons pertain to the workplace. The first one, measured through access to it, is reflected in the patriarchal attitude "Men have more right to a job than women," which has a mean value of 10.45%. As the level of the job position increases, so does the difficulty women face in accessing such positions, as evidenced by the mean value of 12.00% for the patriarchal attitude "Men are better executives than women." Finally, the top echelon refers to politics. Being the highest level of power, the barriers to reaching such positions of representation are greater, as shown by the patriarchal attitude "Men are better political leaders than women," with a mean value of 13.91%. The average firm return (PROF) is approximately 12.47%. In terms of board characteristics, approximately 47% of directors are independent (BOARD_IND), the average board size is 10 persons (BOARD_SIZE) and the proportion of chief executive officers who are also chairpersons (CEO) is approximately 28%.

Table 2 Descriptive statistics

Variables	Mean	Std. Dev	Median	Min	Max
CEP	−0.0713	0.1884	0	−0.80949	0
WOM_BOARD	0.1304	0.1256	0.1111	0	0.5
Men_job_right	0.1045	0.0698	0.0650	0.0200	0.4010
Men_executives	0.1200	0.0450	0.0940	0.0460	0.3220
Men__politicians	0.1391	0.0532	0.1150	0.0520	0.3850
University_boy	0.0552	0.0280	0.0410	0.0110	0.1870
PROF	0.1247	0.0974	0.1171	−1.2181	0.4975
TOBIN	1.5521	3.5252	0.9313	0.0436	59.7483
DEBT	0.2660	0.1705	0.2599	0.0000	0.9256
LIQ	1.5790	2.7549	1.2172	0.1174	90.6923
TANG	0.2736	0.2381	0.2154	0.0000	0.9308
RDA	0.0084	0.0281	0.0000	0.0000	0.2451
NDTS	0.0318	0.0313	0.0255	0.0000	0.2095
AGE	35.9800	29.8760	25.0000	1.0000	184.0000
BOARD_SIZE	10.2944	3.5753	10.0000	3.0000	22.0000
BOARD_IND	0.4650	0.4988	0.0000	0.0000	1.0000
CEO	0.2577	0.4374	0.0000	0.0000	1.0000
GENDER_QUOTA	0.2818	0.4499	0.0000	0.0000	1.0000

Variable definitions in Table 1

Table 3 shows the mean values for carbon performance, gender diversity and patriarchal attitudes by country. The countries with the highest gender diversity are Cyprus, Finland and Greece, all of which are above the average of the sample analysed. In terms of carbon performance, Italy and Poland are the countries with the highest emissions per unit of sales due to the fact that they have the values furthest from 0. In terms of patriarchal attitudes, Cyprus, Greece and the Czech Republic are the countries with the highest patriarchal attitudes, while Sweden, Denmark and Norway show the lowest values.

Table 4 shows the correlations between the dependent and explanatory variables of the models. The values indicate a low correlation, which validates the use of these variables.

In panel A, we find the correlations of the main variables under study, we observe that CEP is positively related to women's board participation, although it is not significant. We highlight the negative sign of the correlation between the four patriarchal attitude variables and gender diversity on boards of directors which confirms the strong effect of societal culture on workplace decisions (Diehl et al., 2020; Hoobler et al., 2018).

Table 3 CEP, gender diversity and patriarchal attitudes by country

Country	CEP	WOM_BOARD	Men_job_right	Men_executives	Men_politicians	University_boy
Austria	−0.0757	0.0165	0.1370	0.1350	0.1910	0.0860
Cyprus	0.0000	0.1991	0.3382	0.2885	0.2967	0.1345
Czech_Republic	0.0000	0.0773	0.2150	0.3020	0.3710	0.1500
Denmark	−0.0978	0.1035	0.0220	0.1160	0.0860	0.0220
Finland	−0.1316	0.1924	0.0511	0.1283	0.1292	0.0333
France	−0.0643	0.1227	0.1291	0.1042	0.1354	0.0599
Germany	−0.1536	0.0552	0.1223	0.1694	0.1475	0.1054
Greece	0.0000	0.1858	0.3770	0.2400	0.2600	0.0800
Hungary	0.0000	0.1653	0.2122	0.2789	0.3223	0.1228
Italy	−0.4044	0.0457	0.2452	0.1573	0.1909	0.1139
Netherlands	−0.0443	0.1065	0.0765	0.1075	0.1248	0.0453
Norway	−0.0660	0.1672	0.0542	0.1544	0.1163	0.0344
Poland	−0.2904	0.1561	0.2498	0.2245	0.3005	0.1111
Portugal	0.1213	0.1209	0.1780	0.1030	0.1560	0.0650
Spain	−0.0890	0.1064	0.1275	0.1248	0.1433	0.0963
Sweden	−0.0380	0.1863	0.0236	0.0683	0.0838	0.0203
United_Kingdom	−0.0422	0.1324	0.0826	0.1051	0.1261	0.0457
Total	−0.0713	0.1304	0.1045	0.1200	0.1391	0.0552

Variable definitions in Table 1

Table 4 Correlation matrix

Panel A. Main variables under study

	CEP	WOM_BOARD	Men_job_right	Men_executives	Men_politicians	University_boy
CEP	1					
WOM_BOARD	0.0020	1				
Men_job_right	−0.0752*	−0.0984*	1			
Men_executives	−0.0851*	−0.0785*	0.7196*	1		
Men_politicians	−0.0868*	−0.1121*	0.8379*	0.8977*	1	
University_boy	−0.1165*	−0.1612*	0.7352*	0.6837*	0.7801*	1

Panel B. Control variables

	CEP	PROF	TOBIN	DEBT	LIQ	TANG	RDA	NDTS	AGE	BOARD_ SIZE	BOARD_ IND	CEO	GENDER_ QUOTA
CEP	1												
PROF	−0.0144	1											
TOBIN	−0.0270	−0.0028	1										
DEBT	0.0149	−0.1156*	0.0301	1									
LIQ	0.0278	−0.0738*	0.0069	−0.1629*	1								
TANG	−0.2643*	0.1123*	0.0769*	0.2159*	−0.0730*	1							
RDA	−0.0243	−0.0202	0.0272	−0.1856*	0.0862*	−0.1354*	1						
NDTS	−0.0243	0.2706*	0.0259	0.1292*	−0.1050*	0.2504*	−0.0790*	1					
AGE	0.0290	−0.0278	0.0911*	−0.0301	−0.0500*	−0.0463*	−0.0365	−0.0702*	1				
BOARD_ SIZE	−0.1275*	−0.0849*	0.0502*	0.1353*	−0.0971*	0.1034*	−0.0864*	0.0827*	0.0859*	1			

(continued)

Table 4 (continued)

Panel B. Control variables													
BOARD_IND	−0.0186	−0.0217	0.0669*	0.1152*	0.0084	0.0490*	0.0641*	0.1361*	−0.0312	0.0981*	1		
CEO	0.0361	−0.0736*	0.0011	0.0748*	−0.0113	−0.0144	0.0096	0.0383	−0.0099	0.2940*	0.0298	1	
GENDER_QUOTA	−0.0479*	−0.0993*	0.0277	0.1371*	−0.0149	−0.0164	−0.0253	0.1294*	−0.0028	0.2331*	0.2008*	0.2846*	1

* denotes significance at the 1% level

In panel B, we find the correlations of CEP with the control variables. We highlight the negative relationship between tangibility (TANG) and carbon performance (CEP). Companies with higher tangibility values are often associated with industrial sectors that, on average, emit higher GHG, which negatively impacts carbon performance (Li & Ramanathan, 2018; Wang et al., 2014).

4.2 Multivariate Results

Table 5 presents the effect of gender diversity on carbon performance, as well as the moderating role of patriarchal attitudes on the previously analysed baseline relationship.

Column (1) of Table 5 addresses Hypothesis 1. Gender diversity on boards has a positive and significant relationship with carbon performance. Therefore, the presence of women on boards enables the implementation of policies that improve environmental performance and favour sustainable behaviour (Fan et al., 2023; Valls Martínez et al., 2022). According to the results obtained, we can confirm hypothesis 1.

To test the second hypothesis, we applied the model [2] in which patriarchal attitudes are included. The results are shown in columns 2–5, each column including gender diversity, a patriarchal attitude and the interaction between the two variables.

In the four regressions we can see that the effect of gender diversity on carbon performance is positive and significant, as we saw in the previous hypothesis. As for the effect of patriarchal attitudes on the base relationship, the coefficients indicate that high patriarchal values affect negatively, lowering or even turning negative the base ratio. That is, the joint effect on carbon performance is positive at lower levels of patriarchal attitudes but is offset and becomes negative as patriarchal attitudes become stronger. If we analyse each patriarchal attitude independently, we can observe that in all cases the positive effect of the presence of women on boards of directors on carbon performance decreases or disappears. In column 2, in mean values, the effect of gender diversity on CEP has a coefficient of 0.0202, while the moderating effect of the patriarchal attitude (Men_job_right) has a coefficient of −0.3402. Thus, taken together, there is a negative effect. We can observe the same effect in column 3, where we obtain coefficients 0.0532 and −0.4741 for the patriarchal attitude (Men_executives). For the patriarchal attitude Men_political_leaders (column 4) or University_boy (column 5) we find the same results: 0.0450 and −0.4241 in the former case, or 0.0292 and −0.8165 in the latter. The effect of patriarchal attitudes is consistently significant and negative (Castro et al., 2023), which reduces and even turns negative the positive effect of the presence of women on the board of directors on firms' carbon performance. Therefore, our second hypothesis is confirmed.

Table 5 Carbon performance, board gender diversity, and patriarchal attitudes

	(1)	(2)	(3)	(4)	(5)
CEP_{t-1}	0.9131***	0.9412***	0.9603***	0.9415***	0.9412***
	[0.0006]	[0.0002]	[0.0002]	[0.0002]	[0.0002]
WOM_BOARD	0.0125***	0.0202***	0.0532***	0.0450***	0.0292***
	[0.0007]	[0.0005]	[0.0008]	[0.0009]	[0.0007]
Men_job_right		0.0792***			
		[0.0013]			
WOM_BOARD*Men_job_right		−0.3402***			
		[0.0039]			
Men_executives			0.0782***		
			[0.0017]		
WOM_BOARD*Men_executives			−0.4742***		
			[0.0062]		
Men_political_leaders				0.0856***	
				[0.0015]	
WOM_BOARD*Men_political_leaders				−0.4241***	
				[0.0066]	
University_boy					0.0568***
					[0.0024]
WOM_BOARD*University_boy					−0.8165***
					[0.0127]
PROF	−0.0107***	−0.0035***	0.0095***	−0.0044***	−0.0034***
	[0.0008]	[0.0003]	[0.0004]	[0.0002]	[0.0003]
TOBIN	−0.0006***	0.0002***	0.0003***	0.0002***	0.0002***
	[0.0000]	[0.0000]	[0.0000]	[0.0000]	[0.0000]
DEBT	−0.0047***	0.0200***	0.0109***	0.0200***	0.0199***
	[0.0007]	[0.0003]	[0.0003]	[0.0003]	[0.0003]
LIQ	0.0000***	0.0004***	−0.0000***	0.0005***	0.0005***
	[0.0000]	[0.0000]	[0.0000]	[0.0000]	[0.0000]
TANG	0.0007	0.0077***	−0.0083***	0.0069***	0.0065***
	[0.0007]	[0.0003]	[0.0003]	[0.0003]	[0.0003]
RDA	0.0112**	−0.0165***	0.0084***	−0.0204***	−0.0208***
	[0.0051]	[0.0010]	[0.0015]	[0.0012]	[0.0014]
NDTS	−0.0288***	−0.0345***	−0.0895***	−0.0302***	−0.0333***
	[0.0040]	[0.0013]	[0.0019]	[0.0013]	[0.0015]
AGE	−0.0001***	−0.0001***	−0.0001***	−0.0001***	−0.0001***
	[0.0000]	[0.0000]	[0.0000]	[0.0000]	[0.0000]
BOARD_SIZE	−0.0004***	0.0000	−0.0003***	−0.0000*	−0.0000***

(continued)

Table 5 (continued)

	(1)	(2)	(3)	(4)	(5)
	[0.0000]	[0.0000]	[0.0000]	[0.0000]	[0.0000]
BOARD_ INDEPENDENCE	−0.0011***	−0.0026***	−0.0002***	−0.0028***	−0.0028***
	[0.0001]	[0.0000]	[0.0000]	[0.0001]	[0.0001]
CEO	−0.0030***	−0.0002**	−0.0032***	−0.0003***	−0.0001
	[0.0002]	[0.0001]	[0.0001]	[0.0001]	[0.0001]
GENDER_QUOTA	−0.0066***	−0.0061***	−0.0060***	−0.0058***	−0.0057***
	[0.0003]	[0.0001]	[0.0001]	[0.0001]	[0.0001]
Constant	0.0051***	−0.0161***	−0.0036***	−0.0197***	−0.0119***
	[0.0008]	[0.0003]	[0.0004]	[0.0004]	[0.0003]
Observations	4361	4361	4361	4361	4361
Number of firms	665	665	665	665	665
Country Dummies	Yes	Yes	Yes	Yes	Yes
Time Dummies	Yes	Yes	Yes	Yes	Yes
Industry Dummies	Yes	Yes	Yes	Yes	Yes
Hansen Test	369	471.8	458.1	464.6	469.9
Freedom degrees	337	477	438	477	477
Sig. Hansen	0.111	0.558	0.244	0.649	0.583
AR2	1.641	1.615	1.617	1.619	1.618
Sig. AR2	0.101	0.106	0.106	0.106	0.106

Variable definitions in Table 1. Standard errors are shown in brackets. ***, **, and * denote significance at the 1%, 5%, and 10% levels, respectively

4.3 Robustness Analysis

4.3.1 Allocation Shortfall

To test the robustness of our results in the previous section, we will use the allocation shortfall (Clarkson et al., 2015) to replace carbon performance as the dependent variable in our model. The allocation shortfall variable (ALLOCSHORT) is calculated as the difference between the firm's total carbon emissions and the firm's allocated carbon allowances designated to them by the European Commission. This variable is a proxy for the level of compliance of the firm with the reduction goals set by the EC, with positive (negative) values indicating emissions over (below) their target. Therefore, firms have to buy allowances in the market, which means incurring higher costs. In the negative case, this tells us that firms are making fewer emissions than they are allowed to, in compliance with regulations. Therefore, we expect the effect of gender diversity on the allocation shortfall to be negative, in line with our first hypothesis.

In column 1 of Table 6 we can observe the negative effect of gender diversity on the allocation shortfall, in line with our first hypothesis. In columns 2–5 we observe again the positive effect discussed above and how patriarchal attitudes moderate this relationship, reducing or even reversing the effect of gender diversity on the allocation shortfall for the four patriarchal attitude variables, in line with our second hypothesis.

4.3.2 Gender Inequality and Gender Development Indices

To check the robustness of the results, following Castro et al. (2023) we replace patriarchal attitudes with development and gender inequality indices as moderating variables. The Gender Inequality Index (GII) quantifies gender inequality, specifically measuring the loss of achievements (within development) in a country due to gender inequality. It relies on three key aspects of human development: reproductive health, empowerment, and women's participation in the labor market. A higher GII value indicates greater disparities in these three dimensions between men and women, thus reflecting higher levels of gender inequality. The Gender Development Index (GDI) represents the ratio of the Human Development Index (HDI) values for women to that of men. It assesses gender inequalities in three fundamental dimensions: health, education, and control over economic resources. Higher GDI values signify greater gender disparities in human development. As can be observed in Table 7, the results using both GII (column 1) and GDI (column 2) are very similar to those found in the main analysis. The positive effect of gender diversity on carbon performance is confirmed, as well as the moderating effect of the two indices. In both cases, the relationship between diversity and carbon performance is reduced or even negative, which would support our second hypothesis.

5 Conclusion

Although there is growing concern about gender equality and climate change, the relationship between the presence of women in top levels of corporate power and the carbon performance of companies is an underdeveloped line of inquiry. Previous papers addressing this relationship primarily focus on examining the direct effect of gender diversity on corporate CO_2 emissions. In our case, we sought to analyze this relationship using all the greenhouse gases emitted by firms, as well as incorporating a new cultural lens that serves to examine the moderating effect that culture may have on any type of corporate relationship.

This study examines the relationship between gender diversity on corporate boards and the carbon performance of companies. Likewise, we have also analyzed the moderating effect of culture, measured through patriarchal attitudes, on the aforementioned base relationship.

Table 6 Allocation shortfall, board gender diversity, and patriarchal attitudes

	(1)	(2)	(3)	(4)	(5)
$ALLOCSHORT_{t-1}$	0.3858***	0.7005***	0.6996***	0.7001***	0.6943***
	[0.0011]	[0.0034]	[0.0035]	[0.0035]	[0.0034]
WOM_BOARD	−0.0018*	−0.0309***	−0.0989***	−0.1065***	−0.0750***
	[0.0010]	[0.0046]	[0.0085]	[0.0081]	[0.0067]
Men_job_right		−0.0339***			
		[0.0100]			
WOM_BOARD*Men_job_right		0.6666***			
		[0.0414]			
Men_executives			−0.0937***		
			[0.0126]		
WOM_BOARD*Men_executives			1.1293***		
			[0.0725]		
Men_political_leaders				−0.0787***	
				[0.0097]	
WOM_BOARD*Men_political_leaders				1.0223***	
				[0.0562]	
University_boy					−0.1712***
					[0.0161]
WOM_BOARD*University_boy					2.0627***
					[0.1251]
PROF	0.0010	−0.0273***	−0.0222***	−0.0252***	−0.0294***
	[0.0010]	[0.0037]	[0.0039]	[0.0038]	[0.0038]
TOBIN	−0.0000***	−0.0011***	−0.0011***	−0.0011***	−0.0011***
	[0.0000]	[0.0000]	[0.0000]	[0.0000]	[0.0000]
DEBT	0.0170***	0.0106***	0.0117***	0.0101***	0.0090***
	[0.0012]	[0.0020]	[0.0022]	[0.0021]	[0.0021]
LIQ	0.0002***	0.0001***	0.0002***	0.0002***	0.0002***
	[0.0000]	[0.0000]	[0.0000]	[0.0000]	[0.0000]
TANG	0.0236***	0.0085***	0.0137***	0.0125***	0.0124***
	[0.0016]	[0.0024]	[0.0024]	[0.0024]	[0.0024]
RDA	0.0037	−0.0682***	−0.0592***	−0.0626***	−0.0927***
	[0.0065]	[0.0141]	[0.0143]	[0.0148]	[0.0151]
NDTS	−0.0120*	0.0365**	0.0466***	0.0430***	0.0630***
	[0.0068]	[0.0148]	[0.0149]	[0.0144]	[0.0146]
AGE	0.0002***	0.0000	0.0001***	0.0000**	0.0000**
	[0.0000]	[0.0000]	[0.0000]	[0.0000]	[0.0000]
BOARD_SIZE	0.0007***	−0.0006***	−0.0008***	−0.0005***	−0.0007***

(continued)

Table 6 (continued)

	(1)	(2)	(3)	(4)	(5)
	[0.0001]	[0.0001]	[0.0001]	[0.0001]	[0.0001]
BOARD_ INDEPENDENCE	0.0040***	0.0071***	0.0080***	0.0077***	0.0092***
	[0.0003]	[0.0006]	[0.0007]	[0.0006]	[0.0005]
CEO	0.0054***	−0.0026***	−0.0025***	−0.0021***	−0.0021***
	[0.0005]	[0.0006]	[0.0007]	[0.0007]	[0.0006]
GENDER_QUOTA	0.0007*	0.0081***	0.0104***	0.0100***	−0.0013**
	[0.0004]	[0.0008]	[0.0008]	[0.0009]	[0.0005]
Constant	−0.0219***	0.0037*	0.0057**	0.0072***	0.0044**
	[0.0016]	[0.0022]	[0.0026]	[0.0023]	[0.0021]
Observations	4361	4361	4361	4361	4361
Number of firms	665	665	665	665	665
Country Dummies	Yes	Yes	Yes	Yes	Yes
Time Dummies	Yes	Yes	Yes	Yes	Yes
Industry Dummies	Yes	Yes	Yes	Yes	Yes
Hansen Test	344.5	311.4	313	311.7	316.2
Freedom degrees	337	296	296	296	296
Sig. Hansen	0.377	0.259	0.238	0.254	0.201
m2	0.0848	0.542	0.544	0.524	0.501
Sig. m2	0.932	0.588	0.586	0.600	0.616

Standard errors are shown in brackets. ***, **, and * denote significance at the 1%, 5%, and 10% levels, respectively

For an international sample of 665 listed firms from 17 European countries for the period 2005–2019, our results show that greater diversity on corporate boards leads to better carbon performance of companies. This is due to women being more environmentally conscious and inclined towards socially responsible and ethical policies (Adams et al., 2011; Brahma et al., 2020). Our results show that gender-related cultural aspects present in countries (Hoobler et al., 2018) can also be observed in workplaces (Diehl et al., 2020). That is, those countries with more patriarchal attitudes will be the ones with the highest barriers and obstacles for women in the workplace. This is because companies adopt the cultural norms and values of the countries in order to be socially accepted (legitimacy theory) and comply with societal expectations and pressures (institutional theory) (Elmagrhi et al., 2019; Scott, 1995). Patriarchal attitudes show a significant negative effect (Castro et al., 2023) that moderates the relationship between gender diversity on boards and firm performance. That is, patriarchal attitudes reduce or turn negative the positive impact of the presence of women on boards. Finally, our results are robust to the use of two alternative measures of cultural gender bias, GII and GDI, as well as for the use of allocation shortfall.

Table 7 Carbon performance, board gender diversity, GII and GDI

	(1)	(2)
CEP_{t-1}	0.9352***	0.9316***
	[0.0003]	[0.0004]
WOM_BOARD	0.0397***	0.1937***
	[0.0005]	[0.0289]
GII	−0.1000***	
	[0.0013]	
WOM_BOARD*GII	−0.1883***	
	[0.0019]	
GDI		0.9159***
		[0.0111]
WOM_BOARD*GDI		−0.1871***
		[0.0294]
PROF	−0.0168***	−0.0068***
	[0.0003]	[0.0003]
TOBIN	−0.0007***	−0.0007***
	[0.0000]	[0.0000]
DEBT	0.0090***	0.0107***
	[0.0002]	[0.0003]
LIQ	0.0001***	0.0001***
	[0.0000]	[0.0000]
TANG	0.0042***	0.0108***
	[0.0003]	[0.0005]
RDA	−0.0117***	−0.0022
	[0.0023]	[0.0029]
NDTS	−0.0169***	−0.0304***
	[0.0011]	[0.0025]
AGE	−0.0000***	−0.0002***
	[0.0000]	[0.0000]
BOARD_SIZE	−0.0007***	−0.0009***
	[0.0000]	[0.0000]
BOARD_INDEPENDENCE	−0.0022***	0.0013***
	[0.0000]	[0.0001]
CEO	−0.0004***	0.0001
	[0.0001]	[0.0001]
GENDER_QUOTA	−0.0052***	−0.0140***
	[0.0001]	[0.0001]
Constant	0.0120***	−0.8953***
	[0.0003]	[0.0108]

(continued)

Table 7 (continued)

	(1)	(2)
Observations	4,361	4,361
Number of firms	665	665
Country Dummies	Yes	Yes
Time Dummies	Yes	Yes
Industry Dummies	Yes	Yes
Hansen Test	489	470.3
Freedom degrees	504	434
Sig. Hansen	0.676	0.111
m2	1.578	1.647
Sig. m2	0.115	0.100

Standard errors are shown in brackets. ***, **, and * denote significance at the 1%, 5%, and 10% levels, respectively

The most significant contribution of the chapter is the use of patriarchal attitudes as moderating variables for the effect of women's participation on corporate boards on carbon performance. This is the first work to use these variables as proxies for a country's culture.

We used four patriarchal attitude variables obtained from the World Values Survey. Each patriarchal attitude assesses how society perceives gender equality at different levels of society (university students, general workers, business executives and political leaders). Although all four variables show a significant and negative effect on the baseline relationship between gender diversity and carbon performance, as the level of power of the status asked about increases, the patriarchal attitude also increases. There is a significant difference between the patriarchal attitude at the top level (men are better politicians than women) and the attitude at the lowest level (university is more important for boys than for girls). Through this work, we provide empirical evidence to support the still scarce theoretical basis that analyses the influence of society's cultural values on gender inequality.

The main practical implications can be addressed to two groups: firstly, to policy makers and legislators in the process of decreasing or eradicating gender inequality and GHG, and secondly, to business leaders in countries with high patriarchal attitudes. In the first case, regulations and incentives should be put in place to encourage both the inclusion of women in management positions and socially responsible behaviour. In the second case, the presence of women on boards of directors brings different perspectives that encourage companies to orient their strategies more towards sustainability and environmental protection.

This work was supported by the Spanish Ministry of Universities (FPU21/01744).

References

Abad, D., Lucas-Pérez, M. E., Mínguez-Vera, A., & Yagüe, J. (2017). Does gender diversity on corporate boards reduce information asymmetry in equity markets? *Business Research Quarterly, 20*(3), 192–205. https://doi.org/10.1016/j.brq.2017.04.001

Adams, R. B., & Funk, P. (2012). Beyond the glass ceiling: Does gender matter? *Management Science, 58*(2), 219–235. https://doi.org/10.1287/mnsc.1110.1452

Adams, R. B., Licht, A. N., & Sagiv, L. (2011). Shareholders and stakeholders: How do directors decide? *Strategic Management Journal, 32*(12), 1331–1355. https://doi.org/10.2139/ssrn.1549482

Aguilera, R. V., Rupp, D. E., Williams, C. A., & Ganapathi, J. (2007). Putting the S back in corporate social responsibility: A multilevel theory of social change in organizations. *Academy of Management Review, 32*(3), 836–863.

Arellano, M. (2003). *Panel data econometrics*. Oxford University Press.

Arellano, M., & Bover, O. (1995). Another look at the instrumental variable estimation of error-components models. *Journal of Econometrics, 68*(1), 29–51. https://doi.org/10.1016/0304-4076(94)01642-D

Atif, M., Hossain, M., Alam, M. S., & Goergen, M. (2021). Does board gender diversity affect renewable energy consumption? *Journal Corporate Finance, 66*, 101665.

Atif, M., Liu, B., & Huang, A. (2019). Does board gender diversity affect corporate cash holdings? *Journal of Business, Finance & Accounting, 46*(7–8), 1003–1029. https://doi.org/10.1111/jbfa.12397

Ben-Amar, W., Chang, M., & McIlkenny, P. (2017). Board gender diversity and corporate response to sustainability initiatives: Evidence from the carbon disclosure project. *Journal of Business Ethics, 142*(2), 369–383. https://doi.org/10.1007/s10551-015-2759-1

Bernardi, R. & Threadgill, V. (2010). Women directors and corporate social responsibility. *EJBO: Electronic Journal of Business Ethics and Organizational Studies,15*(2), 15–21.

Blundell, S., & Bond, R. (1998). Initial conditions and moment restrictions in dynamic panel data models. *Journal of Econometrics, 87*(1), 115–143.

Botta, M. (2020). Financial crises, debt overhang, and firm growth in transition economies. *Applied Economics, 52*(40), 4333–4350. https://doi.org/10.1080/00036846.2020.1734184

Brahma, S., Nwafor, C., & Boateng, A. (2020). Board gender diversity and firm performance: The UK evidence. *International Journal of Finance & Economics*, 1–16. https://doi.org/10.1002/ijfe.2089

Bruckmüller, S., & Branscombe, N. R. (2010). The glass cliff: When and why women are selected as leaders in crisis contexts. *British Journal of Social Psychology, 49*(3), 433–451. https://doi.org/10.1348/014466609X466594

Busch, T., & Lewandowski, S. (2018). Corporate carbon and financial performance. A meta-analysis. *Journal of Industrial Ecology, 22*(4), 745–759. https://doi.org/10.1111/jiec.12591

Cabeza-García, L., Del Brio, E. B., & Rueda, C. (2019). Legal and cultural factors as catalysts for promoting women in the boardroom. *Business Research Quarterly, 22*(1), 56–67. https://doi.org/10.1016/j.brq.2018.06.004

Carter, D. A., D'Souza, F., Simkins, B. J., & Simon, W. G. (2010). The gender and ethnic diversity of US boards and board committees and firm financial performance. *Corporate Governance: An International Review, 18*(5), 396–414. https://doi.org/10.1111/j.1467-8683.2010.00809.x

Castro, P., Tascón, M. T., & Corral, S. (2023). Can patriarchal attitudes moderate the relation between women on boards and firm economic performance? *European Research on Management and Business Economics, 29*(3), 100222.

Clarkson, P. M., Li, Y., Pinnuck, M., & Richardson, G. D. (2015). The valuation relevance of greenhouse gas emissions under the European Union carbon emissions trading scheme. *European Accounting Review, 24*(3), 551–580.

Davis, L. S., & Williamson, C. R. (2019). Does individualism promote gender equality? *World Development, 123*, 1–16. https://doi.org/10.1016/j.worlddev.2019.104627

Dezsö, C. L., & Ross, D. G. (2012). Does female representation in top management improve firm performance? A panel data investigation. *Strategic Management Journal, 33*(9), 1072–1089. https://doi.org/10.1002/smj.1955

Diehl, A. B., Stephenson, A. L., Dzubinski, L. M., & Wang, D. C. (2020). Measuring the invisible: Development and multi-industry validation of the gender bias scale for women leaders. *Human Resource Development Quarterly, 31*(3), 249–280. https://doi.org/10.1002/hrdq.21389

Eagly, A. H., & Wood, W. (2012). Social role theory. *Handbook of Theories of Social Psychology, 2*, 458–476.

Eagly, A. H., Wood, W., & Diekman, A. B. (2000). Social role theory of sex differences and similarities: A current appraisal. *The Developmental Social Psychology of Gender, 12*(174), 9781410605245–12.

Elmagrhi, M. H., Ntim, C. G., Elamer, A. A., & Zhang, Q. (2019). A study of environmental policies and regulations, governance structures, and environmental performance: The role of female directors. *Business Strategy and the Environment, 28*(1), 206–220. https://doi.org/10.1002/bse.2250

European Parliament. (2017). *Resolution of 14 March 2017 on equality between women and men in the European Union in 2014–2015 (2016/2249(INI)).*

European Commission. (2020). *Paris Agreement | Climate Action.*

Fan, P., Qian, X., & Wang, J. (2023). Does gender diversity matter? Female directors and firm carbon emissions in Japan. *Pacific-Basin Finance Journal, 77*, 101931.

Fernández-Gago, R., Cabeza-García, L., & Nieto, M. (2018). Independent directors' background and CSR disclosure. *Corporate Social Responsibility and Environmental Management, 25*, 991–1001. https://doi.org/10.1002/csr.1515

Heilman, M. E. (2001). Description and prescription: How gender stereotypes prevent women's ascent up the organizational ladder. *Journal of Social Issues, 57*(4), 657–674. https://doi.org/10.1111/0022-4537.00234

Herbohn, K., Gao, R., & Clarkson, P. (2019). Evidence on whether banks consider carbon risk in their lending decisions. *Journal of Business Ethics, 158*, 155–175. https://doi.org/10.1007/s10551-017-3711-3

Hillman, A. J., Shropshire, C., & Cannella, A. A. (2007). Organizational predictors of women on corporate boards. *Academy of Management Journal, 50*(4), 941–952. https://doi.org/10.5465/amj.2007.26279222

Hofstede, G. (2011). Dimensionalizing cultures: The Hofstede model in context. *Online Readings in Psychology and Culture, 2*(1), 1–26. https://doi.org/10.9707/2307-0919.1014

Hoobler, J. M., Masterson, C. R., Nkomo, S. M., & Michel, E.J. (2018). The business case for women leaders: Meta-analysis, research critique, and path forward. *Journal of Management*, 2473–2499. https://doi.org/10.1177/0149206316628643

Huffman, M. L., Cohen, P. N., & Pearlman, J. (2010). Engendering change: Organizational dynamics and workplace gender desegregation 1975–2005. *Administrative Science Quarterly, 55*, 255–277. https://doi.org/10.2189/asqu.2010.55.2.255

Inglehart, R., Haerpfer, C., Moreno, A., Welzel, C., Kizilova, K., Diez-Medrano, J., Lagos, M., Norris, P., Ponarin, E., Puranen, B., et al. (Eds.) (2020a) *World values survey: Round five-country-pooled datafile version.* JD Systems Institute. Available at: http://www.worldvaluessurvey.org/WVSDocumentationWV5.jsp

Inglehart, R., Haerpfer, C., Moreno, A., Welzel, C., Kizilova, K., Diez-Medrano, J., Lagos, M., Norris, P., Ponarin, E., Puranen, B., et al. (Eds.) (2020b). *World values survey: Round six-country-pooled datafile version.* JD Systems Institute. Available at: http://www.worldvaluessurvey.org/WVSDocumentationWV6.jsp

Inglehart, R., Haerpfer, C., Moreno, A., Welzel, C., Kizilova, K., Diez-Medrano, J., Lagos, M., Norris, P., Ponarin, E., Puranen, B., et al. (Eds.) (2020c). *World values survey: Round seven-country-pooled datafile version.* JD Systems Institute. Available at: http://www.worldvaluessurvey.org/WVSDocumentationWV7.jsp

Jeong, S. H., & Harrison, D. A. (2017). Glass breaking, strategy making, and value creating: Meta-analytic outcomes of women as CEOs and TMT members. *Academy of Management Journal, 60*(4), 1219–1252. https://doi.org/10.5465/amj.2014.0716

Johnson, J. L., Daily, C. M., & Ellstrand, A. E. (1996). Boards of directors: A review and research agenda. *Journal of Management, 22*(3), 409–438. https://doi.org/10.1016/S0149-2063(96)90031-8

Kim, Y. B., An, H. T., & Kim, J. D. (2015). The effect of carbon risk on the cost of equity capital. *Journal of Cleaner Production, 93*, 279–287. https://doi.org/10.1016/j.jclepro.2015.01.006

Li, H., & Chen, P. (2018). Board gender diversity and firm performance: The moderating role of firm size. *Business Ethics: A European Review, 27*(4), 294–308. https://doi.org/10.1111/beer.12188

Li, R., & Ramanathan, R. (2018). Impacts of industrial heterogeneity and technical innovation on the relationship between environmental performance and financial performance. *Sustainability, 10*(5), 1653.

Liao, Z., Zhang, M., & Wang, X. (2019). Do female directors influence firms' environmental innovation? The moderating role of ownership type. *Corporate Social Responsibility and Environmental Management, 26*(1), 257–263. https://doi.org/10.1002/csr.1677

Liu, C. (2018). Are women greener? Corporate gender diversity and environmental violations. *Journal Corporate Finance, 52*, 118–142. https://doi.org/10.1016/j.jcorpfin.2018.08.004

Loyd, D. L., Wang, C. S., Phillips, K. W., & Lount, R. B., Jr. (2013). Social category diversity promotes premeeting elaboration: The role of relationship focus. *Organization Science, 24*(3), 757–772. https://doi.org/10.1287/orsc.1120.0761

Nguyen, T., Locke, S., & Reddy, K. (2015). Does boardroom gender diversity matter? Evidence from a transitional economy. *International Review of Economics & Finance, 37*, 184–202. https://doi.org/10.1016/j.iref.2014.11.022

Nuber, C., & Velte, P. (2021). Board gender diversity and carbon emissions: European evidence on curvilinear relationships and critical mass. *Business Strategy and the Environment, 30*(4), 1958–1992. https://doi.org/10.1002/bse.2727

Pfeffer, J. (1973). Size, composition, and function of hospital boards of directors: A study of organization-environment linkage. *Administrative Science Quarterly*, 349–364.

Post, C., & Byron, K. (2015). Women on boards and firm financial performance: A meta-analysis. *Academy of Management Journal, 58*(5), 1546–1571. https://doi.org/10.5465/amj.2013.0319

Post, C., Rahman, N., & Rubow, E. (2011). Green governance: Boards of directors' composition and environmental corporate social responsibility. *Business & Society, 50*(1), 189–223. https://doi.org/10.1177/0007650310394642

Rao, K., & Tilt, C. (2016). Board diversity and CSR reporting: An Australian study. *Meditari Accountancy Research, 24*(2), 182–210. https://doi.org/10.1108/MEDAR-08-2015-0052

Romano, M., Cirillo, A., Favino, C., & Netti, A. (2020). ESG (environmental, social and governance) performance and board gender diversity: The moderating role of CEO duality. *Sustainability, 12*(21), 1–16. https://doi.org/10.3390/su12219298

Scherer, A. G., & Palazzo, G. (2007). Towards a political conception of corporate responsibility. *Academy of Management Review, 32*(4), 1096–1120. https://doi.org/10.5465/amr.2007.26585837

Scott, W. R. (1995). *Institutions and organizations*. Sage.

Shaukat, A., Qiu, Y., & Trojanowski, G. (2016). Board attributes, corporate social responsibility strategy, and corporate environmental and social performance. *Journal of Business Ethics, 135*, 569–585. https://doi.org/10.1007/s10551-014-2460-9

Suchman, M. C. (1995). Managing legitimacy: Strategic and institutional approaches. *The Academy of Management Review, 20*(3), 571–610.

Terjesen, S., Aguilera, R. V., & Lorenz, R. (2015). Legislating a woman's seat on the board: Institutional factors driving gender quotas for boards of directors. *Journal of Business Ethics, 128*(2), 233–251. https://doi.org/10.1007/s10551-014-2083-1

Trumpp, C., & Guenther, T. (2017). Too little or too much? Exploring U-shaped relationships between corporate environmental performance and corporate financial performance. *Business Strategy and the Environment, 26*, 49–68. https://doi.org/10.1002/bse.1900

Uribe-Bohorquez, M. V., Martínez-Ferrero, J., & García-Sánchez, I. M. (2019). Women on boards and efficiency in a business-orientated environment. *Corporate Social Responsibility and Environmental Management, 26*, 82–96. https://doi.org/10.1002/csr.1659

Valls Martínez, M. C., Cruz Rambaud, S., & Parra Oller, I. M. (2019). Gender policies on board of directors and sustainable development. *Corporate Social Responsibility and Environmental Management, 26*, 1539–1553. https://doi.org/10.1002/csr.1825

Valls Martínez, M. C., Santos-Jaén, J. M., Román, R. S., & Martín-Cervantes, P. A. (2022). Are gender and cultural diversities on board related to corporate CO_2 emissions? *Journal of Cleaner Production, 363*, 132638.

Wang, L., Li, S., & Gao, S. (2014). Do greenhouse gas emissions affect financial performance?—An empirical examination of Australian public firms. *Business Strategy and the Environment, 23*(8), 505–519. https://doi.org/10.1002/bse.1790

Yarram, S.R. & Adapa, S. (2021). Board gender diversity and corporate social responsibility: is there a case for critical mass? *Journal of Cleaner Production, 278*, 123319. https://doi.org/10.1016/j.jclepro.2020.123319

Zaid, M.A.A., Wang, M., Adib, M., Sahyouni, A., & Abuhijleh, S. T. (2020). Boardroom nationality and gender diversity: implications for corporate sustainability performance. *Journal of Cleaner Production, 251*, 119652. https://doi.org/10.1016/j.jclepro.2019.119652

Zhang, J. Q., Zhu, H., & Ding, H. (2013). Board composition and corporate social responsibility: An empirical investigation in the post sarbanes-oxley era. *Journal of Business Ethics, 114*(3), 381–392. https://doi.org/10.1007/s10551-012-1352-0

Sara Corral Ph.D. student in Business Economics at the University of León. Sara has obtained a teaching and research training grant (FPU) funded by the Spanish Ministry of Universities, attached to the Department of Financial Economics and Accounting of the Department of Management and Business Economics at the University of León, to carry out her teaching and research training.

Graduated in Finance and in the Master in Research in Business Administration and Economics, during her student years, she received the extraordinary end-of-degree prize as well as the prize for the best record in the Degree in Finance awarded by the University of León. Her Bachelor Thesis in Finance on the influence of gender on business failure received an accreditation at an international congress.

Although her first steps in research focused on business failure, her current doctoral research focuses on the relationship between firm performance and gender diversity on boards of directors. Her first work as part of her doctoral thesis has been published in the journal *European Research on Management and Business Economics* and has won several awards in the different international conferences in which it has been presented, among which we can highlight: prize for best paper at the ASEPELT 2023 International Conference, finalist for the prize for best paper at the ACEDE 2023 International Conference, and the diploma of the Spanish Journal of Finance and Accounting (REFC) for the quality and scientific contribution of the paper presented at the AECA 2023 International Conference.

She is a member of the Finance Research Group (GIF) and the Financial Education Teaching Innovation Group (GIDEF), participating as a researcher in two national research projects and in a European teacher training project.

The Influence of Corporate Governance on the Sustainability of American Company Buildings

María del Carmen Valls Martínez, José-María Montero, María Estefanía Sánchez Pacheco, and Fernando José Zambrano Farías

1 Introduction

In 2015, the member countries of the United Nations agreed on 17 Sustainable Development Goals (SDGs) to be met by 2030. The aim is to eradicate poverty, protect the planet, and improve the lives and prospects of people worldwide. In particular, goal 9 concerns industry, innovation, and infrastructure. More specifically, target 9.4 is to achieve the modernization of infrastructures and reconversion of sectors to make them sustainable, which will involve using resources efficiently and promoting clean and environmentally sound technologies. On the other hand, goal 11, which refers to sustainable cities and communities, has as target 11.3 to promote inclusive and sustainable urbanization. Many companies are headquartered in city centers, especially service companies. Therefore, the construction of new buildings or the reconversion of old ones must always be done sustainably. In addition, target 12.6 of objective 12, on responsible production and consumption, establishes that companies should adopt sustainable practices and include information on sustainability in their public reports (https://sdgs.un.org/goals).

Sustainable or green building is a multidimensional concept (Alsulaili et al., 2020; Ding et al., 2018; Samer, 2013). Green buildings permit maintaining or improving their environment's quality of life through construction and equipment. They are

M. C. Valls Martínez (✉)
Mediterranean Research Center on Economics and Sustainable Development, Economics and Business Department, University of Almería, Almería, Spain
e-mail: mcvalls@ual.es

J.-M. Montero
Department of Political Economy and Public Finance, Economic and Business Statistics, and Economic Policy, University of Castilla-La Mancha, Toledo, Spain

M. E. Sánchez Pacheco · F. J. Zambrano Farías
International University of Ecuador, Guayaquil, Ecuador

M. d. C. Valls Martínez and J. M. Santos-Jaén (eds.), *Environmentally Sustainable Production*, https://doi.org/10.1007/978-3-031-52656-5_18

designed with eco-friendly materials, are low-energy and water-consumption buildings, use clean energy, and minimize pollution. Another feature that characterizes green buildings is that they blend in with their environment, integrating into it and going unnoticed. Their location is also strategically chosen, seeking places with access to public transport, reducing pollution caused by the traffic of private vehicles.

Although corporate concern for social and environmental issues has become an essential aspect following the establishment of the SDGs (ElAlfy et al., 2020), it has received increasing attention since the 1990s. Indeed, the disclosure of environmental, social, and governance (ESG) actions in a document issued together with companies' financial statements has become a common practice, especially since this decade's economic, social, and environmental scandals (Fernández-Gago et al., 2018; Sial et al., 2018; Valls Martínez et al., 2022a).

ESG investment and inclusion in sustainability indices have become mainstream, with companies specializing in the valuation of ESG scores, such as Bloomberg or Thomson Reuters (Duque-Grisales & Aguilera-Caracuel, 2019; Machmuddah & Wardhani, 2020; Pyles, 2020) being used as pillars when selecting a portfolio by investment fund managers and individual investors. One of the variables considered in the environmental part of the ESG scores is whether or not the company has green buildings, which is the study aim of this research in relation to corporate governance.

The board of directors determines the company's fundamental strategies, mission, goals, and objectives, epitomizing corporate governance (Pletzer et al., 2015). The structure and characteristics of company boards have been the subject of research in recent years, relating them to several measures of financial performance, both accounting (ROA, ROE, etc.) and market (market-to-book ratio, PER, Tobin's Q, beta, etc.), as well as to the disclosure of corporate social responsibility measures or ESG scores (Ben-Amar et al., 2017; Charumathi & Rahman, 2019; Dienes & Velte, 2016; Kyaw et al., 2017; Liao et al., 2018; Sial et al., 2018; Velte, 2017; Zhang et al., 2013). The ESG score analysis has often been broken down into three pillars: environmental, social, and governance. However, each of these pillars comprises a multiplicity of determinants, and their individualized study is unusual. Recently, CO_2 emissions and their relationship with variables such as gender diversity on the board of directors and cultural diversity have been the subject of some empirical studies (García Martín & Herrero, 2020; Haque, 2017; Nuber & Velte, 2021; Valls Martínez et al., 2022b).

Following this research line, the present research seeks to explore the relationship between the company's adoption of green buildings and the characteristics of the board of directors, such as its size, composition (gender and skills of the directors, as well as the relationship with the company's management), the involvement level of board members (remuneration and duration in the position) or the degree of activity. The empirical study is based on a sample of companies included in the Standard & Poor 500 index from 2011 to 2019. Data analysis is performed using multivariate logistic regression with panel data. The results show that the probability of the company adopting sustainable buildings has a significant positive direct relationship with the percentage of women on the board of directors, the compensation of board members, the longer tenure of board members, the number of board meetings, and the size of the board.

This research contributes to the literature on sustainability and corporate governance since it is the first time that the composition and characteristics of the board of directors are related to the green buildings of companies. Furthermore, the study refers to a market as important as the U.S. and one of the most significant financial indexes worldwide.

The remainder of this chapter is as follows. Section 2 establishes the research hypotheses. Section 3 describes the variables under study and explains the methodology applied. Section 4 presents the results of the investigation. Next, Sect. 5 discusses the results obtained, relating them to previous literature on social responsibility and corporate governance, and offers the conclusions.

2 Research Hypotheses

For some years, the possible relationship between corporate governance and company profitability (ROA, ROE, Tobin's Q, etc.) has been studied (Gallego-Álvarez et al., 2010; Rodríguez Fernández et al., 2013). The link between corporate governance and social responsibility has also been extensively studied, so the literature even offers review papers (Dawar & Singh, 2016; Velte, 2017). However, individualized analysis of particular components of corporate social responsibility has been very limited and is arguably a recent line of research.

In this sense, the analysis of the relationship between CO_2 emissions and gender diversity on the board of directors can be cited, having found a negative relationship between both variables (García Martín & Herrero, 2020; Valls Martínez et al., 2022b; Varrone et al., 2020). However, companies' adoption of sustainable buildings has not been analyzed so far in relation to corporate governance characteristics.

The incorporation of women in positions of responsibility and the current concern for gender equality has led to the fact that the percentage of women on the board of directors is precisely the variable that has aroused the most interest in recent research (for example, the CO_2 studies mentioned above). Most of the investigation concludes that the greater presence of women is related to stronger implementation and disclosure of corporate social responsibility practices (Aslam et al., 2018; Charumathi & Rahman, 2019; Kyaw et al., 2017). Therefore, we formulate the following hypothesis:

Hypothesis 1 There is a positive relationship between the percentage of women on the board of directors and the probability that the company has green buildings.

It is reasonable that the greater presence of board members with a financial background or in the industry itself favors the company's financial performance, which previous studies have corroborated (Defond et al., 2005). However, the relationship of this type of board members with corporate social responsibility practices, which may include green buildings, is a little explored aspect, although the scarce literature has not found a significant relationship (Dienes & Velte, 2016; Jizi et al., 2014). Indeed, and especially since the establishment of the sustainable development goals by the United Nations, the interest for companies to have sustainable buildings is evident. Thus, the market could penalize those that are less environmentally sustainable, so if directors are looking for better financial results, they should advocate the adoption of green buildings. In this regard, we propose the next research hypothesis:

Hypothesis 2 There is a positive relationship between the percentage of board members with experience in the industry or financial background and the probability that the company has green buildings.

Non-executive board members reduce agency costs caused by the clash of interests between executives and shareholders (Fama & Jensen, 1983). Indeed, executives are more concerned with the company's short-term profits, while non-executive directors focus on long-term yields (Aslam et al., 2018; Jizi et al., 2014). Non-executives are also more likely to be responsive to stakeholder demands related to non-financial issues, especially those related to corporate environmental responsibility. Although previous studies found no relationship between the percentage of non-executive directors and the implementation of corporate social responsibility practices (Fernández-Gago et al., 2018; Liao et al., 2018; Sial et al., 2018), other research has found a positive relationship (Aslam et al., 2018; Prado-Lorenzo et al., 2012; Zhang et al., 2013), thus confirming the theoretical foundation. Based on this, we made the third hypothesis:

Hypothesis 3 There is a positive relationship between the percentage of board members who are not executives and the probability that the company has green buildings.

Board members with stronger ties to the company are more likely to be inclined to meet stakeholder demands by protecting the company's stability and survival. The remuneration of board members and their permanence in such a position will be more guaranteed with better positioning of the company in the market, which has been shown to occur in those companies with sustainable practices. Despite the limitation established in some countries regarding the duration of directors because a long-term association would result in a loss of independence (Ratri et al., 2021), empirical evidence has shown that the long-serving of board members is linked to fewer corporate social responsibility controversies (Patro et al., 2018). Likewise, long tenure implies a deeper understanding of the company and a superior ability to deal with negative externalities. Therefore, we establish the following two hypotheses:

Hypothesis 4 There is a positive relationship between the board members' remuneration and the probability that the company has green buildings.

Hypothesis 5 There is a positive relationship between the average length of time directors serve on the board of directors and the probability that the company has green buildings.

Identifying the CEO with the Chairperson is a variable widely used in the literature on corporate social responsibility and on which there are no conclusive results. If one of the tasks of the board of directors is to monitor the executives' actions, the fact that the CEO occupies the position of Chairperson on the board would result in worse control and, therefore, higher agency costs. It seems reasonable to think that the CEO, as the principal director, would not favor the control of managers. In short, the company's commitment to corporate social responsibility could be affected, showing a negative relationship with the CEO-Chairperson identity (Aslam et al., 2018; Ben-Amar et al., 2017; Giannarakis et al., 2014). However, strategy planning by the board of directors and its effective monitoring of the company's activities requires in-depth business knowledge provided by the CEO as the company's primary insider. This would lead to a positive relationship between the CEO's presence on the board as Chairperson and the company's sustainable commitments. Conversely, the duality of positions can reinforce the CEO's willingness to adopt social responsibility and sustainability policies (Jizi et al., 2014; Prado-Lorenzo & Garcia-Sanchez, 2010; Sial et al., 2018). According to the previous reasoning, we formulate the following hypotheses:

Hypothesis 6 There is a positive relationship between the identity of the CEO-Chairperson and the probability that the company has green buildings.

Hypothesis 7 There is a positive relationship between the identity of the Chairperson-Ex-CEO and the probability that the company has green buildings.

From the point of view of agency theory, more board meetings facilitate the control of non-executive board members, allowing them to participate in the company's governance. This increased monitoring leads to lower agency costs. On the other hand, according to the legitimacy theory, more meetings translate into better attention to stakeholders' interests and, consequently, to their demands on social and environmental responsibility. Moreover, in the event of an unfavorable incident, the company's response would be faster, reducing the possible impacts that could damage the company's image. Although the literature has sometimes found no relationship between the number of board meetings and the performance or disclosure of corporate social responsibility practices (Charumathi & Rahman, 2019; Liao et al., 2018; Sial et al., 2018), there is research showing this relationship to be significantly positive (Cuadrado Ballesteros et al., 2015; Kent & Monem, 2008; Martínez-Ferrero et al., 2015). Consequently, the present hypothesis is formulated as follows:

Hypothesis 8 There is a positive relationship between the number of annual board meetings and the probability that the company has green buildings.

According to resource dependence theory, a larger board of directors will have more excellent external relationships, allowing access to a broader range of resources (Dienes & Velte, 2016). It will maintain closer contact with stakeholders and be able to serve their concerns efficiently. In addition, the larger number of members will have the capacity to perform more accurate monitoring of the company's operations.

In short, the larger board will be in a better position to detect the needs of internal and external stakeholders, including those related to corporate social responsibility (Giannarakis, 2014; Liao et al., 2018). The board would also dispose of greater resources to meet these demands. Following the above, we formulate the next hypothesis:

Hypothesis 9 There is a positive relationship between the number of board members and the probability that the company has green buildings.

3 Methodology

The empirical analysis was undertaken with data corresponding to the companies belonging to the Standard & Poor 500 index during the period 2011–2019, extracted from the Bloomberg database, which is a reliable source of data used by finance professionals and by scholars in the study of the sustainable performance of companies (Liao et al., 2019; Nollet et al., 2016; Valls Martínez et al., 2022a). The nine-year analysis period is sufficiently long to draw reliable conclusions. The final year of the study is 2019 to avoid distortion of the results because of the COVID-19 pandemic. Thus, considering only the cases for which there were data for all the variables included in the research, the final sample comprised 3996 observations.

Table 1 shows the variables considered in the study, their definition, and the abbreviations that will be used henceforth. The dependent variable, Green buildings (G.B.), will take the value 1 if the company has sustainable buildings and 0 otherwise. The fact that the variable is dichotomous led to applying the logistic regression methodology.

Nine independent variables referring to different characteristics related to the company's board of directors were considered to cover a broad spectrum of corporate governance. The percentage of women on the board of directors (PWO) was considered a proxy for gender diversity, as is common in the literature (Ben-Amar et al., 2017; Francoeur et al., 2019). Similarly, the percentage of board members with prior industry experience or financial background was regarded as a proxy for board member skills (BSK) (Dienes & Velte, 2016; Valls Martínez et al., 2022b). Another variable widely used in the literature on the board of directors is the percentage of non-executive board members (NEX) (Fernández-Gago et al., 2018; Furlotti et al., 2019). Two variables that can influence the directors' commitment, which was therefore important to include, are their compensation (BCO) (Esa & Zahari, 2016; Sarhan & Al-Najjar, 2023) and the length of time they remain on the board (ABT) (Patro et al., 2018; Ratri et al., 2021). CEO involvement has traditionally been considered fundamental, so two variables related to CEO involvement were included: CEO-Chairperson duality (DUA) (Aslam et al., 2018; Kyaw et al., 2017) and whether the Chairperson is a former CEO (CEO) (E-Vahdati et al., 2023). Finally, two board variables that may influence the board's operability and decisions were introduced: the number of meetings held (NMB) (Charumathi & Rahman, 2019; Liao et al., 2018) and the size of the board (BME) (Sial et al., 2018; Velte, 2017).

Table 1 Definition of variables

Abbreviation	Variable	Definition
GB	Green buildings	Dummy variable, 1 if the company has sustainable buildings and 0, otherwise
PWO	Board gender diversity	Percentage of women on the company's board of directors
BSK	Board specific skills	Percentage of board members with experience in the industry or financial background
NEX	Non-executive board members	Percentage of board members who are not executives
BCO	Board member compensation	Board members remuneration
ABT	Average board tenure	The average length of time directors serve on the board of directors
DUA	Duality	Dummy variable, 1 if the CEO is also the Chairperson of the board of directors and 0 otherwise
CEO	Chairperson is ex-CEO	Dummy variable, 1 if the Chairperson is ex-CEO and 0 otherwise
NBM	Number of board meetings	Number of annual board meetings
BME	Board size	Number of board members
SCO	Sustainability Committee	Dummy variable, 1 if the company has a corporate social responsibility committee and 0, otherwise
SRE	Sustainability reporting	Dummy variable, 1 if the company discloses corporate social responsibility reports and 0 otherwise
LTA	Company size	The logarithm of the company's total assets
NIN	Net income	The logarithm of the net income reported by the company
TBQ	Tobin's Q	Market capitalization plus debt divided by total assets

Two groups of control variables were addressed. The first, related to the company's commitment to sustainability (Isidro & Sobral, 2015; Valls Martínez et al., 2022a; Velte, 2017), includes whether or not the company has a sustainability committee (SCO) and whether or not the company discloses a sustainability report (SRE). The second group comprises three financial variables: the company's size (Boulouta, 2013; Cuadrado Ballesteros et al., 2015), measured by total assets (LTA); the company's net income (NIN), as a measure of accounting performance (Giannarakis, 2014); and Tobin's Q (TBQ) (Valls Martínez & Cruz Rambaud, 2019; Wiggins & Ruefli, 2002), as a measure of market performance.

Table 2 shows the averages by sector of the dependent and independent variables. The industry with the highest percentage of companies (Financials) has one of the lowest percentages of green buildings. On the other hand, the sector with the least representation in the sample (Telecommunications Services) has the highest percentage of sustainable buildings. In general, there are no substantial variations between industries in the different variables studied.

The multivariate analysis methodology used in the empirical analysis was logistic regression. Previously, a univariate analysis was performed using the main descriptive statistics for each of the continuous variables involved in the study; in the case of the dichotomous variables, a frequency analysis was conducted.

Bivariate analysis was then performed by determining the Pearson correlations between the continuous variables. Likewise, considering the two possible dependent variable values, mean difference tests were carried out for the continuous variables and Chi2 tests for the dichotomous variables.

Finally, logistic regression models were calculated. Models 1 and 2 considered the data in pooled form. Model 1 included all variables. In Model 2, only the significant regressors of Model 1 were considered. Subsequently, in Models 3 and 4, the data were evaluated in panel form. Model 3 again included all the study variables, and Model 4 only those significant in Model 3. The Hausman test was applied to select the option of fixed or random effects (Hausman, 1978). The Akaike (AIC) and Bayesian (BIC) tests were used to determine the best-proposed models (Akaike, 1974; Schwarz, 1978).

4 Results

Table 3 summarizes the main descriptive statistics of the continuous variables. It shows that the average percentage of women on the board of directors of U.S. companies included in the S&P 500 index in the period considered is 22.77%, which shows that gender equality is far from being achieved. More than half of the directors have previous experience in the industry or the financial sector, less than 13% of the directors are themselves executives in the company, the average number of years that directors have been in their positions is more than 8, on average there are more than 7 board meetings per year, and the average number of directors is more than 11. Regarding the compensation of the members of the board of directors, there is a wide dispersion.

Table 4 shows the frequency of dichotomous variables. It is observed that more than 68% of the companies report having green buildings, and more than 75% have sustainability committees and disclose sustainability reports. It is usual for the CEO to be the Chairperson of the Board of Directors (in 77.13% of the companies, this is the case), and once they are no longer CEO, they continue on the Board of Directors (in 75.52% of the cases).

Table 5 reports the Pearson correlation coefficients between the continuous variables. No high correlations were observed that could give rise to multicollinearity problems in the subsequent regression. Among the independent variables, the highest correlation is 0.4519 between BME and BCO. Regarding the control variables, the highest correlation is shown by TBQ and LTA with −0.5017. None of these values is high; therefore, all the variables proposed in the regression were maintained.

Table 2 Sample composition

Sector	Companies (%)	GB	PWO	BSK	NEX	BCO	ABT	DUA	CEO	NBM	BME
Basic materials	4.78	0.65	23.28	53.23	88.23	736.85	8.34	0.78	0.73	7.46	11.80
Consumer cyclicals	15.49	0.69	22.45	53.89	86.96	743.19	8.98	0.76	0.72	7.34	11.78
Consumer non-cyclicals	7.33	0.71	23.67	51.55	87.35	824.64	8.59	0.80	0.76	7.46	12.16
Energy	5.91	0.65	21.88	54.69	87.67	827.04	8.43	0.77	0.71	7.51	11.72
Financials	19.67	0.66	22.82	53.82	87.12	796.08	9.03	0.76	0.73	7.41	11.95
Healthcare	12.46	0.70	23.03	52.63	87.47	824.96	8.61	0.78	0.72	7.72	11.61
Industrials	14.59	0.68	22.62	52.23	87.25	786.82	8.61	0.78	0.72	7.39	11.66
Technology	13.19	0.69	22.32	54.85	86.05	739.96	8.76	0.75	0.71	7.09	11.39
Telecommunication services	0.80	0.88	24.74	48.06	88.71	911.00	7.24	0.81	0.81	8.56	11.87
Utilities	5.78	0.75	23.29	50.39	87.75	800.11	8.65	0.79	0.73	7.22	12.12

Number of observations: 3996

Table 3 Descriptive statistics of the continuous variables

Variable	Mean	Median	SD	Minimum	Maximum
PWO	22.7661	22.2222	8.0975	0	57.1429
BSK	53.1993	53.8461	18.5105	0	100.0000
NEX	87.1695	90.0000	6.4945	50.0000	100.0000
BCO	784.9775	835.5000	398.1224	1.0000	1400.0000
ABT	8.7391	8.3571	2.9793	0.2500	27.4444
NBM	7.4059	6.5000	3.6425	1.0000	28.0000
BME	11.7698	12.0000	2.1256	5.0000	24.0000
LTA	23.5855	23.4961	1.4388	18.6952	28.5952
NIN	19.9219	20.6543	5.9467	−21.9868	24.8436
TBQ	2.1441	1.5615	2.2172	0.0447	43.7608

Number of observations: 3996

Table 4 Frequency in the dichotomous variables

	Value 0		Value 1	
Variable	Frequency	Percentage	Frequency	Percentage
GB	1255	31.41	2741	68.59
DUA	914	22.87	3082	77.13
CEO	1098	27.48	2898	72.52
SCO	984	24.62	3012	75.38
SRE	815	20.40	3181	79.60

Number of observations: 3996

Table 6 contains the mean tests performed on the continuous variables and the Chi2 tests on the dichotomous variables as a function of the dependent variable GB. The tests were significant for all the dependent variables, except for CEO, which, nevertheless, it was decided to keep in the analysis due to its theoretical link with DUA, a variable widely used in the literature on corporate governance and its influence on corporate social responsibility proxies or derived from this concept.

It is observed that, on average, when the building is sustainable, the percentage of women, the number of non-executive board members, the compensation of directors, the number of board meetings, and the number of directors are higher. Conversely, the percentage of directors with an industry or financial sector background and the number of years on the board are lower.

Table 7 depicts the logistic regression results when the data are considered a pool. Model 1 includes all the variables in the study, while Model 2 only includes those significant in Model 1. It is observed that the results remain stable, with both models showing similar levels of fit, reaching a concordance percentage of more than 76%.

Table 8 presents the logistic regression results considering the data as a panel. Model 3 contains all the study variables, while Model 4 only keeps the significant

Table 5 Pearson correlations between the continuous variables

	PWO	BSK	NEX	BCO	ABT	NBM	BME	LTA	NIN
BSK	−0.0580*** (0.0002)	1.0000							
NEX	0.1269*** (0.0000)	−0.2319*** (0.000)	1.0000						
BCO	0.0569*** (0.003)	−0.1559*** (0.0000)	0.1271*** (0.0000)	1.0000					
ABT	−0.0714*** (0.0000)	0.0641** (0.0127)	−0.1844*** (0.0000)	−0.1265*** (0.0000)	1.0000				
NBM	0.0487*** (0.0021)	0.0329** (0.0375)	0.1249*** (0.0000)	0.1282*** (0.0000)	−0.2057*** (0.0000)	1.0000			
BME	0.0636*** (0.0001)	−0.1624*** (0.0000)	0.2418*** (0.0000)	0.4519*** (0.0000)	0.0158 (0.3183)	0.1311*** (0.0000)	1.0000		
LTA	0.0669*** (0.0000)	−0.0517*** (0.0011)	0.0398** (0.0118)	0.2131*** (0.0000)	−0.0512*** (0.0012)	0.0940*** (0.0000)	0.1018*** (0.0000)	1.0000	
NIN	0.0181 (0.2521)	−0.0040 (0.8014)	−0.0279* (0.0778)	0.0588*** (0.0002)	−0.0125 (0.0001)	0.0424*** (0.0074)	0.0241 (0.1277)	0.2284*** (0.0000)	1.0000
TBQ	0.0219 (0.1657)	0.0384** (0.0151)	−0.0343** (0.0299)	−0.0754*** (0.0000)	0.0473*** (0.0028)	−0.0363** (0.0218)	−0.0581*** (0.0002)	−0.5017*** (0.0000)	−0.1200** (0.0037)

P-value in parentheses

***, ** and * indicate a significance of less than 1%, 5% and 10%, respectively

Number of observations: 3996

Table 6 Difference of means between the dependent variable and the regressors

Panel A: Continuous variables

Variable	Mean GB = 0	Mean GB = 1	Difference
PWO	21.3193	23.4285	−2.1092*** (0.0000)
BSK	56.7163	51.5890	5.1273*** (0.0000)
NEX	86.0535	87.6804	−1.6269*** (0.0000)
BCO	624.6614	858.3802	−233.7188*** (0.0000)
ABT	8.8799	8.6746	0.2054** (0.0431)
NBM	6.6446	7.4059	−1.1098*** (0.0000)
BME	11.1044	12.0744	−0.9700*** (0.0000)
LTA	23.4109	23.6654	−0.2545*** (0.0000)
NIN	19.8057	19.9750	−0.1693 ns (0.4036)
TBQ	2.1753	2.1299	0.0455 ns (0.5475)

Panel B: Dichotomous variables

	Pearson Chi^2
DUA	5.1424** (0.023)
CEO	2.1395 ns (0.144)
SCO	483.5498*** (0.000)
SRE	372.0626 *** (0.000)

P-value in parentheses
ns denotes not significant
***, ** and * indicate a significance of less than 1%, less than 5% and less than 10%, respectively

regressors in Model 3. In both cases, considering the results of the Hausman test, fixed effects were applied. It is observed that the results remain stable when the non-significant variables are eliminated. On the other hand, considering the AIC and BIC criteria results, the panel data models considerably outperformed the pooled data models.

The results confirmed hypotheses 1, 4, 5, 8 and 9. Thus, it is observed that the probability that buildings are sustainable increases as the percentage of women on the board of directors is higher, the directors have higher remuneration, the number of years that directors remain as such increases, there are more board meetings, and the size of the board is larger.

However, hypotheses 2, 3, 6, and 7 were not confirmed. That is, the likelihood of buildings being sustainable does not seem to be influenced by the fact that board members have a background in industry or finance, there are more members non-executives, and the CEO is himself the Chairperson of the board or is a member of the board after they step down as CEO.

Table 7 Logistic regression with pool data

Variable	Model 1	Model 2
PWO	0.02333*** (0.000)	0.02354*** (0.000)
BSK	−0.00318 ns (0.145)	
NEX	−0.00178 ns (0.786)	
BCO	0.00076*** (0.000)	0.00078*** (0.000)
ABT	0.05133*** (0.000)	0.05075*** (0.000)
DUA	0.02072 ns (0.909)	
CEO	−0.03047 ns (0.857)	
NBM	0.07209*** (0.000)	0.07080*** (0.000)
BME	0.07343*** (0.001)	0.07511*** (0.000)
SCO	0.93692*** (0.000)	0.94378*** (0.000)
SRE	0.79865*** (0.000)	0.80772*** (0.000)
LTA	0.12352*** (0.001)	0.11957*** (0.896)
NIN	−0.00515 ns (0.449)	
TBQ	0.06009*** (0.002)	0.07144*** (0.002)
Intercept	−5.81546*** (0.000)	−6.16517*** (0.000)
Year dummies	Yes	Yes
Sector dummies	Yes	Yes
Number of observations	3996	3996
LR Chi^2	808.33*** (0.000)	805.55*** (0.000)
Wald Chi^2	633.84*** (0.000)	631.51*** (0.000)
Nagelkerke	0.2572	0.2564
Percent concordant	76.58	76.48
AIC	4229.200	4221.979
BIC	4430.578	4391.891

P-value in parentheses

ns denotes not significant

***, ** and * indicate a significance of less than 1%, less than 5% and less than 10%, respectively

5 Discussion and Conclusion

Sustainability, in a broad sense, is a global goal. Indeed, this was established by the United Nations in 2015 when member countries agreed on the 17 SDGs to be achieved by 2030. In companies, green buildings are tied to the following goals: (9) Industry, innovation, and infrastructure; (11) Sustainable cities and communities; and (12) Responsible consumption and production. Thus, factories in industrial areas, as well as offices and facilities that companies have in urban areas, must be designed in a sustainable way so that production respects the conditions of economic, social, and environmental sustainability.

Table 8 Logistic regression with panel data

Variable	Model 3	Model 4
PWO	0.02021*** (0.000)	0.02013*** (0.000)
BSK	−0.00352 ns (0.145)	
NEX	0.00247 ns (0.735)	
BCO	0.00083*** (0.000)	0.00083*** (0.000)
ABT	0.04065*** (0.007)	0.03794*** (0.009)
DUA	0.11357 ns (0.563)	
CEO	−0.15517 ns (0.399)	
NBM	0.07339*** (0.000)	0.07231*** (0.000)
BME	0.08087*** (0.001)	0.08549*** (0.000)
SCO	0.92638*** (0.000)	0.93413*** (0.000)
SRE	0.64209*** (0.000)	0.64792*** (0.000)
LTA	−0.38234*** (0.004)	−0.44833*** (0.000)
NIN	−0.00931 ns (0.256)	
TBQ	0.03326 ns (0.350)	
Number of observations	3586	3586
LR Chi^2	542.37*** (0.000)	537.09*** (0.000)
Wald Chi^2	427.95*** (0.000)	425.96*** (0.000)
Nagelkerke	0.2410	0.2388
Hausman test	47.57*** (0.0000)	46.31*** (0.0000)
AIC	2616.830	2610.119
BIC	2703.417	2659.597

P-value in parentheses
ns denotes not significant
***, ** and * indicate a significance of less than 1%, less than 5% and less than 10%, respectively

It should be remembered that green buildings encompass various perspectives (Alsulaili et al., 2020; Ding et al., 2018; Samer, 2013): eco-friendly construction materials, limited energy and water consumption, water and waste treatment processes, use of renewable energies, location in areas accessible by public transportation, mimicry with the environment to eliminate visual pollution, etc.

In the business world, corporate social responsibility has been considered for several decades as a way for companies to be accountable for their actions (Fernández-Gago et al., 2018). It is a requirement that the different stakeholders (investors, customers, governments, etc.) have imposed as almost necessary. So, companies of a certain size have been disclosing social responsibility reports along with their financial statements (Valls Martínez et al., 2020). The fact that the company's buildings meet the sustainable requirements of green buildings can be considered part of corporate social responsibility actions.

Considering that the highest governing body of companies, the one that establishes the main lines of action and monitors management decisions, is the board of directors, to date, researchers have found the theoretical basis and developed numerous empirical studies that relate the structure and characteristics of the board of directors to corporate social responsibility, whether the latter is considered as a whole or disaggregated into its three pillars: environmental, social, and governance (Giannarakis et al., 2014; Kyaw et al., 2017; Velte, 2017). However, the particular analysis of such an important aspect as green buildings in relation to corporate governance has not been addressed, which is the objective of this work.

The present study was based on the companies listed in the American S&P 500 index in 2011–2019, covering 3996 observations, with 68.59% of the companies having green buildings, applying the multivariate logistic regression methodology. It should be noted that since no prior research has been found that deals with green buildings and corporate governance, the results obtained in this work will be compared with those obtained previously about corporate social responsibility.

It was found that a higher percentage of women on the board of directors is related to a higher probability of the company having green buildings. Studies on gender diversity have increased in recent years, coinciding with the incorporation of women in positions of maximum organizational responsibility and the development of laws favoring gender equality. Most results show that women are more sensitive to social and environmental problems, and their higher weight in direction positions tends to strengthen corporate social responsibility. Some research has even linked the higher proportion of women on the board of directors with lower CO_2 emissions (García Martín & Herrero, 2020; Haque, 2017; Valls Martínez et al., 2022b). Therefore, it seems logical that the relationship found in this research between women board members and green buildings is positive. The theoretical basis for this can be found in the agency, legitimacy, resource dependence, stakeholder, social role, and upper echelons theories, among others (Valls Martínez et al., 2022a; Valls Martínez & Soriano Román, 2022).

A stronger link between directors and the company is related to a higher probability that the company will have green buildings. Today, environmental sustainability issues are highly valued and required by governments, investors, environmental pressure groups, and the general public. Arguably, the long-term success and survival of the company is threatened by social and ecological non-compliance. Since the board members with the most profound ties to the company would be the most adversely affected in the event of business failure, it is logical that they would be the most interested in the company meeting the required explicit or implicit canons. This fact explains the positive and significant relationship found between board members' remuneration and board tenure with green buildings (Patro et al., 2018). It would also explain the absence of a significant connection between the percentage of non-executive directors and the dependent variable.

According to agency and legitimacy theories, results have shown that more board meetings are positively related to the probability that the company has green buildings, enabling directors to monitor management's actions better and steer the company's operations to meet stakeholder requests, in line with previous studies

(Cuadrado Ballesteros et al., 2015; Martínez-Ferrero et al., 2015). These two theories, together with the resource dependence theory, are the basis for the positive relationship found in this research between the number of board members and the probability that the company will have green buildings. Indeed, if the board of directors is larger, it can better perform its role of controlling management, obtaining resources, and attending to stakeholders, in vein with existing literature (Fernández-Gago et al., 2018).

In addition, the empirical study has shown that the financial or industry background of the board members, neither the CEO-Chairperson duality nor the fact that the Chairperson is a former CEO, are significantly related to the company's sustainable buildings. While these variables have sometimes been positively associated with corporate social responsibility (Sial et al., 2018), most previous studies found no relationship between these variables frequently used in research and corporate sustainability (Dienes & Velte, 2016; Kyaw et al., 2017; L. Liao et al., 2018).

In conclusion, the results showed that companies are more likely to have sustainable buildings when the percentage of women on the board of directors, director compensation and tenure, number of board meetings, and board size increase. On the other hand, the percentage of board members with a financial or industry background, the rate of non-executive directors, the CEO-Chairperson duality, and the Chairman being a former CEO did not show a significant relationship with the probability of green buildings.

These results have implications for investors and policymakers. The results provide arguments to be made in favor of gender quota laws and no restrictions on the tenure of the board members. Investors wishing to place their savings in socially responsible investments should look for companies with the characteristics we have highlighted on the board of directors. In addition, shareholders should encourage boards that are not too small in size, active in the number of meetings, with substantial compensation for board members and long tenure, as well as increase the number of women.

This study is not without limitations. Given that this is the first research on green buildings and the structure and characteristics of the board of directors, it would be advisable to extend the time horizon of the sample and the geographical scope, and consider American companies not included in the S&P 500 index. In addition, the methodology could be completed with techniques such as bootstrap and predictive analysis of the multivariate models proposed.

Acknowledgements This research was funded by PPIT-UAL, Junta de Andalucía-ERDF 2021–2027. Programme: 54.A to the Consolidated Research Group on Ethics, Gender, and Sustainability (SEJ-647).

References

Akaike, H. (1974). A new look at the statistical model identification. *IEEE Transactions on Automatic Control, 19*(6), 716–723. https://doi.org/10.1109/TAC.1974.1100705

Alsulaili, A. D., Al-Matrouk, M. F., Al-Baghli, R. A., & Al-Enezi, A. F. (2020). Environmental and economic benefits of applying green building concepts in Kuwait. *Environment, Development and Sustainability, 22*(4), 3371–3387. https://doi.org/10.1007/s10668-019-00352-1

Aslam, S., Abdul, M., Makki, M., Mahmood, S., & Amin, S. (2018). Gender diversity and managerial ownership response to corporate social responsibility initiatives: Empirical evidence from Australia. *Journal of Managerial Sciences, 12*(2), 131–151.

Ben-Amar, W., Chang, M., & McIlkenny, P. (2017). Board gender diversity and corporate response to sustainability initiatives: Evidence from the carbon disclosure project. *Journal of Business Ethics, 142*(2), 369–383. https://doi.org/10.1007/s10551-015-2759-1

Boulouta, I. (2013). Hidden connections: The link between board gender diversity and corporate social performance. *Journal of Business Ethics, 113*, 185–197. https://doi.org/10.1007/s10551-012-1293-7

Charumathi, B., & Rahman, H. (2019). Do women on boards influence climate change disclosures to CDP?—Evidence from large indian companies. *Australasian Accounting, Business and Finance Journal, 13*(2), 5–31. https://doi.org/10.14453/aabfj.v13i2.2

Cuadrado Ballesteros, B., García Rubio, R., & Martínez Ferrero, J. (2015). Effect of the composition of the board of directors on corporate social responsibility. *Spanish Accounting Review, 18*(1), 20–31. https://doi.org/10.1016/j.rcsar.2014.02.003

Dawar, G., & Singh, S. (2016). Corporate social responsibility and gender diversity: A literature review. *Journal of IMS Group, 13*(1), 61–71.

Defond, M. L., Hann, R. N., & Hu, X. (2005). Does the market value financial expertise on audit committees of boards of directors? *Journal of Accounting Research, 43*(2), 153–193. https://doi.org/10.1111/j.1475-679x.2005.00166.x

Dienes, D., & Velte, P. (2016). The impact of supervisory board composition on CSR reporting. Evidence from the German two-tier system. *Sustainability, 8*(1), 1–20. https://doi.org/10.3390/su8010063

Ding, Z., Fan, Z., Tam, V. W. Y., Bian, Y., Li, S., Illankoon, I. M. C. S., & Moon, S. (2018). Green building evaluation system implementation. *Building and Environment, 133*, 32–40. https://doi.org/10.1016/j.buildenv.2018.02.012

Duque-Grisales, E., & Aguilera-Caracuel, J. (2019). Environmental, social and governance (ESG) scores and financial performance of multilatinas: Moderating effects of geographic international diversification and financial slack. *Journal of Business Ethics, 168*(2), 315–334. https://doi.org/10.1007/s10551-019-04177-w

ElAlfy, A., Darwish, K. M., & Weber, O. (2020). Corporations and sustainable development goals communication on social media: Corporate social responsibility or just another buzzword? *Sustainable Development, May*, 1–13. https://doi.org/10.1002/sd.2095

Esa, E., & Zahari, A. R. (2016). Corporate social responsibility: Ownership structures, board characteristics & the mediating role of board compensation. *Procedia Economics and Finance, 35*, 35–43. https://doi.org/10.1016/S2212-5671(16)00007-1

E-Vahdati, S., Wan-Hussin, W. N., & Mohd Ariffin, M. S. (2023). The value relevance of ESG practices in Japan and Malaysia: Moderating roles of CSR award, and former CEO as a Board Chair. *Sustainability, 15*(3), Article 3. https://doi.org/10.3390/su15032728

Fama, E. F., & Jensen, M. C. (1983). Separation of ownership and control. *Journal of Law and Economics, 26*(2), 301–325.

Fernández-Gago, R., Cabeza-García, L., & Nieto, M. (2018). Independent directors' background and CSR disclosure. *Corporate Social Responsibility and Environmental Management, 25*(5), 991–1001. https://doi.org/10.1002/csr.1515

Francoeur, C., Labelle, R., Balti, S., & El Bouzaidi, S. (2019). To what extent do gender diverse boards enhance corporate social performance? *Journal of Business Ethics, 155*, 343–357. https://doi.org/10.1007/s10551-017-3529-z

Furlotti, K., Mazza, T., Tibiletti, V., & Triani, S. (2019). Women in top positions on boards of directors: Gender policies disclosed in Italian sustainability reporting. *Corporate Social Responsibility and Environmental Management, 26*(1), 57–70. https://doi.org/10.1002/csr.1657

Gallego-Álvarez, I., García-Sánchez, I. M., & Rodríguez-Dominguez, L. (2010). The influence of gender diversity on corporate performance. *Revista De Contabilidad-Spanish Accounting Review, 13*(1), 53–88. https://doi.org/10.1016/S1138-4891(10)70012-1

García Martín, C. J., & Herrero, B. (2020). Do board characteristics affect environmental performance? A study of EU firms. *Corporate Social Responsibility and Environmental Management, 27*(1), 74–94. https://doi.org/10.1002/csr.1775

Giannarakis, G. (2014). Corporate governance and financial characteristic effects on the extent of corporate social responsibility disclosure. *Social Responsibility Journal, 10*(4), 569–590. https://doi.org/10.1108/SRJ-02-2013-0008

Giannarakis, G., Konteos, G., & Sariannidis, N. (2014). Financial, governance and environmental determinants of corporate social responsible disclosure. *Management Decision, 52*(10), 1928–1951. https://doi.org/10.1108/MD-05-2014-0296

Haque, F. (2017). The effects of board characteristics and sustainable compensation policy on carbon performance of UK firms. *The British Accounting Review, 49*(3), 347–364. https://doi.org/10.1016/j.bar.2017.01.001

Hausman, J. A. (1978). Specification tests in econometrics. *Econometrica, 46*(6), 1251–1271. https://doi.org/10.2307/1913827

Isidro, H., & Sobral, M. (2015). The effects of women on corporate boards on firm value, financial performance, and ethical and social compliance. *Journal of Business Ethics, 132*(1), 1–19. https://doi.org/10.1007/s10551-014-2302-9

Jizi, M. I., Salama, A., Dixon, R., & Stratling, R. (2014). Corporate governance and corporate social responsibility disclosure: Evidence from the US banking sector. *Journal of Business Ethics, 125*(4), 601–615. https://doi.org/10.1007/s10551-013-1929-2

Kent, P., & Monem, R. (2008). What drives TBL reporting: Good governance or threat to legitimacy? *Australian Accounting Review, 18*(4), 297–309. https://doi.org/10.1111/j.1835-2561.2008.0036.x

Kyaw, K., Olugbode, M., & Petracci, B. (2017). Can board gender diversity promote corporate social performance? *Corporate Governance: the International Journal of Business in Society, 17*(5), 789–802. https://doi.org/10.1108/CG-09-2016-0183

Liao, L., Lin, T. P., & Zhang, Y. (2018). Corporate board and corporate social responsibility assurance: Evidence from China. *Journal of Business Ethics, 150*(1), 211–225. https://doi.org/10.1007/s10551-016-3176-9

Liao, Z., Zhang, M., & Wang, X. (2019). Do female directors influence firms' environmental innovation? The moderating role of ownership type. *Corporate Social Responsibility and Environmental Management, 26*(1), 257–263. https://doi.org/10.1002/csr.1677

Machmuddah, Z., & Wardhani, R. (2020). *Environmental social governance (ESG) disclosure score rating of bloomberg* (pp. 48–52). https://doi.org/10.2991/ahsr.k.200723.011

Martínez-Ferrero, J., Vaquero-Cacho, L. A., Cuadrado-Ballesteros, B., & García-Sánchez, I. M. (2015). Corporate governance and corporate social responsibility in banking: The role of the board of directors. *Investigaciones Europeas De Dirección y Economía De La Empresa, 21*(3), 129–138. https://doi.org/10.1016/j.iedee.2015.01.001

Nollet, J., Filis, G., & Mitrokostas, E. (2016). Corporate social responsibility and financial performance: A non-linear and disaggregated approach. *Economic Modelling, 52*(Part B), 400–407. https://doi.org/10.1016/j.econmod.2015.09.019

Nuber, C., & Velte, P. (2021). Board gender diversity and carbon emissions: European evidence on curvilinear relationships and critical mass. *Business Strategy and the Environment, 30*(4), 1958–1992. https://doi.org/10.1002/bse.2727

Patro, S., Zhang, L. Y., & Zhao, R. (2018). Director tenure and corporate social responsibility: The tradeoff between experience and independence. *Journal of Business Research, 93*, 51–66. https://doi.org/10.1016/j.jbusres.2018.08.033

Pletzer, J. L., Nikolova, R., Kedzior, K. K., & Voelpel, S. C. (2015). Does gender matter? Female representation on corporate boards and firm financial performance—a meta-analysis. *PLoS ONE, 10*(6), 1–20. https://doi.org/10.1371/journal.pone.0130005

Prado-Lorenzo, J. M., & Garcia-Sanchez, I. M. (2010). The role of the board of directors in disseminating relevant information on greenhouse gases. *Journal of Business Ethics, 97*(3), 391–424. https://doi.org/10.1007/s10551-010-0515-0

Prado-Lorenzo, J. M., García-Sánchez, I. M., & Gallego-Álvarez, I. (2012). Effects of activist shareholding on corporate social responsibility reporting practices: An empirical study in Spain. *Journal of Economics, Finance and Administrative Science, 17*(32), 7–16.

Pyles, M. K. (2020). Examining portfolios created by Bloomberg ESG scores: Is disclosure an alpha factor? *The Journal of Impact and ESG Investing*. https://doi.org/10.3905/jesg.2020.1.001

Ratri, M. C., Harymawan, I., & Kamarudin, K. A. (2021). Busyness, tenure, meeting frequency of the CEOs, and corporate social responsibility disclosure. *Sustainability, 13*(10), Article 10. https://doi.org/10.3390/su13105567

Rodríguez Fernández, M., Fernández Alonso, S., & Rodríguez Rodríguez, J. (2013). Estructura del consejo de administración y rendimiento de la empresa española cotizada. *Revista Europea De Dirección y Economía De La Empresa, 22*(3), 155–168. https://doi.org/10.1016/j.redee.2012.10.002

Samer, M. (2013). Towards the implementation of the green building concept in agricultural buildings: A literature review. *Agricultural Engineering International: CIGR Journal, 15*(2), Article 2.

Sarhan, A. A., & Al-Najjar, B. (2023). The influence of corporate governance and shareholding structure on corporate social responsibility: The key role of executive compensation. *International Journal of Finance & Economics, 28*(4), 4532–4556. https://doi.org/10.1002/ijfe.2663

Schwarz, G. (1978). Estimating the dimension of a model. *The Annals of Statistics, 6*(2), 461–464. https://doi.org/10.1214/aos/1176344136

Sial, M. S., Zheng, C., Cherian, J., Gulzar, M. A., Thu, P. A., Khan, T., & Khuong, N. V. (2018). Does corporate social responsibility mediate the relation between boardroom gender diversity and firm performance of Chinese listed companies? *Sustainability, 10*(10), 1–18. https://doi.org/10.3390/su10103591

Valls Martínez, M. C., Martín Cervantes, P. A., & Miralles Quirós, M. M. (2022a). Sustainable development and the limits of gender policies on corporate boards in Europe. A comparative analysis between developed and emerging markets. *European Research on Management and Business Economics, 28*(1), 1–15. https://doi.org/10.1016/j.iedeen.2021.100168

Valls Martínez, M. C., & Soriano Román, R. (2022). Women in monitoring positions and market risk. Are the stocks of companies with gender diverse boards less volatile? *Frontiers in Psychology, 13*(November), 1–19. https://doi.org/10.3389/fpsyg.2022.1049175

Valls Martínez, M. C., & Cruz Rambaud, S. (2019). Women on corporate boards and firm's financial performance. *Women's Studies International Forum, 76*(102251), 1–11. https://doi.org/10.1016/j.wsif.2019.102251

Valls Martínez, M. C., Martín Cervantes, P. A., & Cruz Rambaud, S. (2020). Women on corporate boards and sustainable development in the American and European markets: Is there a limit to gender policies? *Corporate Social Responsibility and Environmental Management, 27*(6), 2642–2656. https://doi.org/10.1002/csr.1989

Valls Martínez, M. C., Santos Jaén, J. M., Soriano Román, R., & Martín Cervantes, P. A. (2022). Are gender and cultural diversities on board related to corporate CO_2 emissions? *Journal of Cleaner Production, 363*, 1–17. https://doi.org/10.1016/j.jclepro.2022.132638

Varrone, N., D'Angelo, E. D., Gangi, F., & Daniele, L. M. (2020). Does cultural diversity of board of directors affect corporate environmental performance? Evidence from the energy sector. *European Scientific Journal ESJ, 16*(28). https://doi.org/10.19044/esj.2020.v16n28p287

Velte, P. (2017). Does board composition have an impact on CSR reporting? *Problems and Perspectives in Management, 15*(2), 19–35. https://doi.org/10.21511/ppm.15(2).2017.02

Wiggins, R. R., & Ruefli, T. W. (2002). Sustained competitive advantage: Temporal dynamics and the incidence and persistence of superior economic performance. *Organization Science, 13*(1), 82–105. https://doi.org/10.1287/orsc.13.1.81.542

Zhang, J. Q., Zhu, H., & Ding, H. (2013). Board composition and corporate social responsibility: An empirical investigation in the post Sarbanes-Oxley era. *Journal of Business Ethics, 114*(3), 381–392. https://doi.org/10.1007/s10551-012-1352-0

María del Carmen Valls Martínez is a Ph.D. in Economics and Business. In 1995, she joined the Economics and Business Department at the University of Almería (Spain), where she is an Associate Professor in Financial Economics and Accounting and leads the *Ethics, Gender and Sustainability Research Group.* Her research interests include Financial Operations, Ethical Banking, Gender Economics Studies, Health Economics and Sustainability. She has published numerous books and book chapters in reputable publishers such as Springer, Dykinson, Pirámide, McGraw-Hill, etc. She is the author of a large number of articles in prestigious scientific journals, such as the *Journal of Cleaner Production, Corporate Social Responsibility and Environmental Management, Risk Management, Physica A: Statistical Mechanics and its Applications, European Research on Management and Business Economics, Mathematics, Plos One, Women's Studies International Forum, International Journal of Environmental Research and Public Health, Frontiers in Psychology, Oeconomia Copernicana,* etc. She is also an editor at *Plos One, Frontiers in Public Health* and *Mathematics.*

José-María Montero is a Full Professor in Applied Economics at the University of Castilla-La Mancha (Spain). He is director of the Exponential Education International Academic Council, President of the International Association of Applied Economics (ASEPELT), Honorary Life Member of the International Atlantic Economic Society (USA), Vice-President of the International CESTIRA Association, and an advisor to the Social Sciences and Humanities Research Council of Canada, in addition to serving on the editorial boards of scientific journals in economics, finance, and statistics. Professor Montero has joined the board of the Philippe Cousteau Foundation 'Union des Océans,' whose work is focused on conserving the environment and marine biodiversity, sustainable development, international cooperation, and the fight against marine pollution. He is the author of numerous books on statistics. He has published in prestigious journals such as *Frontiers in Public Health, Atmospheric Pollution Research, Mathematics, Review of Income and Wealth, The European Journal of Finance, Environmetrics, Journal of Cleaner Production, Mathematical and Statistical Methods for Actuarial Sciences and Finance,* etc.

María Estefanía Sánchez Pacheco is a Ph.D. student in Economics, Business, and Law at the University of Almería. She is an associate researcher in business and innovation at the International University of Ecuador (UIDE). She holds a Master's Degree in Financial Management and another in Human Talent Management. In addition, she holds a Marketing Degree and a Bachelor's degree in International Business. This background has allowed her to collaborate on publishing articles of regional and global impact in prestigious international journals. Professionally, she has worked in Human Resources as Coordinator of the Department at PDVSA Ecuador S.A. and is currently a Management Advisor at the Public Company of the State University of Milagro.

Fernando José Zambrano Farías holds a Ph.D. in Economics, Business and Law. He is a research professor at the International University of Ecuador UIDE and the University of Guayaquil. He has taught Corporate Finance, International Finance, and Research Methodology at the Faculty of Administrative Sciences since 2015. He holds a Master's Degree in Accounting and Corporate Finance, a Master's Degree in Financial Management, and a Master's Degree in Economics with a major in Corporate Finance and Projects. In addition, he holds a degree in Economics with a major in Business Management and a Bachelor of Science in Education. He has published regional and global impact articles in prestigious international journals where he is also a reviewer. Professionally, he has worked in the financial area as Financial Manager of some companies in Ecuador and is currently a member of the Board of Directors of the company Bicert Ecuador S.A.S.